W9-BHM-749

DATE DUE

GAYLORD   #3523PI   Printed in USA

The Hummingbirds of North America

Also by Paul A. Johnsgard

The Avian Brood Parasites: Deception at the Nest (in press)

The Baby Bird Portraits of George Miksch Sutton: Watercolors in the Field Museum (in press)

Ruddy Ducks and Other Stifftails: Their Behavior and Biology (with M. Carbonell) (1996)

This Fragile Land: A Natural History of the Nebraska Sandhills (1995)

Arena Birds: Sexual Selection and Behavior (1994)

Cormorants, Darters, and Pelicans of the World (1993)

Ducks in the Wild: Conserving Waterfowl and Their Habitats (1992)

Bustards, Hemipodes, and Sandgrouse: Birds of Dry Places (1992)

Crane Music: The North American Cranes (1991)

Hawks, Eagles, and Falcons of North America: Biology and Natural History (1990)

Waterfowl of North America: The Complete Ducks, Geese, and Swans (1989) (species accounts)

North American Owls: Biology and Natural History (1988)

The Quails, Partridges, and Francolins of the World (1988)

Diving Birds of North America (1987)

Birds of the Rocky Mountains (1986)

The Pheasants of the World (1986)

Prairie Children, Mountain Dreams (1985)

The Platte: Channels in Time (1984)

The Cranes of the World (1983)

The Hummingbirds of North America (1983)

The Grouse of the World (1983)

Dragons and Unicorns: A Natural History (with Karin Johnsgard) (1982)

Teton Wildlife: Observations by a Naturalist (1982)

Those of the Gray Wind: The Sandhill Cranes (1981)

The Plovers, Sandpipers, and Snipes of the World (1981)

A Guide to North American Waterfowl (1979)

Birds of the Great Plains: Breeding Species and Their Distribution (1979)

Ducks, Geese, and Swans of the World (1978)

The Bird Decoy: An American Art Form (editor) (1976)

Waterfowl of North America (1975)

American Game Birds of Upland and Shoreline (1975)

Song of the North Wind: A Story of the Snow Goose (1974)

Grouse and Quails of North America (1973)

Waterfowl: Their Biology and Natural History (1968)

Animal Behavior (1967)

Handbook of Waterfowl Behavior (1965)

The Hummingbirds of North America

Second edition

Paul A. Johnsgard

**Color Plates by Mark Marcuson,
James McClelland, and Sophie Webb**

SMITHSONIAN INSTITUTION PRESS
Washington, D. C.

Copy editor and typesetter: Princeton Editorial Associates
Production editor: Duke Johns
Designer: Janice Wheeler

Library of Congress Cataloging-in-Publication Data

Johnsgard, Paul A.
   The hummingbirds of North America / Paul A. Johnsgard. — 2nd ed.
     p. cm.
   Rev. ed. of: The hummingbirds of North America. 1983.
   Includes bibliographical references (p.   ) and index.
   ISBN 1-56098-708-1 (alk. paper)
   1. Hummingbirds—North America. I. Johnsgard, Paul A.
Hummingbirds of North America. II. Title.
QL696.A558J63 1997
598.8'99—dc20   96-44168

British Library Cataloguing-in-Publication Data is available

Manufactured in the United States of America

04 03 02 01 00 99 98 97 5 4 3 2 1

Contents

Preface to the Second Edition

During the past few years I have been told repeatedly by friends and my Smithsonian Institution Press acquisitions editor, Peter Cannell, that my *Hummingbirds of North America* should be revisited and revised, but I have always found it more interesting to undertake new books than to wrestle with updating old ones. However, during a chance meeting in the summer of 1995, my old friend Luis Baptista urged me not only to revise my original book but also to add the strictly Mexican species of hummingbirds. That sounded like an exciting idea, so I immediately began to determine what was needed to bring it about.

I soon decided that the Isthmus of Tehuantepec represents a much more meaningful and biologically defensible terminus for a book on North American hummingbirds than the Mexican–Guatemalan border, and by limiting my coverage to species that breed north of that region I would need to add only 25 new species accounts to my text coverage. The earlier book had described 23 species, although the inclusion of one of these (the Antillean crested hummingbird) was based on very questionable evidence as to its historical continental occurrence, and I have eliminated it from the present edition. Thus, the present book includes a minimum of 47 so-called "species," although several of these populations are admittedly complex species groups that are regarded as two or more species by various authorities.

I have followed the comprehensive world checklist by Sibley and Monroe (1990) for recognizing both species limits and taxonomic sequence of species, but have not always followed that reference for recognizing generic limits. Nor have I followed the accepted taxonomy of the American Ornithologists' Union (AOU) in several cases, but wherever my terminology differs from that of the AOU I have provided the currently accepted AOU nomenclature, based on the most recent *Check-list of North American Birds* (AOU, 1983) including relevant supplements. A new edition of this checklist should appear soon, and it may include additional taxonomic changes appearing too late to incorporate into this text.

To illustrate the 25 newly included species, I obtained the help of Mark Marcuson, who on short notice produced four wonderful watercolor plates, depicting males of each newly included species. James McClelland kindly allowed me to reprint the 16 stunning watercolors of 22 hummingbird species that he had painted for my original volume. I also was fortunate to obtain the permission of Sophie Webb to reproduce four of her splendid paintings of 56 taxa of North American and Central American hummingbirds from *A Guide to the Birds of Mexico and Northern Central America* (Howell and Webb 1995). Her paintings not only illustrate male, female, and immature plumages of these species, but also include several additional forms that are either peripheral Central American species or represent recognizable endemic Mexican races if not biologically distinct sibling species. I also relied heavily on Howell and Webb's wonderfully pioneering field guide in drawing my range maps for these many Mexican species, and gratefully acknowledge its importance in facilitating this component of my book. I also used Howell and Webb's vernacular nomenclature for most English names of hummingbirds, including those several forms here considered as races but which may be recognized as distinct species after more is learned about their ranges and biologies.

In addition to the help of these persons, I am especially grateful to the Field Museum in Chicago, and in particular to Dr. Daniel Willard of the Bird Division, who arranged for the loan of hummingbird specimens. Dr. Willard also provided me with valuable museum specimen data on hummingbird weights. Associate Librarian Benjamin Williams also very kindly allowed me to examine references in the museum's rare book collection.

Preface to the First Edition

It is frequently difficult to remember just when the germ of an idea for a book emerges, but the one for this book certainly arose when an impudent female calliope hummingbird constructed her nest near my cabin in Grand Teton National Park while I was doing fieldwork there in the summer of 1975. Watching the progression of her incubation and brooding, often from only a meter or two away, was a fascinating exercise in ecology and behavior. Thus, I resolved to investigate the whole group of hummingbirds more carefully someday.

That day came in the spring of 1980, when I found the time to write a fairly short and not-too-technical book on an appealing subject. After only a little thought, I settled on hummingbirds. My first idea was simply to produce a series of essays on the breeding biologies and natural histories of the North American hummingbirds. Almost immediately, however, I realized that limiting my coverage to the strictly North American forms would eliminate the vast majority of hummingbirds and their biology. Thus, I decided to incorporate a series of introductory chapters on comparative hummingbird biology, not restricted to North America. This approach, in turn, supported the inclusion of a complete synopsis of the hummingbirds of the world, together with a limited amount of information on their ranges and identification.

From the outset I believed that the book should contain suitable illustrations, inasmuch as hummingbirds are among the most spectacularly plumaged of all birds. Since my own photographic files were almost totally lacking in hummingbird subjects, I asked James McClelland, a highly gifted watercolor artist of Lincoln, Nebraska, if he would consider doing a series of plates for me. This he immediately agreed to do, and his fine efforts have greatly assisted me in illustrating this book. The University of Nebraska Herbarium loaned specimens to me as needed, and whenever possible Glen Drohman kindly provided fresh plant material for me. C. F. Zeillemaker and Richard Zusi undertook critical readings of the entire manuscript. I have received valuable assistance from many other persons, including David Willard of the Field Museum of Natural History, Chicago, and Robert Mengel of the University of Kansas Museum of Natural History, for loan of specimens. Specimen information was also provided me by Lloyd Kiff, Western Foundation of Vertebrate Zoology, Los Angeles; David Niles, Delaware Museum of Natural History, Greenville; Fernando I. Ortiz-Crespo, Museum of Vertebrate Zoology, Berkeley; Dwain Warner, James Ford Bell Museum of Natural History, Minneapolis; and George Watson, National Museum of Natural History, Washington, D.C. Other information or assistance was provided by James Bond, John A. Crawford, Crawford Greenewalt, Lester Short, Jr., and Charles G. Sibley. A grant from the Frank Chapman Fund of the American Museum of Natural History allowed me to visit that museum and obtain necessary specimen and library information available there. I am also indebted to the School of Life Sciences, University of Nebraska–Lincoln, for providing me with the facilities necessary for undertaking this project and for secretarial assistance in manuscript preparation.

Introduction

"Of all the numerous groups into which the birds are divided there is none other so numerous in species, so varied in form, so brilliant in plumage, and so different from all others in their mode of life." Thus did the American ornithologist Robert Ridgway (1890) introduce the hummingbirds in a monograph a century ago, and no writer before or since has been able to refer to this incredible group of birds without resorting to superlatives. John James Audubon described a hummingbird as a "glittering garment of the rainbow," and John Gould called them "wonderful works of creation."

The all-too-human tendency initially to admire such beauty and then to demand to possess it has been responsible for an impassioned interest in hummingbirds, which has lasted for several hundred years. At its heyday in the mid-1800s, the market for hummingbirds was so great that hundreds of thousands were killed in South America to supply the upper-class Europeans with specimens for collections and for ornamental uses. A Brazilian port once shipped 3000 skins of a single hummingbird species on one consignment, and public sales in London for one month in 1888 included more than 12,000 hummingbird skins. London auctions sold some 400,000 bird skins that year, a good percentage of which were doubtless hummingbirds. During that period ornithologists discovered numerous new hummingbird species from among the vast number of specimens flowing into England and the Continent. Many of these came from unspecified parts of northern South America, and a few of them have never been found again.

The period of maximum importation of hummingbirds, occurring at the peak time for animal monographs with magnificent hand-colored plates, spawned many sumptuous volumes on the subject. One of the first of significance was that by the Frenchman R. P. Lesson, who published one volume in 1829, added a "supplement" in 1830–1831, and completed a third and last volume in 1832. Although the three volumes were discrete works with separate titles, they formed a comprehensive treatment of the roughly 110 species known to him and contained many colored plates. But Lesson contributed early to the confusion of hummingbird nomenclature by indiscriminately renaming many previously described species.

Between 1849 and 1861, John Gould produced a five-volume monograph on hummingbirds. This magnificent folio production, one of the most beautiful of all bird monographs, contained 360 hand-colored lithographic plates and collectively catalogued some 416 "species," many described for the first time. But the chaotic state of hummingbird taxonomy at that time is evident from the frequent divergence between the names on the plates and those in the accompanying letterpress. Nevertheless, the monograph set a level in book production and in bird art that has rarely been surpassed (if ever).

During the latter decades of the nineteenth century a host of new hummingbird species were described, and many new names were applied to species already known. Nevertheless, in 1878, D. G. Elliot produced a landmark monograph of the whole group. This work made a serious effort at redefining all genera, describing all the known species, and providing an identification key as well as detailed illustrations of representatives of all the genera accepted by Elliot.

About a decade later—in 1890—Robert Ridgway provided his equally outstanding contribution on the subject, including an extended survey of the characteristics of the hummingbird family as a whole, as well as individual treatments of the 17 species that had been found within the boundaries of the United States. In 1911 he published a larger work, which included all of the nearly 150 species that Ridgway accepted as occurring north of the Isthmus of Panama and in the West Indies. This is still the most valuable and definitive single volume on the hummingbirds of North and Central America, particularly for its keys and plumage descriptions.

Surprisingly few monographic approaches to the entire family appeared during the 1900s, despite the tremendous interest of the preceding century. One of the few—an *Histoire Naturelle* of the hummingbirds, published by E. Simon in 1921—was little more than a revised taxonomy of the group. It had none of the strengths of Elliot's earlier approach and burdened the hummingbird literature with what proved to be the ultimate in generic and specific names. But Simon's classification clearly affected the subsequent taxonomies of the group, especially James Peters' classification, which has been the standard authority since its appearance in 1945.

Although the London Zoo acquired its first living hummingbird as early as 1905, the bird died within two weeks. It was not until the development of rapid transatlantic air travel and suitable diets that hummingbirds began to appear regularly in European zoos. The English aviculturalist Alfred Ezra established the importance of adding proteins, fats, minerals, and vitamins to the basic honey diet of hummingbirds, and later Walter Scheithauer in Germany experimented extensively to establish specific optimum requirements for these food components. Scheithauer's popular book *Hummingbirds*, published initially in German in 1966 and in English translation the following year, not only provided this information, but also included superb color photographs of 31 species as well as a detailed discussion of the flying abilities of these birds. A few years earlier in North America, Crawford Greenewalt produced a similar volume, also titled *Hummingbirds*, which contained 69 outstanding color photographs of 59 species, a relatively technical discussion of flight characteristics, and detailed analyses of the physics of flight and iridescent coloration.

Comprehensive summaries of the biology of hummingbirds are almost as scarce as works dealing with other aspects of these birds. A. C. Bent described the life histories of 18 North American species in 1940, and J. Berlioz (1944) and A. Martin and A. Musy (1959) provided more general approaches. One of the most attractive introductions to hummingbird biology is the 1973 volume by Alexander F. Skutch, *The Life of the Hummingbird*, with splendid color illustrations by Arthur Singer.

Over the past decades there has been a major resurgence of interest in the hummingbirds, particularly in their physiology, foraging ecology, and coevolutionary relationships with bird-pollinated flowers. The last of these areas of research was marked by the appearance in 1968 of *Hummingbirds and Their Flowers*, by Karen and Vernon Grant, but the other topics have yet to be summarized in book form.

Because Bent's materials on the natural history and breeding biology of North American hummingbirds are now nearly 60 years old, and because of the great advances in most areas of knowledge of hummingbirds, I thought a new approach to the hummingbirds of North America would be appropriate. Initially I envisioned the major emphasis as a group of up-to-date species accounts of the North American species. However, I soon recognized the need for summarizing certain comparative information and realized that a few preliminary chapters on comparative ecology, behavior, physiology, and the like might best serve this end. Moreover, these sections could contain information for the non–North American species, which would not be included in the species accounts. Lastly, it seemed desirable to provide a new synopsis of the entire family of hummingbirds, including both common and scientific names and a limited amount of information on the species not given separate descriptive accounts.

The Hummingbirds of North America

PLATE 1. Adult males of (clockwise from lower left) wedge-tailed sabrewing, little hermit, violet sabrewing, long-tailed hermit, and white-necked jacobin visiting *Alpinia purpurea*. *Painting by Mark Marcuson.*

PLATE 2. Adult males of (clockwise from lower left) green-breasted mango, black-crested coquette, dusky hummingbird, fork-tailed emerald, emerald-chinned hummingbird, and rufous-crested coquette (long-crested taxon), perched on *Tillandsia leiboldia*. *Painting by Mark Marcuson.*

PLATE 3. Adult males of (clockwise from lower left) blue-throated goldentail, cinnamon hummingbird, white-bellied emerald, crowned woodnymph, Xantus hummingbird, and azure-crowned hummingbird visiting *Razissa spicata*. *Painting by Mark Marcuson.*

PLATE 4. Adult males of (clockwise from lower left) amethyst-throated hummingbird, sparkling-tailed hummingbird, garnet-throated hummingbird, long-billed starthroat, Mexican sheartail, and stripe-tailed hummingbird visiting *Aechmea nudicaulus*. *Painting by Mark Marcuson.*

PLATE 5. Green violet-ear male and lucifer hummingbird pair visiting *Salvia henryi*. *Painting by James McClelland.*

PLATE 6. Bahama woodstar pair and Cuban emerald male (lower left) visiting *Tillandsia fasiculata*. *Painting by James McClelland.*

PLATE 7. White-eared hummingbird pair (right) and broad-billed hummingbird pair (left) visiting *Bouvardia ternifolia*. *Painting by James McClelland.*

PLATE 8. Violet-crowned hummingbird female (above) and two males visiting *Castilleja miniata*. *Painting by James McClelland.*

PLATE 9. Buff-bellied (above), berylline (middle), and rufous-tailed hummingbird (below) males visiting *Erythrina flabelliformes*. *Painting by James McClelland.*

PLATE 10. Blue-throated hummingbird pair visiting *Zauchernia californica*. *Painting by James McClelland.*

PLATE 11. Magnificent hummingbird male and two females visiting *Mimulus cardinalis*. *Painting by James McClelland.*

PLATE 12. Plain-capped starthroat male and bumblebee hummingbirds (two males, one female) visiting *Agave lechuguilla*. *Painting by James McClelland.*

PLATE 13. Ruby-throated hummingbird pair visiting *Lobelia cardinalis*. *Painting by James McClelland.*

PLATE 14. Black-chinned hummingbird pair visiting *Castilleja chromosa*. *Painting by James McClelland.*

PLATE 15. Anna hummingbird pair visiting *Ribes speciosus*. *Painting by James McClelland.*

PLATE 16. Costa hummingbird pair visiting *Fouquierria splendens*. *Painting by James McClelland.*

PLATE 17. Calliope hummingbird pair visiting *Ipomopsis aggregata*. *Painting by James McClelland.*

PLATE 18. Broad-tailed hummingbird pair visiting *Penstemon barbatus*. *Painting by James McClelland.*

PLATE 19. Rufous hummingbird pair visiting *Ribes sanguineum*. *Painting by James McClelland.*

PLATE 20. Allen hummingbird pair visiting *Aquilegia formosa*. *Painting by James McClelland.*

PLATE 21. "Mexican" variant of long-tailed hermit (1), long-tailed hermit (2), little hermit (3), wedge-tailed sabrewing (4), "long-tailed" variant of wedge-tailed sabrewing (5), plain-capped starthroat (6), long-billed starthroat (7), green-breasted mango (8), white-necked jacobin (9), scaly-breasted hummingbird (10), rufous sabrewing (11), violet sabrewing (12), purple-crowned fairy (13). *Painting by Sophie Webb.*

PLATE 22. Golden-crowned emerald, male, female, and immature (1); broad-billed hummingbird, "doubleday" form (2), and typical form (3); dusky hummingbird (4); Xantus hummingbird (5); crowned woodnymph (6); blue-throated goldentail (7); stripe-tailed hummingbird (8); "blue-tailed" hummingbird (9); berylline hummingbird (10); cinnamon hummingbird (11); rufous-tailed hummingbird (12); buff-bellied hummingbird (13); "white-tailed" hummingbird (14); "blue-capped" hummingbird (15). *Painting by Sophie Webb.*

PLATE 23. Violet-crowned hummingbird (1), "cinnamon-sided" hummingbird (2), green-fronted hummingbird (3), azure-crowned hummingbird (4), white-bellied emerald (5), white-eared hummingbird (6), green violet-ear (7), amethyst-throated hummingbird (8), blue-throated hummingbird (9), magnificent hummingbird (10), garnet-throated hummingbird (11), green-throated mountain-gem (12). *Painting by Sophie Webb.*

PLATE 24. Rufous hummingbird (1), broad-tailed hummingbird (2), calliope hummingbird (3), Costa hummingbird (4), ruby-throated hummingbird (5), black-chinned hummingbird (6), bumblebee hummingbird (7), wine-throated hummingbird (8), sparkling-tailed hummingbird (9), slender sheartail (10), lucifer hummingbird (11), beautiful hummingbird (12), Mexican sheartail (13), short-crested variant of rufous-crested coquette (14), black-crested coquette (15), emerald-chinned hummingbird (16). *Painting by Sophie Webb.*

Plate 1

MARK E. MARCUSON

Plate 2

Plate 3

MARK E. MARCUSON

Plate 4

Mark E. Marcuson

Plate 5

J. McClelland

Plate 6

J. McClellan

Plate 7

J. McClelland

Plate 8

J. McClellan

Plate 9

J. McClelland

Plate 10

J. McClelland

Plate 11

J. McClelland

Plate 12

J. McClelland

Plate 13

J. McClelland

Plate 14

J. McClelland

Plate 15

J. McClelland

Plate 16

J. McClelland

Plate 17

J. McClelland

Plate 18

J. McClelland

Plate 19

J. McClelland

Plate 20

J. McClelland

Plate 21

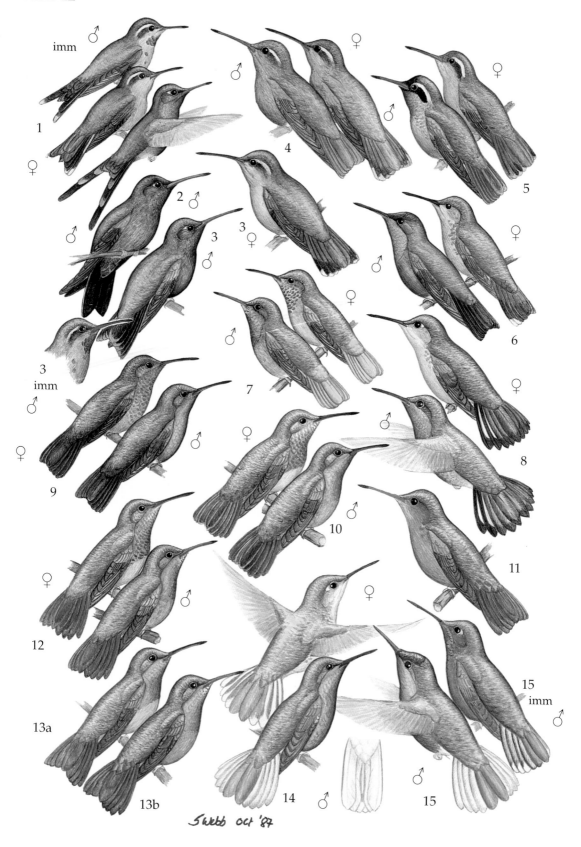

Plate 22

Plate 23

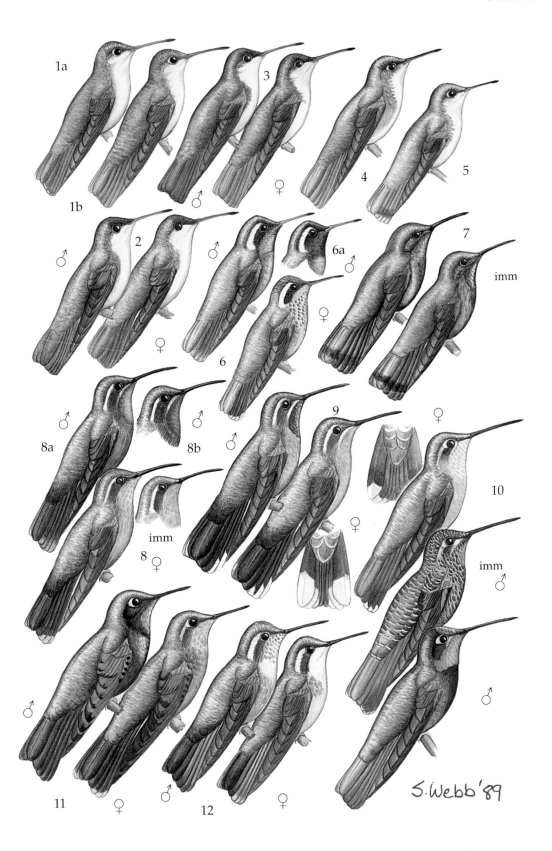

S.Webb '89

Plate 24

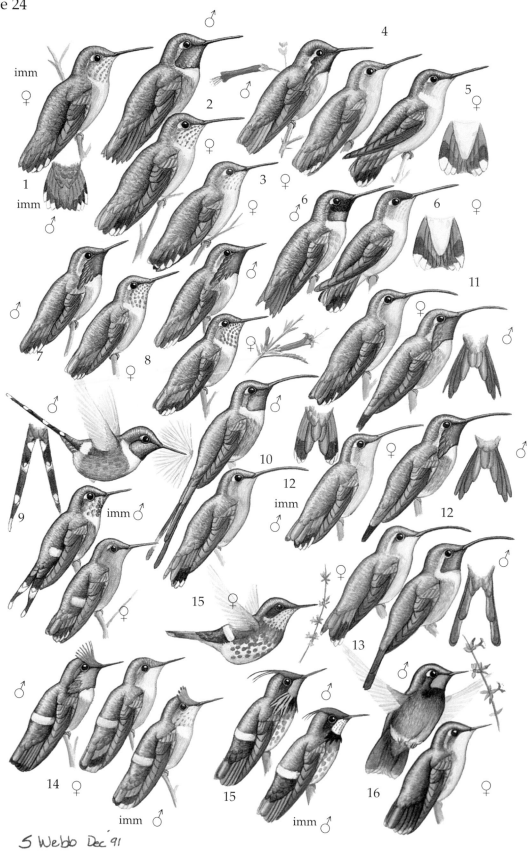

S Webb Dec '91

PART ONE

COMPARATIVE BIOLOGY
OF HUMMINGBIRDS

CHAPTER 1

Classification, Distribution, and General Attributes

Although most bird lovers readily recognize hummingbirds on sight because of their unusually small size, distinctive "humming" flight, and attraction to nectar-producing flowers or feeders, an introduction to the family Trochilidae should begin with a fairly formal definition. In brief, hummingbirds are small (2 to 20 grams) nectarivorous and insectivorous members of the avian order Apodiformes, but differing from others of that group (the swifts) in having long, slender bills and extensile bitubular tongues. They always have 10 primary feathers, 6 or 7 secondaries, 10 rectrices (rarely fewer), and an extremely large sternum. Their feet are small and rather unsuited for walking, but the toes are well developed for perching, with three directed forward and one posteriorly. Their wings are moderately long and pointed, with unusually short arm and forearm bones but relatively long hand bones. They fly by a unique method of rotating the entire wing with little or no wrist and elbow flexing, which makes them capable of forward, backward, and prolonged hovering flight. Adults virtually lack down feathers, and their body feathers are often scale-like and highly iridescent. Sexual dimorphism in adult plumages is common, and non-monogamous mating systems are typical. The entire family is confined to the New World and is mostly tropical in distribution.

This relatively detailed and technical description totally fails to portray the beauty and attractiveness of hummingbirds. Nevertheless, their common names, such as woodstar, starfrontlet, sapphire, emerald, topaz, ruby, sylph, mountain-gem, sunbeam, firecrown, fairy, and sunangel, portray some of the romance that has traditionally endowed the group. In France they are often referred to as *colibri* or by the less attractive name *oiseau-mouche* (fly-birds); in Spanish-speaking countries they are usually called *picaflores* (flower-peckers); and the Portuguese sometimes use the delightful term *beija flor* (flower-kissers). In the Antilles and Guiana they are sometimes called *murmures* (murmurers), *bourdons*, and *frou-frous* by the Creoles; likewise, their native names in parts of Central and South America include a variety of terms of descriptive or metaphorical nature, such as "rays of the sun," "tresses of the day-star," "murmuring birds," and the like (Ridgway, 1890).

Even if hummingbirds were not so beautiful, they would be able to draw our attention simply because probably no other group of birds can lay claim to so many unique or extraordinary characteristics, such as the following:

- They include perhaps the smallest of warm-blooded vertebrates, and have the greatest relative energy output of any warm-blooded animal.
- They are the largest nonpasserine family of birds, and the second largest family of Western Hemisphere birds in number of living species.
- The smaller species have the most rapid wingbeat of all birds (reportedly to 200 per second during courtship, and reliably to 80 per second in forward flight) and are among the fastest fliers of small birds (50 to 65 kilometers per hour in forward flight to 95 kilometers per hour in dives).
- The ratio of their heart size to their body size is the largest of all warm-blooded animals, and their heart rate reaches 1260 beats per minute (second only to some shrews).
- They have the relatively largest breast muscles of all birds (up to 30 percent of total weight), and they are the only birds whose upstroke provides as much power as their downstroke.
- Their plumage is among the most densely distributed of all birds, and their feather structure is among the most specialized; but they have the fewest total feathers of all birds (often less than 1000).
- Their brain size is among the relatively largest of all birds (up to at least 4.2 percent of body weight).
- They have a unique flight mechanism, capable of prolonged hovering and rapid backward flight.
- They are the only birds that regularly become torpid at night, with a drop in body temperature of as much as 19° C; however, their normal body temperature (40° C) is among the highest of all birds.
- Individual hummingbirds often consume more than half their total weight in food and may drink twice their weight in water per day.

Although various regions of the Old World possess nectar-sucking birds, such as the similar sunbirds, these groups do not approach the hummingbirds in their degree of specialization for nectar feeding, nor do they exhibit the great array of color and species diversity for which the hummingbirds are well known. Thus, hummingbirds are perhaps the most notable of all New World bird groups, and, like the birds of paradise of

Australia and New Guinea, they were the source of great interest and folklore by the early explorers of the New World. French colonists of Brazil mentioned them as early as 1558, and in 1671 the nest and eggs of a ruby-throated hummingbird were first described in the *Philosophical Transactions* of Great Britain as a "curiously contrived Nest of a Humming Bird, so called from the humming noise it maketh whilst it flies." The same journal published a description of the bird itself in 1693, in which the author noted that "they feed by thrusting their Bill and Tongue into the blossom of Trees, and so suck the sweet juice of Honey from them; and when he sucks he sits not, but bears up his Body with a hovering Motion of his Wings" (Ridgway, 1890).

By the time of Linnaeus, several species of hummingbirds were known; the 10th edition of his *Systema Naturae* (1758) included descriptions of 18 species known to him, which have since been reallocated to 11 currently recognized species. Thereafter relatively few species were described, until a veritable plethora were discovered in the middle of the 19th century (Table 1). By 1945, when James Peters provided what is still the most widely accepted classification of hummingbirds, the golden age of hummingbird discoveries was essentially over. Nevertheless, in the 52 years since Peters' classification appeared, 13 species of hummingbirds were described (4 more than were described in the previous 35 years). Six of these species were discovered in the 1970s alone, although most of them are of doubtful validity.

Partly because of the great taxonomic problems associated with their classification, it is virtually impossible to state with certainty how many hummingbird species actually exist. Table 2 provides a summary of the total numbers of genera and species of hummingbirds recognized by various authorities since Linnaeus.

Although hummingbirds are usually associated with tropical rainforests, their actual distribution is somewhat at variance with this conception. Certainly the largest number of hummingbird species reside near the equator, but it is the Andes rather than the Amazon basin that support the greatest hummingbird diversity. As may be determined from Table 3, the country that supports the largest number of hummingbird species is Ecuador, followed closely by Colombia. By comparison, Brazil supports only about half as many as either one, in spite of its far greater surface area. If adjustments are made for differences in land areas, then such small Central American countries as Belize, El Salvador, Costa Rica, and Panama all exhibit a diversity of hummingbirds as great or greater than that of most South American areas. Yet, from the vicinity of Panama northward there is a progressive decline in actual species of hummingbirds along a general latitudinal gradient, with the diversity remaining greatest in areas supporting montane forests and other diverse vegetational habitats (see Figures 1 to 3).

A summary of the types shown in Table 3 fails to distinguish the kinds

Table 1

Historical Sequence of Discovery of
Hummingbird Species
(based on taxonomy of Peters, 1945)

Time of Discovery	Number of Species Described	Cumulative % of Total
1758	11*	3
1761–1770	5	5
1771–1780	1	5
1781–1790	17	10
1791–1800	1	10
1801–1810	3	12
1811–1820	15	16
1821–1830	29	25
1831–1840	54	42
1841–1850	79	66
1851–1860	43	79
1861–1870	18	84
1871–1880	14	89
1881–1890	9	91
1891–1900	12	95
1901–1910	7	97
1911–1945	9	100

*The 18 species of Linnaeus represent 11 recognized by Peters.

Table 2

Summary of Genera and Species Accepted by
Various Authorities

Authority	Genera*	Species
Linnaeus, 1758	1	18
Brisson, 1760	2	36
Gould, 1849–61	123	416
Gray, 1861–71	163[†]	469
Elliot, 1878	127 (37)	327
Simon, 1921	188 (84)	489
Peters, 1945	123 (73)	327
Morony et al., 1975	116 (64)	338
Sibley and Monroe, 1990	109 (53)	319

*Monotypic genera in parentheses.

[†]Including subgenera.

of hummingbirds in various regions, as is especially illustrated by the seemingly more "primitive" hermit hummingbirds. This group of generally plain-colored birds with relatively long and often decurved bills are largely associated with equatorial regions and are most abundant in the lowland forests of the Amazon basin. They seem to be more insectivorous than the hummingbird types that have colonized North America, and their bill shapes appear to be highly adapted for extracting insects and nectar from the flowers of such tropical forest plants as *Heliconia*, *Centropogon*, and the like.

Although the hermits comprise a distinctive and reasonably isolated group, which Gould designated as a separate subfamily as long ago as 1861, the remainder of the family is much less readily characterized. Ridgway (1911) listed three subfamilies—the hermits (Phaethornithinae), the coquettes and thorntails (Lophornithinae), and all the remaining species (Trochilinae)—which he distinguished primarily on the basis of the location and relative development of the nasal operculum. Most recent classifications, such as that of Peters (1945), do not contain subfamilies, although Zusi (1980) supported Gould's original recognition of two subfamilies, Phaethornithinae and Trochilinae, with the former more specialized for gleaning arthropods and the latter for nectar-feeding. He

Table 3
Hummingbird Diversity in Various Regions

Country or Region	Hummingbird Species		Area (1000 km^2)	Land Area per Species*	Reference
	Total	No. Endemic			
Argentina	24	0	2,808	177.0	Olrog, 1963
Belize (British Honduras)	19	0	23	1.3	Russell, 1964
Bolivia	63	2	1,079	16.9	de Schauensee, 1966
Brazil	84	22	8,549	101.8	Sick, 1993
Canada	5	0	10,108	2,022.8	Godfrey, 1986
Chile	7	2	744	106.6	Johnson, 1967
Colombia	135	12	1,144	8.3	de Schauensee, 1964
Costa Rica	52	3	42	0.8	Slud, 1964
Ecuador	163	5	276	1.6	Greenewalt, 1960b
El Salvador	21	0	34	1.6	Dickey and van Rossem, 1938
Guatemala	37	0	117	3.1	Land, 1970
Guyana	36	0	216	5.9	Synder, 1966
Honduras	40	1	153	3.6	Monroe, 1968
Mexico	50	8	1,976	39.0	Peterson and Chalif, 1973
Panama	54	3	73	1.3	Wetmore, 1968
Paraguay	12	0	408	33.8	Olrog, 1968
Peru	100	15	1,253	12.5	Zimmer, 1950–53
Surinam	27	0	140	5.2	Haverschmidt, 1968
United States[†]	19	0	5,972	312.0	This work
Uruguay	4	0	187	46.8	Olrog, 1968
Venezuela	97	5	915	9.4	de Schauensee and Phelps, 1978
West Indies	17	13	212	12.5	Bond, 1971

*1000 km^2/species.

[†]Excluding Alaska and Hawaii.

listed five genera (*Glaucis, Threnetes, Ramphodon, Phaethornis,* and *Eutoxeres*) in the subfamily Phaethornithinae, excluding *Doryfera* and *Androdon,* although the latter is convergent with *Ramphodon.* However, Ruschi (1965) noted that, although *Doryfera* fits with the Trochilinae in having a typical ventrally supported cup-like nest, *Androdon* (like *Ramphodon* and the other hermit-like forms) builds a nest attached laterally to the underside of a palm leaf, with hanging streamers below. Thus, the form of the nest seems to be a very good trait for separating the two subfamilies—if one believes that *Androdon* actually belongs with the hermits and *Doryfera* with the typical hummingbirds.

Although the hermits seem to be more specialized for insect-feeding than are the typical hummingbirds, all the species consume a certain proportion of animal food. Indeed, ancestral hummingbirds were probably exclusively insectivorous, probing vegetation and especially flower blossoms for tiny insects. During evolution, their bill and tongue structure likely became progressively more suitable for obtaining nectar as well as

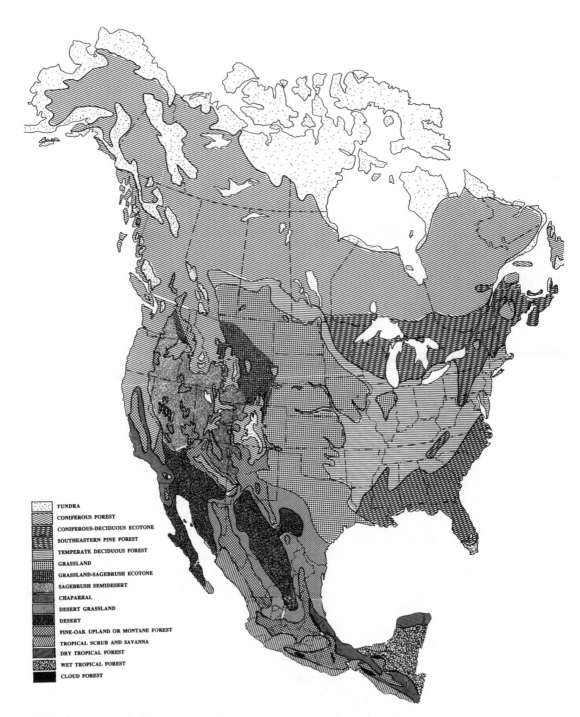

TUNDRA
CONIFEROUS FOREST
CONIFEROUS-DECIDUOUS ECOTONE
SOUTHEASTERN PINE FOREST
TEMPERATE DECIDUOUS FOREST
GRASSLAND
GRASSLAND-SAGEBRUSH ECOTONE
SAGEBRUSH SEMIDESERT
CHAPARRAL
DESERT GRASSLAND
DESERT
PINE-OAK UPLAND OR MONTANE FOREST
TROPICAL SCRUB AND SAVANNA
DRY TROPICAL FOREST
WET TROPICAL FOREST
CLOUD FOREST

1. Major climax plant formations of North America. *(After various sources)*

2. Species-density map of hummingbirds, showing total numbers of breeding or resident species within the enclosed areas. In part after Cook (1969); South American map based on incomplete information and subject to modification. Areas enclosed by lines represent the numbers of species and hummingbirds breeding within each area. Increasing species density toward the tropics is indicated by four levels of overlay patterning.

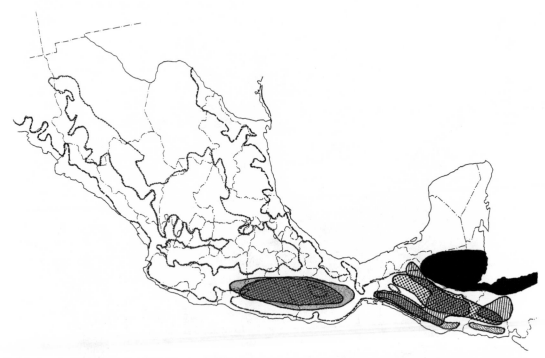

3. Mexican topography and selected hummingbird ranges, including two northern montane species, the beautiful (horizontal hatching) and dusky (cross-hatching); and three Central American species, the rufous sabrewing (light stippling), green-throated mountain-gem (heavy stippling), and slender sheartail (vertical hatching). The common ranges of two Caribbean lowland species that also terminate near the Isthmus, the purple-crowned fairy and scaly-sided hummingbird, are also shown (inked). Contour intervals of 180 meters and 1800 meters are indicated by dashed and dotted lines, respectively. *(Original, by author)*

solid foods. Thus, reciprocal evolution began to occur between plants and birds, as the plants evolved pollination adaptations through developing nectar production and floral structures that both attracted the birds and tended to restrict the range of potential pollinators. Eventually, a special class of bird-pollinated (ornithophilous) plants emerged in the Western Hemisphere, and the hummingbirds were the principal beneficiaries. (The floral adaptations associated with pollinating adaptations such as distinctive coloration, high nectar production, tubular blossoms, and lack of insect-attracting devices are described in Chapter 4.) This process has clearly played an important role in shaping the blossom types in many families of Western Hemisphere plants, as well as affecting the structure of the bill and tongue in nearly all hummingbirds.

The major evolutionary forces that produced the astonishing array and beauty of hummingbird plumages are quite different. Scarcely a single species of hummingbird is not iridescent in at least some feather areas, most generally on the dorsal surface and, in the more highly specialized forms, in the neck region, where an often highly differentiated "gorget" occurs. In some relatively dull-colored forms, such as the hermits, males and females frequently are almost identical in appearance; in the more sexually dimorphic groups, the male invariably exhibits the elaborate crests, trains, and highly colorful plumages.

Apart from the unpigmented white areas, hummingbird feathers possess only two fairly simple melanin pigments, producing blackish or rufous coloration. Light refraction by specialized feathers generates the metallic colors, which range the entire visible spectrum from ruby red to intense violet. Further, the precise angle at which the light strikes the feathers alters the color, so that a feather may shift in apparent coloration with only slight changes in angle relative to the light source. In hummingbirds, the barbules of the iridescent feathers are highly flattened and twisted, so that the microscopic feather surface faces the observer. The melanin granules within the barbules are themselves responsible for the light refraction; in some other bird groups the interference is caused by the outer keratin layer, and the melanin simply prevents the reflection of other colors. The melanin granules of hummingbirds are distinctively shaped into rows of oval platelets, with each platelet containing numerous small air bubbles. The specific colors reflected depend on the thickness of the platelets and the sizes of the enclosed bubbles (Greenewalt, 1960a).

One of the most highly developed attributes of hummingbirds is their tongue (see Figure 7, p. 23). Besides being extremely long, it is capable of considerable extension because of the elongated hyoid apparatus that curls backward and upward around the eye sockets in species so far studied. Although long known to be split, forming a double tubular structure at the tip, the tongue was once believed to be hollow as well. Research (Weymouth et al., 1964) has effectively disproved that contention, although the terminal portions of the tongue are probably able to take up liquids rapidly, perhaps by capillary action. A foraging hummingbird does not keep its tongue continuously in the fluid, as a moth does, but rather rapidly inserts and removes it. Insects do not appear to be sucked up directly, although they may well be caught up in the sticky material adhering to the fringed edges of the tongue. Further, although a few hummingbirds have distinctly serrated edges on the bill, which may facilitate the holding of insects, generally hummingbirds seem incapable of manipulating insects with the bill or even holding them in it for more than a few seconds (Thompson, 1974).

For almost two centuries we have known that hummingbirds feed at least in part on insects, although the actual proportion of animal foods in their diet continues to be debated. For one thing, nectar is very rapidly assimilated and thus is usually absent from the stomach contents of dead individuals. On the other hand, often their stomachs are almost entirely filled with minute insects and spiders, especially among the insect-adapted hermits and their allies. Furthermore, nestling hummingbirds no more than two days old are frequently fed almost exclusively on these items (Ridgway, 1890). The unusually well-developed crop of young hummingbird nestlings nonetheless seems to be equally adapted for holding large volumes of arthropods and fluid foods. The crop is retained into adulthood and, at least in some species, can be expanded to fill the area above the neck vertebrae. Moreover, the entrance and exit openings of the stomach are very close together, so that the gizzard, when filled with arthropods, will not interfere with the passage of nectar.

If the hummingbirds have any deficiencies, one may be their general lack of vocalizations. The voice-producing organ (the syrinx) is located very far forward in the windpipe or trachea, and thus the bronchi are unusually long. Further, the major muscles that originate on the sternum and insert on the trachea of most birds are completely lacking, although two pairs of special intrinsic muscles are present. Thus, hummingbird vocalizations are generally simple and of a twittering and high-pitched nature. One of the few species that has anything like a true song is the tiny vervain hummingbird of the West Indies, which warbles in a weak but sweetly melodious manner for as long as 10 minutes (Ridgway, 1890). Further, the male Anna hummingbird has a fairly elaborate courtship vocalization, as compared with that of most other North American species.

Although hummingbirds may have vocal limitations, their ability to generate instrumental sounds through feather vibrations is certainly substantial, as indeed the name *hummingbird* implies. For example, the attenuated outer primaries of such genera as *Selasphorus, Lafresnaya,* and *Aglaectis* probably are adapted chiefly for sound production, and the unusually narrowed outer rectrices of some species of *Archilochus* and the typical "woodstars" of the genus *Acestrura* may serve similar functions.

In their mating systems the hummingbirds are surprisingly uniform, yet it is exactly this uniformity that may be responsible for the incredible diversity of male plumage patterns and elaborate displays for which the group is noted. In general, the sexes live apart, with little or no association until the breeding season; then the males begin their advertisement displays. They often are dispersed in widely separated breeding territories, but may at times gather in loose courtship groups. "Singing assemblies" composed of a few birds to about 20 or more may develop, with each male defending a small but discrete territory within a larger "lek" area. These assemblies seem most prevalent among the hermits and other essentially forest-dwelling species, where poor illumination requires vocalizations rather than visual displays for the greatest attraction powers. On the other hand, most of the more temperate-climate species tend to display solitarily in well-illuminated areas, which enables the males to exhibit their magnificent gorgets and other iridescent colorations effectively; thus, their weak vocalizations contribute little or nothing to the total display valence.

The male plays virtually no role in the reproductive process beyond fertilization; male participation in incubation or rearing of young has been reported for very few species. The female builds the nest alone, which is normally cuplike and incorporates spider webbing or similar silken materials to bind it to the substrate. She almost invariably lays two eggs, which are white and unusually long and elliptical, and, with rare exceptions, she alone incubates them for 14 to 21 days. The young hatch virtually naked (psilopaedic), helpless, and blind (altricial); they must stay in the nest (nidicolous) until they fledge 18 to 40 days later, depending on food supplies and weather conditions.

CHAPTER 2

Evolution and Speciation

The evolutionary history of hummingbirds is essentially conjectural because there are no fossil remains to use as guidelines. A few fragments of probable swift fossils from the earliest Oligocene strata in France provide us with a rough approximation of the temporal origin of the order Apodiformes. Most scientists believe that hummingbirds and swifts evolved from a common ancestral nucleus, principally because (a) their wing bones have similar proportions; (b) they both lay long, white elliptical eggs, usually in small clutches; and (c) their young are altricial, psilopaedic, and nidicolous.

In contrast to the hummingbirds, swifts feed only on insects, which they capture while in flight. Moreover, swifts have extremely weak legs and feet suitable for perching or clinging, but not for walking. Some swift species build nests on the undersides of palm leaves, making tubular nests of feathers and plant fibers in a manner similar to that of hermit hummingbirds, but swifts secure the nest with sticky saliva rather than with spider webbing. Further, among swifts both sexes typically assist in nest building, incubating the eggs, and brooding. Finally, some swifts and hummingbirds are capable of entering a state of torpidity.

In spite of these considerable similarities, some investigators are uncertain whether the swifts and hummingbirds are truly closely related or whether they have simply undergone convergent evolution toward a similar flight mechanism. After close examination of this question, Cohn (1968) determined that the major skeletal similarities (enlarged sternum, strong wing bones, a short humerus and forearm but greatly elongated hand bones) are somewhat superficial and that actually no linear measurements exist that have the same relationship to body weight in both groups. Thus, the ulna and carpometacarpus of hummingbirds are relatively shorter than the corresponding bones in swifts, but the digits are relatively longer. The trunk and humerus of hummingbirds are also relatively longer than those of swifts, but the hummingbird wings are relatively shorter. In most aspects of the wing and associated skeletal girdle, hummingbirds resemble the perching birds (Passeriformes) and related orders, whereas the swifts resemble the goatsuckers (Caprimulgiformes).

Hummingbirds are unique in that they are able to reverse their primaries during hovering flight through rotary movements at the shoulder and the wrist and in the bones supporting the outer primaries. The swifts, however, which are larger, fly in the same wing-flexing manner as more typical birds, and their wrists are reinforced against rotation (Cohn, 1968). The forearm musculature of both groups achieves a strong fanning action during flight, but in swifts this fanning is apparently associated with speed and maneuverability rather than with development of hovering capabilities. Because of these differences, Cohn suggested that the phyletic relationships of the hummingbirds might be best reflected by erecting a special order, Trochiliformes, for them.

Whether or not hummingbirds are closely related to swifts, their ancestral home must have been South America. Not only is the largest number of species still found there, but also the more insectivorous types (the hermit group) are still essentially confined to that region. A bill type of the form currently found in various hermit species, which combines a moderate degree of elongation for probing into the corollas of flowers with a limited ability for grasping and extracting insects from them, would seem to be the generalized hummingbird bill type. From this beginning, progressively more specialized types have emerged, as co-evolution between plants and particular species of hummingbirds has gradually refined the relationship of bill and tongue structure to specific food sources in nectar-producing plants. In many species of hummingbirds the tip of the tongue is essentially bitubular and probably highly effective in nectar-gathering; in others it is more brush-tipped and evidently effective both in obtaining nectar and in entrapping small insects.

The evolution of a very small clutch-size may have provided the first steps in the shift to a non-monogamous mating system. Selection favoring small body size probably imposed this system and thus improved hovering abilities and survival on limited food resources as the birds became progressively more nectarivorous. Quite possibly the earliest hummingbirds, like the swifts, were monogamous, with both sexes participating in incubation and brooding. Although some swifts lay clutches of up to six eggs, the reduction of the clutch size in hummingbirds to two eggs (or only one, in rare cases) has reduced both the energy drain on the female during egg-laying and the amount of foraging required to feed the developing young during their fledging period. The gradual emancipation of the male from nesting duties may thus have fostered tendencies for successive matings with additional females during the long tropical nesting periods, eventually almost eliminating all pair-bonding tendencies in both sexes.

The origins of flower groups specifically adapted to foraging and pollination by hummingbirds also remain speculative, but they may at least in part have emerged from flower groups previously pollinated largely by bees and butterflies. Plants adapted for pollination by these insects probably already had some of the important attributes, such as daylight blossoming, large showy flower parts, considerable quantities of nectar, and (perhaps) sufficient odor to attract not only bees or butterflies but also tiny insects that exploit the nectar without achieving pollination. To the extent that hummingbirds attracted to these flowers for their insect fauna may have inadvertently achieved pollination, it became progressively more advantageous for the flowers gradually to shift from insect- to hummingbird-pollination mechanisms (by reducing odor production, shifting away from the blue and violet end of the color spectrum for blossom parts, etc.). Thus, the flowers adapted in ways that reduced nectar loss to "illegitimate" insect foragers and (perhaps) increased the total diversity of available pollinators, thereby reducing interspecific pollination possibilities.

The earliest hummingbird forms may well have been largely confined to relatively wooded environments rich in insect life that would thrive among the dense forest vegetation. However, as the birds became progressively more nectarivorous, the hummingbirds probably began to move into edge environments. There, flowering shrubs, vines, and herbs grew abundantly in sunny areas, and individual flowers became more noticeable through large blossom sizes and colors that effectively contrasted the background substrate. At about this time, the males might have become more prone to promiscuous mating systems, and selective pressures began to favor increased male conspicuousness. Thus, greater nectarivory may well have fostered sexual selection for male advertisement devices such as brilliant plumages and conspicuous behavior, especially visual displays like aerial posturing in a well-lighted environment.

In Trinidad, the flowers of trees that are visited by hummingbirds have only a low incidence of red blossoms (4 of 19 species), but those of herbaceous plants and vines have a high incidence of red coloration (9 of 14 species) (Snow and Snow, 1972). Further, the smaller species of hummingbirds in that area tend to forage on plants with fairly small corollas and pale colors; the larger species more often forage on the larger and more colorful blossom types. As Snow and Snow (1972) suggested, the larger species of nectar-adapted hummingbirds may have evolved in parallel with certain flowers, as the latter evolved mechanisms of size or structure that effectively excluded most insects. Even the more insectivorous hermits studied by the Snows apparently preferred to forage at red flowers; therefore the association between red coloration and hummingbird attraction is probably a very ancient one, which apparently is not specifically associated with advanced groups or well-lighted environments. Rather, as has been suggested by various other investigators, it may simply be that red is one of the most effectively contrasting colors against a green background for daylight-foraging vertebrates. It also is essentially the "leftover" portion of the visible spectrum that previously has been unutilized by plants adapted to pollination by bees, butterflies, and other diurnal insects, whose visual spectrums scarcely reach the red region. Red thus became an ideal device for ornithophilous flowers to achieve maximum visibility to hummingbirds, while at the same time making the flowers relatively inconspicuous to rival nectar-feeders such as bees and butterflies.

Trying to discern trends in the plumage patterns of hummingbirds can be dangerous, but some knowledge has been gained. In comparison with the hermit group, the "typical" hummingbirds of the subfamily Trochilinae exhibit not only a greater tendency toward sexual dimorphism, but also a greater probability of having reddish iridescence in their plumage. Indeed, males of nearly all of the species that have colonized in North America can be considered "flame-throated" to varying degrees, with a well-differentiated iridescent gorget that varies from red to violet or purple in its predominant coloration. Given the obvious visual impact of red on the sensory system of hummingbirds, it is curious indeed that the distribution of this "high-valence" color is so restricted in the group. Thus, many hummingbirds are mostly emerald or sapphire-green in coloration, but almost none has an area of red greater than that of the gorget itself. One of the few species that seems to fall in this category is the crimson-topaz hummingbird, in which the gorget itself is red only in the female. Similarly, hummingbirds lack the bright red carotinoid pigments commonly found in sunbirds and various other passerine groups. Perhaps red is such a powerfully attractive color in hummingbirds that it must be restricted to only a few specific areas, such the gorget. There the bird can effectively expose or hide it at will, thus avoiding constant aggression among the males and also providing directed, periodic exposure to others at appropriate times.

The evolutionary diversity of the hummingbirds is so great, and the species that have been closely studied anatomically or behaviorally are so few, that detailed presentations of probable evolutionary relationships below the subfamily level are impossible at present. There are approximately 300 species within the subfamily Trochilinae alone, with a still uncertain number of genera that should be recognized. This book covers 47 species and about 20 of these genera, thus providing only the slightest taste of evolutionary diversity within the entire hummingbird group. However, here I present (Figures 4 and 5) a highly tentative and greatly oversimplified diagrammatic representation of the possible relationships among them, in my judgment, to point out some of the obvious ways in which the groups seem to differ from one another and may be partially characterized at the generic levels. Much more detailed descriptions of

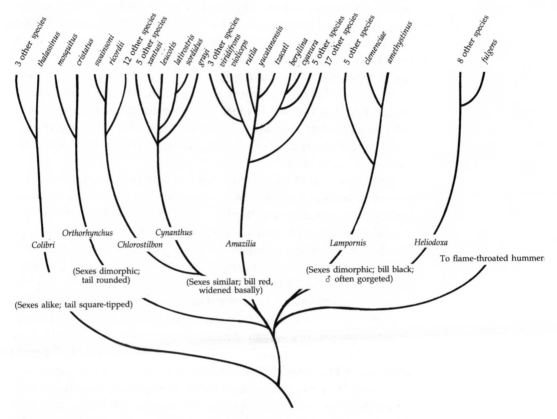

4. Dendrogram of presumptive relationships among North American hummingbirds, exclusive of the flame-throated species.

each genus are provided in Part Two of this book; the diagrams here are for general information only.

In Figure 5 are the presumed affinities of 7 genera and 11 species of predominantly Central and South American trochiline hummingbirds having populations that usually only barely reach the North American borders or are only accidental visitors to the United States. Excluded are more than 60 genera and well over 100 species, which occur in linear taxonomic sequence between *Colibri* and *Heliodoxa* but have not been reported within the U.S. boundaries. Thus, it is a very simplified dendrogram indicating highly speculative relationships.

Figure 6 illustrates the seemingly more advanced groups that have the center of their breeding distributions in Central America, with a few species reaching South America and many breeding regularly within the United States. This dendrogram includes all of the genera and species occurring in linear taxonomic sequence between *Heliomaster* and *Selasphorus*, even if they do not reach the U.S. boundaries. Nevertheless, this too is a quite hypothetical and tentative diagram, not to be interpreted literally.

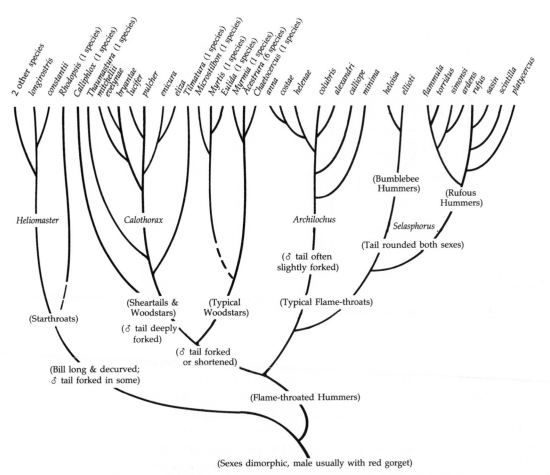

5. Dendogram of presumptive relationships among North American flame-throated hummingbirds.

Although Part Two discusses the probable evolutionary affinities of
each species described in a separate account, coverage of some general
questions of evolution and speciation patterns in North American hummingbirds is appropriate here. If the suggested phylogeny in Figure 5 is
acceptable, then North America has been colonized by a group of hummingbirds with a distributional center primarily in the highlands of Central America. They are generally smaller than the more southern-adapted
forms, and they have fairly short beaks that are relatively straight and
allow for foraging from a wide diversity of nectar-producing plants.
There is a rather surprising uniformity in bill length of the North American species of both *Selasphorus* and *Archilochus*, corresponding with a
great apparent potential for different species to forage on the same food
plants, as well as for a single species to forage from a wide diversity of
food plants. Thus, Austin (1975) noted that the ruby-throated hummingbird has been observed foraging on at least 31 plant species that represent 18 different families, and that at least 19 species of eastern North

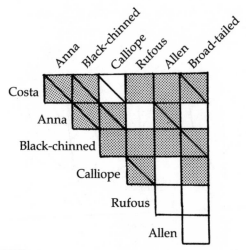

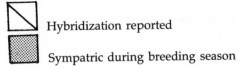

Hybridization reported

Sympatric during breeding season

6. Summary of hybridization records of North American flame-throated hummingbirds.

American plants have undergone evolutionary changes favoring pollination by hummingbirds. This is a remarkable statistic, since in the course of hummingbird/flower evolution a fairly specific plant–bird relationship was probably favored to minimize interspecific pollination probabilities. Part of the answer to this anomaly may be that over much of eastern North America the only available bird pollinator is the ruby-throated hummingbird, and thus numerous plants have adapted to this single species. In turn, it has retained a rather generalized bill shape that allows exploitation of all of these flower types, thus maximizing the potential geographic breeding range of the species.

The western states contain a much larger total number of ornithophilous plant species as well as a much greater diversity of hummingbirds, as documented by Grant and Grant (1968). Here, speciation in the hummingbirds has resulted in the evolution of seven endemic species of breeding hummingbirds of the "flame-throated" group (broad-tailed, rufous, calliope, black-chinned, Costa, Allen, Anna). These species exhibit limited but varying amounts of geographic overlap during the breeding season and exploit the same or similar food plants in at least some parts of their ranges. Correspondingly, their bill lengths differ only by a few millimeters in average length, and all possess essentially identical bill shapes. Interspecific territoriality is fairly common in situations where more than one species coexist (Pitelka, 1951a); likewise hybridization among the species is relatively common. At least 9 interspecific hybrid combinations occur among these 7 species, which represents almost half of the total 21 mathematically possible combinations. Further, at least 6 of the undescribed combinations are impossible or unlikely on the basis of breeding range configurations; 6 other possible hybrids may exist on the basis of known breeding ranges, but have yet to be described (Figure 6). The single reported hybrid specimen involving the calliope and Costa hummingbirds is an unexplained anomaly. Although it was reportedly taken in California (Banks and Johnson, 1961), there are no areas of probable breeding overlap either in that state or elsewhere, except possibly in Nevada. This surprisingly high incidence of hybridization implies that the differences in male plumage do not always serve as absolute isolating mechanisms; rather, the close similarities in the plumages of the females are likely to be a much better index to their actual relationships. Furthermore, given this potential for hybridization and interspecific competition for limited resources, each of the western hummingbirds should show a tendency for ecological isolation from its relatives. This indeed seems to be the case (see Chapter 4), although local contacts between two species are certainly prevalent in some areas, and rarely as many as three hummingbirds may breed in fairly restricted areas (Pitelka, 1951b).

CHAPTER 3

Comparative Anatomy and Physiology

From their iridescent feathers to their tiny skeletons, the hummingbirds offer an amazing number of specializations that can be matched by few if any other bird groups. On the basis of their wings alone, the hummingbirds are preeminent among birds. The remarkable elongation of the hand bones and the associated length of the ten primary feathers are examples. The secondary feathers, however, are greatly reduced in both length and number. There typically are only six feathers, but sometimes a rudimentary seventh is present. The rectrices are also almost invariably ten, although one species—the marvelous spatuletail—has only four, two of which are highly specialized.

The skeletal characteristics (Figure 7) of hummingbirds exhibit many obvious specializations for their unique mode of flight. The sternum, which is greatly enlarged in comparison with more typical flying birds (Figure 8), is also deeply keeled, and the eight pairs of ribs (versus six in most land birds) protect it during the great stresses of hummingbird flight. Further, the unusually strong coracoid is attached to the sternum by a shallow ball-and-cup socket, in a manner that is unique to hummingbirds and swifts (Ridgway, 1890).

The tongue of a hummingbird is essentially as long as the species' bill, yet, by virtue of the elongated hyoid bones, it can be greatly extended from the tip of the bill, thus increasing its effective length for foraging in deeply tubular flowers. For about the anterior half of its total length, the tongue is divided into two separate units, which are fringed along the membranous outer edges, probably in conjunction with the amount of insect foods usually consumed (Figure 7). The bill itself varies in length from less than a centimeter in *Ramphomicron* to sometimes more than 10 centimeters in *Ensifera;* it is often slightly decurved and is very rarely recurved. In a few genera, such as *Heliothryx* and *Schistes,* the bill is greatly compressed laterally, forming a fine tip when viewed from above. In nearly all hummingbirds the nostril is to some extent covered from

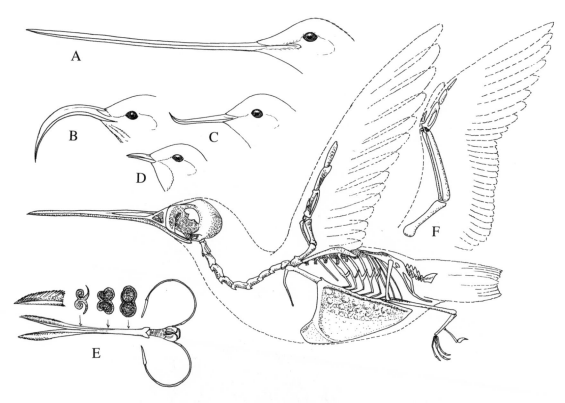

7. Skeleton, bill, and tongue characteristics of hummingbirds, including bills of (A) Andean swordbill, (B) white-tipped sicklebill, (C) avocet-bill, and (D) thornbill. The (E) tongue of the green-throated carib is also diagramed. The wing of a typical flying bird (duck)—drawn to the same scale—is included for comparison (F). *(Partly after Ridgway, 1890)*

above by a distinct shelf-like scale, or operculum. Whether this is related to protection of the nasal cavity from inhaling pollen and the like or has some other unrelated function is unknown.

In nearly all species, the primary feathers are progressively longer from the inside outwardly, and in all but a very few the outermost primary is the longest. The exceptions have an outermost primary that has obviously been modified in structure to generate sound, yet in the streamertail it is shorter than the second primary but not specialized in shape. In some genera, particularly *Campylopterus* and *Aphantochroa*, the shafts of one to three of the outermost primaries are greatly thickened in males, possibly related to a general strengthening of the outer primaries or perhaps for some unknown display purposes.

Although there is remarkable consistency in the number of tail feathers, the shape of the tail varies substantially among hummingbirds, with differential lengthening of the feathers resulting in forked, scissor-like, wedge-shaped, rounded, graduated or pointed tails, or other variations (Figure 9). In several species the outer tail feathers are appreciably narrower than the more centrally located ones, and these probably are set into vibration during aerial display; the Costa and Anna are examples

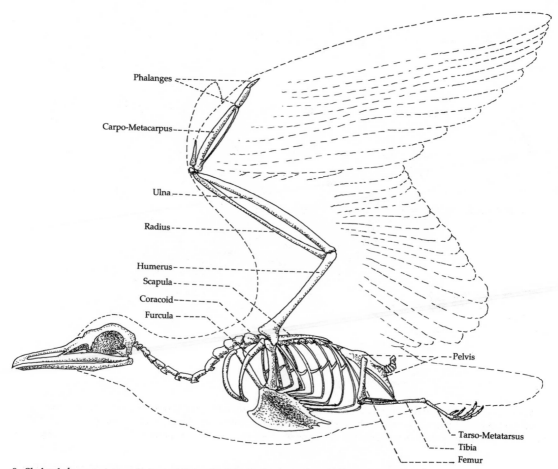

8. Skeletal characteristics of a typical flying bird (gull), for comparison with Figure 7.

among North American species. On the other hand, in the rufous male
the pair of rectrices adjacent to the middle pair are curiously notched
near the tip. They possibly also generate special sounds, although Baptista and Matsui (1979) suggested that, in the case of the Anna, the dive-
noise is largely or entirely vocal in origin.

Generally, neither the remiges nor rectrices of hummingbirds are
extensively iridescent, perhaps because the structural specializations in
the barbules responsible for this coloration seem to inhibit the ability of
the separate feather barbs to "knit" together and maintain an unbroken
airfoil when subjected to stresses. However, with the exception of these
feathers, nearly all of the other feather regions of hummingbirds are relatively iridescent. Their body feathers are also extremely small and closely
packed. Aldrich (1956) reported that a male Allen hummingbird had
1459 feathers and a female had 1659. By comparison, a brown thrasher
(*Toxostoma rufum*) had 1920, but this species has a skin surface area about
ten times larger than a ruby-throat; thus, the density of feathers in the
hummingbird is approximately five times greater (Greenewalt, 1960a).

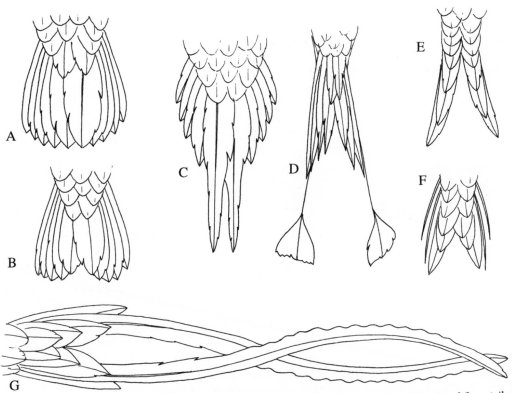

9. Variations in tail shapes of hummingbirds, including (A) rounded, (B) double-rounded, (C) pointed (long-tailed hermit), (D) racket-like (racket-tailed coquette), (E) scissor-like (amethyst woodstar), (F) forked (white-bellied woodstar), and (G) streamer-like (streamertail). *(After Ridgway, 1890, 1911)*

The highly iridescent feathers of the hummingbird gorgets are among the most specialized of all bird feathers. But even in the male's gorget of a species such as the Allen hummingbird, only about the distal third of each feather is modified for iridescence; the close overlapping of adjacent feathers thus generates the unbroken color effect. The iridescence is produced by the proximal parts of the barbules, which are smooth, flattened, and lack hook-like barbicels or hamuli. Beyond the color-producing portion, the barbule is strongly narrowed and curved toward the distal tip of the feather (Figure 10). The barbicels in this area help to hold together the barbules on one side of the barb, but do not unite the barbules of adjacent barbs (Aldrich, 1956).

Greenewalt (1960a) reviewed at length the aspects of optical theory required for an adequate understanding of iridescence in hummingbird feathers; the following is a brief overview: First, the colors do not directly depend on selective pigment absorption and reflection, as do browns and blacks produced by the melanin pigments of non-iridescent feathers. Rather, they depend on "interference coloration," such as that resulting from the colors seen in an oil film or a soap bubble. Basically, the colors

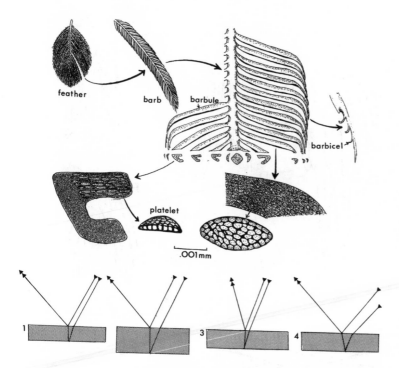

10. Iridescent feather structure of a hummingbird feather, based on photographs in Greenewalt (1960a). Below, diagrams of light pathways of optical films having a higher refractive index than that of air, showing effects of variations in film thickness (1 and 2) and in angle of viewing (3 and 4) on reinforcement of particular wavelengths. (*In part after Greenewalt, 1960a*)

depend on light being passed through a substance with a different refractive index than that of air (1.0), and being partially reflected back again at a second interface. The percentage of light that is reflected back increases with the difference in the refractive indices of the two media; in addition, the thickness of the film through which the light is passing strongly influences the wavelengths of light that are reflected back. Put simply, red wavelengths are longer than those at the violet end of the spectrum and generally require films that are thicker or have higher refractive indices than those able to refract bluish or violet light. Thus, the optimum refractive index for red feathers is about 1.85; for blue feathers it is about 1.5.

Hummingbird feathers may attain any refractive index within this range because the iridescent portions of the barbules are densely packed with tiny, tightly packed layers of platelets. These platelets are only about 2.5 microns in length and average about 0.18 microns in thickness, but they vary in thickness and are differentially filled with air bubbles. The platelet's matrix, probably of melanin, evidently has a refractive index of about 2.2, whereas the air bubbles inside have a refractive index of 1.0. Varying the amount of air in the platelets provides a composite refractive index that ranges from the red end of the spectrum (1.85) to the blue (1.5). An analysis by Greenewalt (1960a) indicated that the theoretical structure of the plates should have an "optical thickness" equal to one-half of the particular color's peak wavelength, or an actual thickness equal to the optical thickness divided by the average refractive index. If, then, half the wavelength of red light (0.6 to 0.7 micron) should be divided by the average refractive index, "red" plates should be about 0.18 micron in average thickness, or the same as the actual thickness indicated by electron microscope examination.

Thus, the actual thickness of the platelets not only significantly determines the quality of the perceived light, but it also affects the amount of air held within the pigment granules and the consequent variations in interference effects. Further, a single pigment granule can produce different color effects according to the angle at which it is viewed. When an optical film is viewed from above, it reflects longer wavelengths than when viewed from angles progressively farther away from the perpendicular (Figure 10). Thus, a gorget may appear ruby red when seen with a beam of light coming from directly behind the eye, but as the angle is changed the gorget color will shift from red to blue and finally to black, as the angle of incidence increases (Greenewalt, 1960a).

In hummingbirds, the color-producing pigment platelets are closely packed into a mosaic surface, and 8 to 10 such layers are then tightly stacked on top of one another in typical iridescent feathers. Far from confusing the visual effects, such stacking actually tends to intensify and purify the resulting spectral color, which is probably why hummingbirds have possibly the most intensively iridescent feathers known in birds (Greenewalt, 1960a).

Iridescence is only one of the many respects in which hummingbirds outdo other birds. When flying, all birds expend a high amount of energy. In most strongly flying birds, the two pairs of muscles originating on the sternum's keel occupy about 15 to 25 percent of total body weight, but in hummingbirds these muscles comprise 25 to 30 percent. Further, the muscle that elevates each wing is approximately half the weight of that responsible for the downstroke; it is usually only 5 to 10 percent in more typical birds (Greenewalt, 1960a, 1962).

The unusually large size of the muscles that elevate the wings is related to the fact that hummingbirds generate power during the upstroke as well as the downstroke, with the wings operating like a variable-pitch rotor. The pitch during the upstroke can modify the thrust generated on the downstroke, so that hummingbirds are readily capable of forward, hovering, and backward flight. These remarkable abilities are achieved by a combination of rotary movements of the outer portion of the wing as well as changes in the plane of the wing movement. Thus, during forward flight the tips of the wings describe a vertical oval in the air, not very different from that of other flying birds. However, during hovering the wings are moved in a manner approaching a horizontal figure eight, with the plane of the movement essentially parallel to the horizon (Figure 11). By a slight backward tilting of the plane the bird can easily move upward and backward, and likewise by tilting the plane downward the bird can begin to proceed slowly forward, much in the manner of a helicopter (Greenewalt, 1960a).

11. Stages in wing action during hovering flight in a hummingbird, in side and dorsal views. Numbered points on the central diagram indicate location of wingtip at successive stages. *(Modified from Greenewalt, 1960a)*

The wings of hummingbirds therefore act like aerial "oars," lacking the flexing at the wrist and forearm joints typical of other birds and operating like mechanical oscillators that have constant-speed motors. Thus, they have a close relationship between wing length and the rate of wing beats, as do other flying birds as well as insects. In fact, hummingbirds appear between insects and other groups of birds when their wing length is plotted against their rate of wing beats, or when their wing length is plotted against average body weight (Greenewalt, 1960a). Their wing length is relatively long relative to body size, and on the average body weight is proportional to the 1.5 power of the wing length (Greenewalt, 1962). Average adult hummingbirds weigh from about 2.1 grams in the vervain species (Lack, 1976) to a maximum of about 22 grams in the giant hummingbird (Lasiewski et al., 1967). The wingbeat rate of the former (as well that of the slightly smaller bee hummingbird) is still undetermined, but in the very slightly larger amethyst woodstar it averages 80 per second (Greenewalt, 1960a). However, that of the giant humming-

bird ranges from about 10 to 15 per second. In spite of this species'
remarkably large size, it is perfectly capable of controlled hovering and
backward flight (Lasiewski et al., 1967).

Plots of frequency distributions of average body weights and average
adult wing lengths of hummingbirds reveal some interesting relationships. Table 4 presents the range of adult weights of 166 species as determined by Carpenter (1976), as well as the average adult wing lengths of
nearly 300 hummingbirds, as obtained from museum specimens and
extracted from literature. The latter total represents about 85 percent of
the known species of hummingbirds (data on the remaining species
were not readily available to me). According to the table, hummingbirds
exhibit an apparent "adaptive aerodynamic peak" in weight at about
4 grams. Likewise there is a distinct peak for wing length between 55 and
60 millimeters, although ecologic factors may dictate other size constraints. Near the lower end of the scale, the average weights of female
hummingbirds are somewhat greater than those of males, but among the
larger species of hummingbirds the reverse is true. This may be because
the effects of sexual selection in the larger species favor male dominance,
whereas in the relatively small species the energy drain of egg-laying
has fostered selection favoring larger female weights as compared with
males. Thus, the female vervain hummingbird lays a clutch of two eggs
of about 0.37 grams each, and their combined weight equals 34 percent
of the average adult weight (Lack, 1976). In the larger species the relative
egg weight is considerably less, and the females spend less energy in
caring for the relatively smaller young (Brown et al., 1978).

Table 4
Average Wing Lengths and Body Weights in the Family Trochilidae

Wing Length (millimeters)	Number of Species*	Percentage of Species	Body Weight (grams)	Number of Species[†]	Percentage of Species
29–34	10	3.4	1.5–2.5	13	7.8
35–39	22	7.4	2.5–3.5	31	18.6
40–44	26	8.8	3.5–4.5	34	20.5
45–49	29	9.8	4.5–5.5	24	14.5
50–54	36	12.2	5.5–6.5	27	16.2
55–59	48	16.3	6.5–7.5	14	8.4
60–64	39	13.2	7.5–8.5	16	9.6
65–69	33	11.2	8.5–9.5	5	3.0
70–74	27	9.2	> 9.5	2	1.2
75–79	15	5.1			
80–84	4	1.3			
85–90	2	0.7			
> 90	3	1.0			

*Based on a survey of 294 species sampled from the literature and museum specimens.

[†]Based on a survey of 166 species sampled by Carpenter (1976), numbers estimated from graph.

Thus, the hummingbirds seem to have evolved along a very narrow evolutionary corridor, which has kept their body weights low enough to meet the energy requirements of their unique flight and associated hovering abilities and has prevented them from becoming larger than their limited food supplies would support. They have also been limited by constraints at the lower end of their potential body size by problems associated with regulating body heat; limits on miniaturization of body parts such as structural strength of bones and minimum brain size; and increasing vulnerability to all sorts of predators, including large insects.

Hummingbirds consume large amounts of oxygen, especially during flight, which often places their circulatory systems under special strains. They have the largest known relative heart size of all birds—up to 2.4 percent of their body weight (in the rufous-tailed)—and likewise the most rapid heartbeat in birds—1260 beats per minute (in the blue-throated hummingbird). Similarly, the density of their erythrocytes is the highest known among birds (6.59 million per cubic milliliter), perhaps as a result of the cells' unusually small size, which is associated with efficient gas-transport capabilities. Body temperatures of active hummingbirds are generally close to 40° C, occasionally reaching as high as 43° C when struggling but restrained (Lasiewski, 1964; Morrison, 1962). The breathing rate in hummingbirds is also very high: approximately 250 per minute for a 3-gram hummingbird at rest.

The rate of metabolism of even a resting hummingbird, as measured by its oxygen consumption, is about 12 times greater than that of a pigeon (*Columba livia*) and 25 times greater than that of domestic fowl (Welty, 1975). A human, metabolizing energy at the rate of a hummingbird, would have to consume roughly double his weight in food such as meat every 24 hours, or about 45 kilograms of pure glucose, and his body temperature would rise to more than 400° C (Scheithauer, 1967). Hummingbirds and the comparably sized shrews among mammals have evidently reached the smallest sizes that are physiologically possible for warm-blooded animals; smaller animals simply could not eat enough food to avoid starvation.

Small birds endure a greater rate of heat loss from the body than do large birds; likewise, small birds are relatively more subject to overheating when placed in an environment that is warmer than their body temperature. Even temperate-zone hummingbirds lack down feathers; therefore, they are relatively unable to increase their insulation effectiveness by feather fluffing during exposure to cold. Thus, they can maintain body temperature only by increasing their heat production. Working on marginal energy balances, hummingbirds cannot accomplish such maintenance over a prolonged period; so, rather than make the attempt, many enter a state of torpor. The metabolic rate may then drop to about one-fiftieth of the basal rate at normal body temperature, and the rate of water loss by evaporation decreases by one-third to one-tenth as compared with that at normal body temperature; thus torpidity may be important as a water-conservation device. The rates of entry into and emergence from torpor are inversely related to the bird's body size: The smaller the hummingbird, the more rapidly it may enter or emerge from a torpid state.

During torpidity, the heartbeat rate varies with body temperature, from 50 to 180 per minute, and breathing becomes irregular, with long periods of nonbreathing at lower temperatures (Lasiewski, 1964). In many torpid birds the body temperature tends to approximate that of the environment, and in species such as the poorwill (*Phalaenoptilus nuttallii*) lowering the external temperature does not stimulate regulatory processes that prevent the body temperature from either falling further or from arousing. However, a few hummingbird species maintain a minimum body temperature of 18 to 20° C below that of the normal resting temperature in spite of even lower air temperatures. Apparently the regulated level of body temperature during torpidity is related to the minimum environmental conditions that are encountered under normal conditions in the wild (Wolf and Hainsworth, 1971). A study of several species of South American hummingbirds found that average body temperatures during daytime activities were about 39° C, with an increase of 2.2° C during maximum activity and a decrease of only 4° to 5° C during deep sleep or torpor conditions. Tropical species of hummingbirds thus seem more sensitive to a lowering of body temperature than are more temperate-adapted species (Morrison, 1962).

We have long wondered how species such as the ruby-throated hummingbird store enough energy to allow them to migrate across the Gulf of Mexico, a minimum overseas distance of about 800 kilometers, given their limited capacities for fat storage and the high rate of energy utilization during flight. Although earlier calculations suggested that the maximum flight range of this species might be about 616 kilometers, or far below the necessary minimum to make the flight, more recent estimates by Lasiewski (1962) have prompted new conclusions. He estimated that the average adult ruby-throat can store and use 2 grams of fat, which would be enough for about 26 hours flight in males and 24.3 hours in the larger females. Given an average air speed of 40 kilometers per hour, the maximum flight range of a male would thus be 1040 kilometers, and that of a female 975 kilometers—more than enough to make a nonstop flight across the gulf. This was based on the estimate that the birds would burn their energy reserves at the rate of 0.69 and 0.74 Calories per hour for the male and female sexes, respectively, and the fact that a gram of fat has an energy content of 9.0 Calories.

A hummingbird must spend a good proportion of its waking hours gathering food simply to stay alive. One wild male Anna hummingbird requires a minimum of 7.55 Calories (assuming torpidity at night) to 10.32 Calories (assuming sleep at night) during a 24-hour period. An average daily period of 12 hours and 52 minutes of activity would require an energy expenditure of 3.81 Calories in perching, 2.46 Calories in nectar flights, 0.09 Calories in insect-catching, and 0.30 Calories in defense of territory. The nectar production of about 1022 fuschia flowers could supply this daily need (Pearson, 1954). For the hummingbird to spend the night without becoming torpid, it must consume a substantial surplus of energy during the hours of activity; one study on the rufous-tailed hummingbird estimated that an excess daily intake of 4.07 Calories would enable the bird to survive the night without going into torpor (Schuchmann et al., 1979). Studies on a few species of hummingbirds indicated daily energy budgets of 7.78 to 12.4 Calories per 24 hours. These energy budgets are 3.1 to 3.6 times the standard ("basal") metabolic rates of the individual species, or considerably greater than figures obtained for comparable nectar-feeders among passerine birds (MacMillen and Carpenter, 1977). For example, Powers and Nagy (1988) estimated the daily field metabolic rate of an Anna's hummingbird weighing 4.48 grams to be about 32 kilojoules (or 7.65 Calories), which amounts to 5.2 times the estimated basal metabolic rate. During daytime hours the rate averaged 6.8 times the basal metabolic rate, but this dropped to 2.1 times greater during nighttime hours. The latter rate is similar to what might be expected in a resting bird experiencing normal body temperatures, but might higher than would be expected if the bird were in a state of torpor. Estimated daily intake of water was 164 percent of body mass (not eight times greater, as has been suggested in early literature). It is likely that enough sucrose is stored in the crop at night to provide a supplemental "energy storage depot" for the birds to draw upon for their nocturnal energy requirements (Powers, 1991).

CHAPTER 4

Comparative Ecology

An enormous number of ecological factors are associated with several hundred species of hummingbirds and their foraging adaptations, as well as related adaptations of the hundreds of species of plants that have coevolved with hummingbirds with respect to mechanisms favoring efficient pollination. The relationship between plants and hummingbirds is an old and close one; in western North America alone about 130 species of plants exhibit features apparently modified through evolution for foraging and pollination by hummingbirds (Grant and Grant, 1968). Another 20 species or more of eastern North American plants have been similarly affected (Austin, 1975), so that at least 150 species of North American flowering plants exhibit an "ornithophilous syndrome" (van der Pilj and Dodson, 1966).

Plants exhibiting these adaptations normally bear large flowers that are solitary or loosely clustered in a horizontal or pendant position, usually at the tip of flexible pedicels. The flowers are often red or red and yellow, holding large quantities of nectar at the base of a long, stout floral tube; and the corolla is often thickened or otherwise modified to protect it from accidental piercing by the bird's beak or from nectar thievery by nonpollinators. The plants typically bloom during daylight hours, have little or no scent, and have projecting stamens and pistils that are likely to intercept the crown of the visiting pollinator. They also lack "landing platforms" suitable for nectar-drinking competitors such as bees, and may have other devices that tend to exclude visits by these and other nonpollinating nectar-drinkers, such as butterflies and moths (Grant and Grant 1968).

Van der Pilj and Dodson (1966) have listed some of the typical features of ornithophilous flowers as compared with those adapted for butterfly and bee pollination. As shown in Table 5, there are greater similarities between blossoms pollinated by hummingbirds and butterflies than between those adapted to birds and bees.

Table 5

Typical Characteristics of Flowers Pollinated by Birds, Butterflies, or Bees (after various sources)

Characteristic	Bird-Pollinated	Butterfly-Pollinated	Bee-Pollinated
Flowering time	diurnal	diurnal	diurnal
Flower shape	weakly zygomorphic or radial	radial or zygomorphic	zygomorphic, usually with landing platform
Blossom color	vivid, often red	vivid, sometimes red	variable but not red
Odor	none	sweet and fresh	sweet and fresh
Nectar	very abundant, in broad tubes	abundant, in narrow tubes	sparse, usually hidden
Flower position	horizontal or hanging	erect	horizontal
Petal position	often recurved	seldom recurved	not recurved, thus allowing for landing

According to Grant and Grant (1968), most of the hummingbird-adapted plants of western North America are either perennial herbs or softwood subshrubs, with a few trees and very few annual herbs as well. Many of the plants are dicotyledons with fused corollas, and the vast majority are red or at least partially red. The Grants concluded that the most common ancestral condition for the hummingbird flowers of North America is a bee-pollinated system, with only a few genera from ancestral groups where lepidopteran (butterfly or moth) pollination typically occurs. Further, a few ornithophilous genera are centered in subtropical or tropical America and probably have been associated with birds for long periods. These presumably have followed the hummingbirds northward to occupy their current ranges in western North America.

In a review of the effect of the evolution of ornithophily on flowers, Stiles (1978) noted that, because this syndrome is energetically expensive for plants, it should occur only when birds provide the optimum vehicle for pollen flow. Pollinating adaptations exhibited by plants all revolve around nectar secretion and the specific manner of presenting it to birds or other pollinators. In dark habitats, plants may produce enough nectar to make bird pollination profitable; similarly, annual plants would benefit less from long-lived pollinators such as birds than would perennials. Likewise, trees benefit relatively little from hummingbirds, since such a rich nectar source would be divided up into small foraging territories, thus reducing pollen flow from tree to tree. On the other hand, many epiphytes are hummingbird pollinated; the small plant sizes associated with the epiphytic habit tend to ensure outcrossing even in territorial species of hummingbirds. Generally, hummingbird-pollinated flowers tend to bloom over a greater proportion of the year than do insect-pollinated flowers, a tendency which may help to stabilize the presence of the pollinator in the community. Many plant species have flowers that are specialized for hummingbird pollination, but usually not by specific bird species. This general rather than precise structural correspondence and codependence may help to buffer the ecosystem against sudden population fluctuations of any single bird or flower species (Stiles, 1978).

In a study area in the White Mountains of Arizona, nine species of hummingbird flowers coexist, all strongly convergent in flower color, size, and shape. Competition among them for pollination is evidently reduced by differences in orientation of anther and stigma location, so that different parts of the bird transport the pollen. Many of the flower species secrete nectar at similar rates, but the local population of cardinal flower produces no nectar at all, and instead attracts hummingbirds by mimicking the more abundant nectar-producing species (Brown and Kodric-Brown, 1979).

As Grant and Grant (1968) noted, the usual relationship between corolla length and bill length of the typical hummingbird visitor is only a general one; nevertheless, there is often a close relationship in both corolla length and corolla shape in these structures (Figures 12 to 14). Brown and Kodric-Brown (1979) noted that seven of the nine species of hummingbird-pollinated flowers they studied had very similar corolla length (20 to 25 millimeters), and the average culmen lengths of the three most common hummingbird visitors ranged from 15.7 to 20.2 millimeters.

A striking case of coevolution between bill length and corolla length in hummingbirds is that of the swordbill, an Andean species of hummingbird that has a culmen length substantially longer than any other known species. The culmen length in adults averages 83 millimeters, ranging to 105 millimeters in some individuals. In their study in Colombia, Snow and Snow (1980) observed that this measurement is apparently specifically dependent upon the blossoms of *Passiflora mixta,* a species of passion flower with a remarkably long corolla tube (about 114 millimeters). The long bill and extensile tongue of the swordbill species enable it to obtain the nectar from the long corolla tube, which is probably inaccessible to all other hummingbirds (Figure 12). The swordbill is evidently nonterritorial, and forages as a "trapliner" by visiting a large number of plants over a fairly wide area. Likewise, the giant hummingbird in Ecuador is heavily dependent upon the flowers of *Agave americana,* and its distribution seems to have spread with that of the plant (Ortiz-Crespo, 1974).

Many other species of plants, though perhaps not so specifically adjusted to bill length as *Passiflora* may be, nonetheless have remarkable adaptations that ensure cross-pollination. Pickens (1927) described the situation in *Macranthera flammea,* which has the typical hummingbird corolla shape (Figure 13) and color (bright orange), but bears an erect flower. As the corolla begins to open, the pistil quickly emerges and reaches its full length. A day or two later the pistil begins to wither, but by then the stamens have grown to the length of the drooping pistil, with the pollen-bearing anther surfaces tilted toward the center of the flower. By this device, a visiting hummingbird is bound to have its crown intercept either the anthers or the stigma, thus assuring cross-pollination

12. Andean swordbill
and *Passiflora mixta*.

when it visits another blossom at a slightly different stage of floral development, while at the same time avoiding self-pollination.

It is apparent that adaptations such as unusually long and curved stamens are common adaptations in hummingbird-adapted flowers. Many of these species have stamens that extend out well beyond the corolla, resulting in pollen deposition on the visiting hummingbird's bill, forehead, or throat, depending on the particular plant species (Figure 14; see also Figure 21).

As hummingbirds are busily extracting nectar and pollen from such plants, they also are frequently receiving uninvited guests in the form of nasal mites (*Rhinoseius* and *Proctolaelaps* species) (Figure 14G). These are almost microscopically small mites that spend most of their lives in the

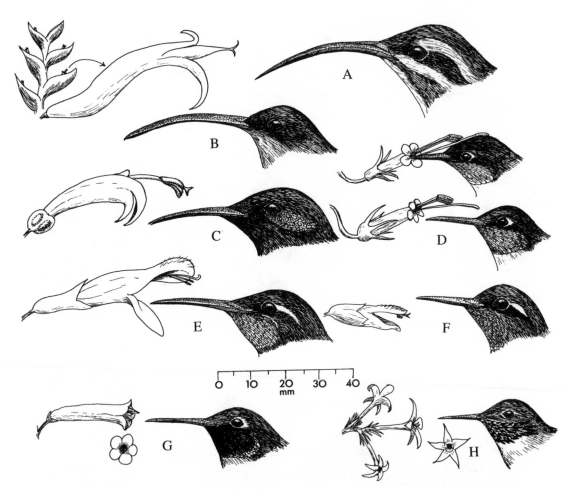

13. Comparisons of corolla shapes and hummingbird bills, showing coevolved characteristics. *Heliconia bihai,* associated with hairy hermit (A) and green hermit (B); *Centropogon valerii,* with green violet-ear (C); *Macranthera flammea,* with ruby-throated hummingbird (D); *Salvia cardinalis,* with blue-throated hummingbird (E); *Salvia mexicana,* with white-eared hummingbird (F); *Penstemon eatoni,* with black-chinned hummingbird (G); and *Ipomopsis aggregata,* with calliope hummingbird (H). *(After various sources)*

blossoms of flowering plants, and compete significantly with humming-birds for the plant's nectar and pollen supplies. Adult mites often move from blossom to blossom in a single flower cluster, but in order to move from plant to plant they need the assistance of a visiting hummingbird as it pokes its bill into the flower's corolla. During the few seconds of such a visit, the mites are able nimbly to climb aboard the hummingbird's bill and quickly run to its nasal opening. There the mites remain, causing no apparent damage to the host, but instead simply waiting until they are provided with a "deplaning" opportunity when the hummingbird again visits the host plant. Experiments by Robert Colwell (1985, 1995) have shown that the mites are attracted to the nectar from their own host plant species in preference to others, and apparently they can identify their

14. Some comparative flower shapes and pollination adaptations with straight-billed North American hummingbirds, including white-eared and *Loesellia mexicana* (A), berylline and *Cinna barina* (B), ruby-throated and *Monarda didyma* (C), Costa and *Fouqueirria splendens* (D), black-chinned and *Trichostemia lanatum* (E), and rufous and *Penstemon labrosus* (F). Also shown is a hummingbird flower mite (*Rhinoseius*) within a hummingbird's nostril (G). (*After various sources*)

specific host species by its odor. They reproduce in the host plant's blossoms, taking about a week to pass from egg to adulthood, then the mites begin their patient wait for another free ride.

Wagner (1946a) described an interesting pollination adaptation in the Mexican plant *Centropogon cordifolius*. In the early stages of floral development, the stamens mature first and tilt downward. After the stamens have withered, the stigmas occupy essentially the same position in the blossom, where they are likely to strike the top of the head of a foraging hummingbird (Figure 15). In a similar case, the stigma of *Lamourouxia exserta* develops earlier than the anthers, so only cross-pollination is possible. Even more remarkable is the highly specialized structure of *Marcgravia*, which has a group of nectaries directly below a series of developing flowers located horizontally above on long, thick stalks around the upper part of the main flower axis. The blossoms face downward toward the nectary below and exhibit "protandry," with the pollen-bearing anthers developing first and the stigmas later. As the

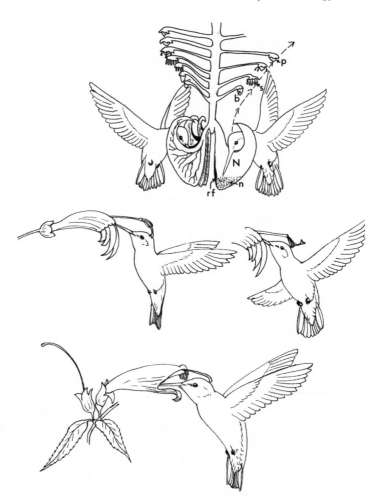

15. Pollination adaptations of
Marcgravia picta (top), *Centropogon*
(middle), and *Lamourouxia exserta*
(bottom), after drawings by
H. Wagner (1946a). See text for
explanation. In top drawing,
b = bud, N = nectary, n = nectar,
p = pistil, rf = rudimentary flower,
and s = stamens.

hummingbird finishes drinking at the nectary and rises backward and
upward, its head comes into contact first with the clustered stamens and
later with the pistil (Figure 15), again assuring cross-pollination (Wagner,
1946a). Similarly, in the southern Andes, the Andean hillstar pollinates a
species of oranged-blossomed composite (*Chuquiraga spinosa*) that
exhibits protandrous development of the stamens. As the growing style
passes through the corolla tube, it pulls the mature pollen upward with
it, coating the inside of the corolla tube and the style itself with pollen.
The lobes of the stigma remain closed until the style's elongation has
lifted the stigma well above the level of the corolla tube, thus avoiding
self-pollination (Carpenter, 1976).

A number of investigators have reviewed the usual association of
bird-pollinated flowers and red coloration. Grant and Grant (1968)
suggested that a single coloration used by a group of hummingbird-
adapted species serves as a common advertisement for food sources;
thus each plant species benefits from becoming a part of a pool of similar

nectar-producing species. Moreover, many hummingbird-pollinated forms probably evolved from bee-pollinated ancestors that were usually blue. The shift from blue to red, which is not attractive to bees, may have augmented the development of a bird-adapted flower form. Thus, in the genus *Penstemon*, which is commonly bee-pollinated, the species *P. centranthifolius* is red, with flowers that are trumpet-shaped and attractive to hummingbirds. The closely related form *P. grinnelli*, which has pale blue, widely bilabiate flowers, is bee-pollinated; and a third species, *P. spectabilis*, has smaller blue, somewhat bilabiate flowers and is wasp-pollinated (Straw, 1956). Grant (1952) described a similar case of floral isolation by flower color, shape, and position in two closely related species of *Aquilegia* found in the Sierra Nevada mountains; these differences reproductively isolate a hummingbird-adapted species from a hawk-moth-adapted species.

Hummingbirds do not exhibit any innate preference for red coloration, but they certainly can learn to associate particular colors with nectar sources. Thus the development of red coloration in bird-adapted flowers provides a convenient, uniformly recognized "flag" for the birds, which is conspicuous against a green background. It is also unlikely to attract bees, which have color vision ranges that barely reach the red portion of the spectrum.

Studies by Stiles (1976) indicated that hummingbirds respond more strongly to energetic aspects of fluid solutions (concentration of sugar, rate of nectar flow) than they do to taste considerations (composition of sugar), and to taste in turn more strongly than to the color of feeder or flower. Among sugars, they select sucrose over glucose, and glucose over fructose. The preferred colors tend to be near the long-wavelength or red end of the spectrum for both tropical and temperate-zone flowers; thus the presence or absence of hummingbird migratory behavior is probably insignificant in fixing the colors of plants.

Besides individualized specialization between birds and plants, specialist foragers such as hummingbirds tend to exhibit a considerable degree of ecological segregation into habitats where interspecific competition from related species are minimal. Of the few quantitative investigations in this area on North American species, studies such as those of Des Granges (1979) have indicated a high degree of interspecific organization of hummingbird guilds in tropical environments. Table 6 presents an approximation of the tendencies for ecological segregation in the species of hummingbirds that breed predominantly in western North America, and Table 7 comprises a similar ecological and geographical tabulation of the primarily Mexican species discussed individually in this text. Neither table allows for a precise estimate of ecological overlaps or interspecific competition between any two species, but both tables provide some indication of the most probable cases of ecological contact between species while on their breeding grounds. Further segregation of habitat between

Table 6
Climatic and Vegetational Affinities of Breeding Hummingbirds in Western North America

Species	Preferred Climate	Preferred Habitat	
		Region	Dominant Plants
Allen	moist	coastal woodlands	deciduous trees
Anna	mesic	chaparral woodlands	deciduous trees
Black-chinned	xeric	scrub lowlands	deciduous trees
Broad-tailed	xeric	open woodlands	conifers
Calliope	mesic	montane coniferous	conifers
Costa	xeric	desert scrub	shrubs
Rufous	moist	moist coniferous	conifers

Note: Excludes species centered primarily in Mexico.

Table 7
Altitudinal and Ecological Affinities of Hummingbirds That Breed in Mexico
and Are Occasionally Found in the United States

Species	Altitude (meters)	Ecological Preference	
		General Area	Particular Habitat
Berylline	900–3000	interior highland	arid pine, pine/oak or fir forest edge
Blue-throated	1800–3900	interior highland	montane meadow, woodland edge
Buff-bellied	0–1200	Atlantic lowland	coastal scrub, forest edge
Bumblebee	1500–3000	interior highland	open pine/oak woodland, cloud forest edge
Green violet-ear	1800–3000	interior highland	humid pine/oak woodland, fir forest
Lucifer	1200–2400	interior highland	arid brush
Plain-capped starthroat	0–1500	Pacific lowland	thorn forest, scrub desert
Magnificent	1500–3300	interior highland	humid pine/oak woodland, cloud forest edge
Rufous-tailed	0–1700	Atlantic lowland	high deciduous forest, rainforest
Violet-crowned	90–2100	interior highland	dry forest, riparian scrub
White-eared	1200–3300	interior highland	scrub oak, pine/oak woodland, arid brush

Note: Excludes Mexican species not recorded from United States; altitudes and habitats mainly after Edwards (1973).

the sexes exists for at least six of the seven species of North American hummingbirds presented in Table 6 (Pitelka, 1951b; Stiles, 1972b), as well as in Mexican species such as the white-eared hummingbird (Des Granges, 1979) and the blue-throated hummingbird (Wagner, 1952). At least one species of hermit (the saw-billed) is sexually dimorphic in bill shape (Selander, 1966).

In a study area in Mexico on the border of Colima and Jalisco, Des Granges (1979) found a foraging guild of some 21 species of hummingbirds, comprising three different groups: (a) the resident group of tropical species inhabiting particular habitats throughout the year; (b) wandering species that visit several habitats during the year and follow seasonal blooms of flowering plants; and (c) migrants present only during the winter. Most of the resident species are territorial, feeding preferentially on tubular flowers. The wanderers are typically "trapliners"—nonterritorial birds that move about, foraging on a variety of blossom types. The migrant species are territorial, defending flowers that provide nectar in excess of the resident and wanderer's requirements, and supplementing their diets with insects. There were several types of ecological segregation, including special segregation of species (dominant and territorial birds defending the tops of trees and shrubs, or areas of tightly packed flowers; and subordinate or nonterritorial birds defending lower areas and often more scattered flowers), seasonal segregation of some species, and a limited degree of sexual segregation in two species. There was no definite indication of diurnal segregation.

At least in the more tropical areas, the species in hummingbird communities tend to fall into one of four foraging modes. "High-reward trapliners" have relatively specialized bills, which have coevolved with particular blossom types. Such species effectively exploit the nectar sources by repeated visits but do not defend specific foraging territories. "Low-reward trapliners" are similar but usually have smaller, straighter bills and are more generalized foragers, visiting more dispersed and less specialized flower types. Typical "territorialists" defend foraging territories, visiting all the suitable flower types within them. Finally, the "territory parasites" are either large species that can feed with impunity in the territories of smaller and less dominant species, or relatively small and fugitive forms that can effectively infiltrate the territories of other species (Feinsinger and Colwell, 1978). Territorial species often have greater flight acceleration and maneuverability than do traplining species, but they hover less effectively and usually have a higher wing–disc loading (the ratio of body weight to the area covered by the outstretched wings) than do trapliners.

In a review, Pyke (1980) noted that honeyeaters of Australia show convergences with hummingbirds in that both groups feed on a combination of nectar and insects; both have long, curved bills and tongues adapted to nectar foraging; and both feed at long red flowers. He summarized studies which indicate that 7 nonhermit species of hummingbirds spent about 84 percent of their time gathering nectar and the remainder of their time catching insects. Studies of 15 species indicated that 86 percent of the observations of foraging were associated with nectar-gathering; 14 percent with insect-catching. Hummingbirds catch insects in many ways, including hawking in flight, gleaning from recesses, variants of gleaning, and sometimes even running or walking (Mobbs, 1979).

CHAPTER 5

Comparative Behavior

The behavior of complex animals such as hummingbirds generally can be organized into three very broad categories: (a) those concerned with the survival and maintenance of the individual (egocentric activities), (b) those essentially self-directed but tending to bring about aggregations of individuals in common habitats or areas of common activity (quasi-social actions), and (c) those directed toward and dependent upon the presence of other organisms for their expression (social behaviors).

Egocentric behaviors of hummingbirds include fundamental features of individual survival such as respiration, ingestion, defecation, and the like, as well as more complex activities such as preening, oiling, shaking, and stretching—all of which might fall under a collective heading of "comfort activities." Unlike most birds, hummingbirds sleep with the neck retracted, head directed forward, bill pointed upward at a distinct angle, and body feathers variably fluffed. Essentially the same posture is typical of fully torpid individuals (Figure 16F). Upon awakening, the bird arches its neck and raises its partly closed wings. It then opens one or both wings fully and stretches them down alongside the body, but not to the rear as in most birds (Figure 16A and C). The tail may be fanned simultaneously. Although hummingbird feet are very small, perching on a single foot has been observed in several genera (Mobbs, 1971).

Apparently hummingbirds never engage in mutual preening, but instead spend a good deal of time in self-preening. For most of this, in common with other birds, they use the bill (Figure 17B), but hummingbirds are remarkably adept at preening themselves with their claws (scratch-preening) in areas of the head and neck that cannot be reached by the bill. Sometimes the claws are used to preen the wing coverts, but not the primaries themselves. Most hummingbirds scratch by raising the foot up and over the wing, as do typical perching birds (Figure 17C), but some long-billed species scratch by bringing the foot forward under the wing (Mobbs, 1973).

16. Hummingbird behavior patterns, including (A) unilateral wing-stretching, (B) "courtship feeding" in Andean emerald, (C) bilateral wing-stretching, (D) feeding of young by female, (E) fighting by sparkling violet-ear, and (F) nocturnal torpor of Andean hillstar. *(After various sources)*

Hummingbirds seem to enjoy bathing very much, and will often bathe in the water film that has accumulated on a large, flattened leaf. At other times they may fly into the flowing water of a small waterfall or other spray and bathe while hovering in the air. When flying around in the jet of a water spray coming from below, the birds will allow themselves to be drenched from beneath and will sometimes catch individual drops of water in the bill with great skill. When bathing on a leafy surface, they will rub their abdomens against the leaves and move

17. Hummingbird behavior patterns, including (A) leaf-bathing by rufous-tailed hummingbird, (B) preening by blue-tailed sylph, (C), over-the-wing preening by horned sungem, (D) threat display by green thorntail, (E) singing with aggressive tail-wagging by long-tailed hermit, and (F) copulation by long-tailed hermit. *(After various sources)*

backwards and forwards over the wet leaf surface (Figure 17A), occasionally sliding off into the air only to fly back and begin the activity again (Scheithauer, 1967).

Quasi-social behaviors include investigative, shelter-seeking, and similar activities that bring individuals into social contact, even though this is not the chief purpose of the activity. Thus, aggregations may develop around limited foraging areas, in localized bathing or roosting sites, and even in favorable nesting sites (albeit rarely). Examples of nesting aggregations include a clustering of five Andean hillstar nests in a small cave in Ecuador (Smith, 1949) and the presence of six Costa hummingbird nests within a 30-meter radius in a cocklebur thicket (Bent, 1940).

The remarkable tendency of hummingbirds to investigate unusual features of their environment probably is related to their constant need to find new and rich sources of food. Thus, they are likely to examine almost anything that is brightly colored, from a red tin can on a camp table to a bright-colored cap. I have seen rufous hummingbirds closely investigate the red stripes of canvas that support the poles of my tent, and of course they will visit red hummingbird feeders almost immediately after one installs them in areas the birds frequent. Such curiosity occasionally can be disastrous, as, for example, when a bird gets caught in the sticky head of a purple thistle and is unable to escape, or is otherwise trapped in an unfamiliar situation.

Closely related to a hummingbird's curiosity is its apparently excellent memory, which enables it to locate food sources perhaps remembered from previous years. Hummingbirds have a remarkable capability of associating food sources with location and color (Miller and Miller, 1971; Scheithauer, 1967), which fosters foraging success. Fitzpatrick (1966) recounted an amazing example of a hummingbird's memory and capabilities for detailed human recognition: He placed a hummingbird feeder outside his bedroom window while he was recuperating from tuberculosis in a California sanitarium. Soon a rufous hummingbird took possession of the feeder, and thereafter Fitzpatrick watched it closely for several months. When he was finally able to go outside in a wheelchair, Fitzpatrick was immediately "greeted" by the hummingbird, which careened around his head and hovered in front of his eyes. After almost a year, when Fitzpatrick returned to his home some 13 kilometers away, the rufous somehow managed to follow him and took up residence near his house. Later, the bird usually accompanied him on his daily walks. It sometimes called his attention to the presence of other animals that he might have otherwise overlooked—once noting a half-hidden rattlesnake—and eventually rode on the rawhide lace that served as a rifle sling. When Fitzpatrick had fully recovered from his illness, he left his house for a month. Yet only moments after he returned and got out of his car, the hummingbird was there, zooming about his head and hovering in front of his eyes!

This remarkable story introduces the area of social behavior, comprising all aspects concerned with individual interactions within and between species. Social behavior includes such altruistic responses as care-giving and care-soliciting behavior, which in hummingbirds is essentially limited to relationships between parents and offspring, although there are a few cases of adult hummingbirds feeding youngsters other than their own. Thus, parental nurturing (Figure 16D) is probably the only type of altruistic behavior among hummingbirds; there is no evidence of succoring behavior between adults, although Weydemeyer (1971) reported seeing a possible example of this. Adults rarely even touch each other, although Poley (1976) photographed contact behavior between two adult female hummingbirds and also photographed what seemed to be typical courtship feeding behavior (Figure 16B). However, the existence of true courtship feeding in hummingbirds is still unproved and unlikely.

The other major aspects of social behavior are agonistic interactions (attack–escape behavior) and sexual activities. In hummingbirds the two components are extremely difficult to separate, for a good deal of what passes for sexual behavior is probably little different from agonistic responses (Figures 16E and 17D). For example, male territoriality in most hummingbirds centers on a supply of food for itself, rather than encompassing a nesting site or available food resources for the female and any offspring. Thus, except in lek-forming species, only secondarily does the territory serve as a mating station, and the display flights of male hummingbirds are probably essentially an intimidation device (Pitelka, 1942). Thus, the bright coloration exhibited by males during territorial advertisement and defense may be essentially agonistic rather than sexual in function. Yet, to the degree that females can discriminate among individual males, and possibly tend to mate with those that are relatively more dominant or conspicuous, the role of sexual selection in the evolution of male plumages and displays cannot be overlooked.

Perhaps the most complete attempt to understand the diversity and significance of hummingbird sexual behavior is that by Wagner (1954), based on his long experience with numerous species of Mexican hummingbirds. A brief summary of his observations can serve as a basis for further discussion. According to Wagner, the female hummingbird searches for a mate only after nest completion. She is likely to mate with the first conspecific male that she meets, and their period of union lasts for only a few hours.

The courtship of the male has two phases: "luring" the attention of females that are ready to mate by species-specific plumage display and associated behavior, followed by the nuptial flight, performed by both sexes immediately before copulation. The luring phase is generally associated with not only posturing but also sound production effected either mechanically or by feather vibration during display flights, and vocalizations. Vocalizations in turn include short-note "songs" of a single bird, group singing in "song-assemblies," and sounds produced during display flights.

For the courtship phase, or nuptial flight, the male must take the initiative, although the female's actions determine the locality of the activity. In different species there are widely varying degrees of sexual dimorphism and coloration associated with precopulatory behavior. In Wagner's (1954) experience, the degree of sexual dimorphism in plumage is closely correlated with the degree of differentiation of the nuptial flight; the more complicated the latter, the greater the degree of sexual dimorphism. In some species this phase of courtship consists only of a high intensity of activities typical of the luring phase; in others it is made up entirely of different instinctive movements.

However, Pitelka's (1942) study regarded male display flights, as well as singing assemblies, essentially as devices for territorial proclamation and intimidation of conspecifics, including females, and courtship or invitation roles were not apparent in their performance. In his opinion, displays in which both the male and the female participated did not indicate the height of courtship, but the contrary, with the female's displays perhaps only an effort to resist the male. He believed that most descriptions of hummingbird copulation were actually simply examples of the usual aggressive clashes between sexes, and that very few credible descriptions of copulation actually exist.

Observations by Scheithauer (1967) tend to confirm the idea that males probably attain copulation primarily by intimidation rather than by courtship. For example, a male blue-tailed emerald displayed for weeks in front of a female, hovering in front of her only 12 millimeters from the tip of her bill as she perched on a branch, following his rapid movements with her head, so that their bills pointed at each other like needles. Invariably, before he was able to take the decisive step toward mating she would dart away and elude him, so that the two birds never paired. Yet, in another case when a female brown inca completed a nest but had not yet secured a mate, she courted a male fawn-breasted brilliant by "dancing" up and down in front of him with a piece of cotton-wool in her bill.

Similarly, Stiles (1982) reported that the dive displays (Figure 18) of North American hummingbirds are essentially aggressive displays associated, in most cases, with the defense of the breeding territory. However, they may also play some role in the initial phases of courtship. The most important displays in courtship per se are close-range, back-and-forth flights by the male, above or in front of the perched female. These "shuttle-flights" are accompanied by species-specific sounds (song in *Calypte* species, wing strokes in other genera), and they are highly species specific in terms of rate, direction, and amplitude of the "shuttles." The displays immediately precede copulation in nearly all cases, but have

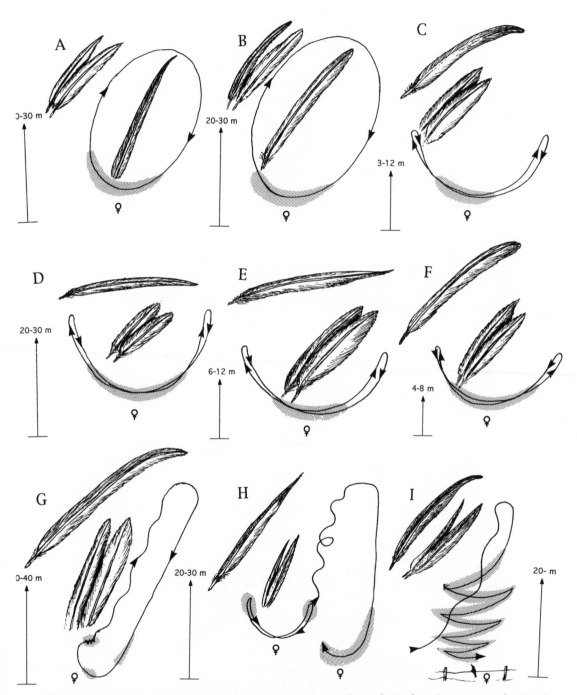

18. Male dive and shuttle-flight displays of nine hummingbirds, together with sketches of outermost male primaries and two outermost rectrices. Species shown are rufous (A), Costa (B), ruby-throated (C), calliope (D), broad-tailed (E), black-chinned (F), Anna (G), Allen (H), and lucifer (I). Stippled areas indicate points of associated vocal or mechanical sounds. *(After various sources)*

never been explicitly described, probably because they usually occur in dense vegetation (as does copulation itself) (Figure 19).

Among the most interesting of the social behaviors of hummingbirds are the song assemblages of territorial males, especially the hermits. Such singing assemblies are present in Trinidad little hermits from about November until the post-breeding molt in July. During this eight-month period each male is at its singing perch for a high proportion of the daylight hours—in one case, 70 percent of the entire time. While on its perch, the male sings about once every 2 seconds (or 12,000 times a day). The songs of individual males vary considerably, but males with neighboring perches tend to have similar song-types. The singing assemblies apparently function as leks, which the females visit for the sole purpose of

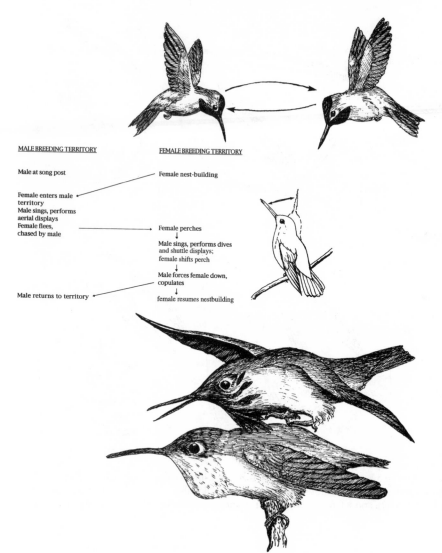

MALE BREEDING TERRITORY

Male at song post

Female enters male
territory
Male sings, performs
aerial displays
Female flees,
chased by male

Male returns to territory

FEMALE BREEDING TERRITORY

Female nest-building

Female perches
↓
Male sings, performs dives
and shuttle displays;
female shifts perch
↓
Male forces female down,
copulates
↓
female resumes nestbuilding

19. Shuttle-flight display of Anna hummingbird (above), with a generalized sequence of sexual behavior in this species. (*Adapted from sketch and diagram of Stiles, 1982*) Copulation by the calliope hummingbird (below). (*After a photo by C. W. Schwartz*)

mating. However, observation of actual copulation in the species has not yet been reported according to Snow (1968). A study of song-types in the little hermit has tended to confirm Snow's finding that similarities exist among the song patterns of different singing assemblages, perhaps around a founding individual that performed an imperfect imitation of a previously existing song pattern. The surprisingly elaborate songs of the species may be related to the fact that the birds display in relatively dark locations, unsuited for visual displays, and also may facilitate the differentiation of song dialects (Wiley, 1971).

Stiles and Wolf (1979) completed a more thorough study of lek behavior for the long-tailed hermit. Of the approximately 30 species of hermits, at least 3 form leks throughout their ranges (long-tailed, green, and little), and a fourth (reddish) forms leks in part of its range. Others, such as the planalto hermit, evidently do not form leks at all. Lek behavior has been observed in the genera *Threnetes* and *Eutoxeres*, but seems to be especially typical of the genus *Phaethornis*. In the long-tailed hermit, leks have been active on the same site for as long as 12 years, with as many as 25 males occurring in a single lek. Territories in the leks evidently serve as mating stations only and are never sufficiently rich in flowers to affect the energy budgets of the resident males.

Song display (Figure 17E) was the major type of territorial advertisement noted by Stiles and Wolf, and visual displays were observed only between birds close to each other. The sexes are identical in this species, and the same displays were given to females as to male intruders. Females apparently signal their sex simply by sitting still long enough for the male to mount, so there are both heterosexual and homosexual mating copulations (Figure 17F). Mating sequences apparently always begin within the male's territory, although actual copulation may take place elsewhere.

Most of the displays of the long-tailed hermit seem to serve both as aggressive signals (toward other males) and as sexual signals, and include aerial activities such as the "float" display (Figure 20A), the "gape and bill-pop" display (Figure 20B), "perch-exchange" behavior (Figure 20C), and "side-by-side" display (Figure 20D).

In the leks of this species, at least, the most dominant males occupy central territories, and subordinate individuals are restricted to more peripheral areas. These central territories were the most stable over time, and the most strongly contested. Resident males returned to the same territories in subsequent years or moved into vacant territories closer to the center of the lek. Most movements toward the lek center occurred after the deaths of central residents, but the dominance status of individual males seemed to change little with age, even over several years. The rate of turnover of lek residents was about 50 percent annually—a high mortality rate which surpasses that of the green hermit in Trinidad, where nearly all resident males survived more than one year (Snow, 1974). In

20. Display behavior of long-tailed hermit, including float display (A), aerial gaping and bill-popping (B), perch-exchange sequence (C), and side-by-side with tail-wagging (D). (Based on sketches of Stiles and Wolf, 1979.) Also shown is gorget display by a male black-capped coquette (E), and sketches of a female calliope hummingbird (F) and a male calliope in normal non-display posture (G), versus during male gorget display (H). Corresponding postures of the Costa hummingbird are also shown (I–K). *(Original, by author)*

this species mating takes place on the male's territorial perch. Moreover, "false matings" by males with leaves or other objects occur frequently in this and other hermit hummingbirds. Male green hermits have visited active nests as well, but nest defense by males has not been recorded.

As is the case with leks of grouse and other lek-forming species such as the ruff (*Philomachus pugnax*), females may cue on activity centers in lek-forming hummingbirds; unlike these species, however, there is no sexual dimorphism in plumage and very little weight dimorphism. In the hermits, the peripheral males apparently have at least some chance of mating, and as a result there is seemingly a less steep dominance/fitness gradient in these birds than in grouse or other species with highly structured leks and extremely localized mating opportunities. This is perhaps partly a result of the rapid turnover rate in resident males, the dense vegetation of the lek that restricts effective widespread dominance by a single male, and the fact that even dominant males must frequently leave the lek to forage, thus increasing mating opportunities for all the remaining individuals. Since nearly all hummingbirds possess the essential prerequisites for lek behavior—male emancipation for breeding participation and extensive available nonmaintenance time and energy for territorial advertisement and defense as well as courtship—it is perhaps surprising that so few species have adopted this breeding strategy. Lek behavior seems to have evolved more frequently in the hermit group than in the more widespread trochiline hummingbirds, because the former have concurrently evolved a high degree of morphological specialization for exploiting, but not defending flowers that have coevolved with these species. However, feeding-resource-centered, territorial, defensive behavior seems to be the most efficient use of male activities for most of the trochiline species, including the North American forms (Stiles and Wolf, 1979).

Among the trochiline hummingbirds, substantial sexual dimorphism in adult plumage is the general rule, and during sexual or aggressive displays males exaggerate their already marked differences from females by erection of usually iridescent crown and throat feathers to produce spectacular visual effects (Figure 20E–K). Clearly the displaying bird is aware of the importance of visual angle in its orientation toward the recipient of the visual signaling, and even seems aware of the importance of direct sunlight in achieving the maximum visual effects. It is impossible for a human to judge how these colors are perceived by another hummingbird, but if ultraviolet perception is within the visual range of birds (for which there is now increasingly good evidence), then the overall effect may be quite different from our own color perception (Goldsmith, 1980; Bowmaker, 1988).

CHAPTER 6

Comparative Reproductive Biology

Among hummingbirds, virtually all of the activities associated with
nesting and the rearing of young are the sole responsibility of the female.
No other major avian family seems to have adopted so overwhelmingly
this trend toward male emancipation from nesting responsibilities and
consequent promiscuous mating tendencies. J. J. Audubon ironically
believed that the beauty of hummingbirds must cause one to "turn his
mind with reverence toward the Almighty Creator." Yet, were it not for
their remarkable mating system and a high degree of associated territorial advertisement behavior, these birds might well have been no more
esthetically attractive than their drab relatives the swifts, which have
consistently held to a monogamous mating system. In adopting an
adventurous and specialized life-style, involving a high degree of nectar-
dependency, a prodigal expenditure of energy during flight, and a
seemingly devil-may-care mating system, the hummingbirds epitomize
a unique kind of high-risk but potentially high-reward strategy for
survival.

Female hummingbirds are among the most tenacious and persistent
of mothers. They often build or rebuild their nests in the most vulnerable
locations, and audaciously attack any human or animal that ventures too
near, including large hawks that might easily consume the bird in a
single swallow. For their part, the males are no less admirable in their
stalwart defense of foraging or mating territories, and on a few rare occasions have been observed incubating eggs or helping to feed the young.
The most notable examples of male incubation were reported by two
independent observers of wild individuals of the sparkling violet-ear
(Moore, 1947; Schäfer, 1952). More recent investigators have also studied
the species, however, and have been unable to confirm participation in
incubation or parental feeding as a regular pattern of male behavior
either in the wild or in captivity.

There have been reports of males of a few other tropical species of hummingbirds helping in incubation or feeding the young, including members of the genera *Glaucis* and *Phaethornis*. A few scattered observations of male incubation have been reported in North America, as for example in the ruby-throated (Welter, 1935) and the rufous hummingbirds (Bailey, 1927). Both of these species are most northerly of all hummingbirds in breeding distribution; in such a climate, with relatively cold environmental temperatures and limited food resources, a monogamous mating system with male participation in incubation and brooding would be most advantageous. There is also a single reported observation of an adult male Anna hummingbird feeding young (Clyde, 1972).

Although the male fiery-throated hummingbird does not defend or feed the young fathered by him, he allows females with which he has mated to forage within his territory, probably because of the considerable sexual segregation in foraging behavior exhibited by this species (Wolf and Stiles, 1970). This species does not exhibit a definite pair bond, but does show this remarkable cooperation of the sexes in their reproductive biology. Perhaps the pair-bonding system in this species should be considered polygynous rather than promiscuous.

The first step in hummingbird nesting is construction, which normally occurs well before fertilization and produces some of the most remarkable of all avian structures. Almost invariably the nest contains extensive wrappings of spider webbing or similar silken materials, which are used to bind it together and to lash it to a solid substrate. In addition, nests of all species contain a very soft inner lining, usually made of a cottony seed material, the wooly surface material of some leaves, or soft bird feathers as may be locally available. Finally, in most species the nests are "decorated" (camouflaged) on the outside with fragments of lichens, bark, moss, or other similar materials, which blend them almost perfectly with the immediate environment.

Although the hummingbird nests are relatively similar in composition, they are placed in a wide variety of locations and substrates. They may be saddled on horizontal branches, partially suspended in a fork or in crotches of trees, adhered to the walls of rock faces, or suspended from above by pendant strands (Figure 21). In the subfamily Phaethornithinae, the typical substrate consists of a hanging leaf, such as that of a palm, with the nest supported on its underside along the leaf and with a long "trailer" of leafy matter hanging downward from the nest. Such seemingly precarious locations are probably quite secure, and protect the nest from rain and from most terrestrial predators.

A few hummingbirds, including the sooty-capped hermit, enhance the equilibrium of the nest by incorporating small bits of clay or pebbles into its bottom and sides to counterbalance the weight of the sitting female (Figure 21). Similarly, the Andean hillstar increases the nest

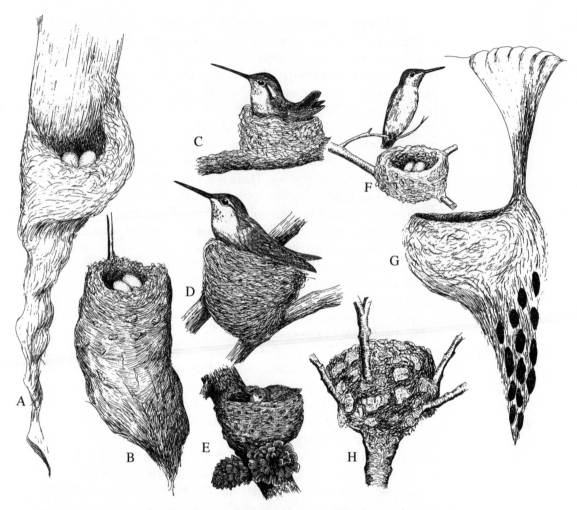

21. Nesting sites and nests of hummingbirds, including palm-leaf nest of long-tailed hermit (A), hanging nest of blue-throated hummingbird (B), saddled nest of black-chinned hummingbird (C), doubly supported nest of Costa hummingbird (D), pine-cone clump nest of calliope hummingbird (E), fork nest of vervain hummingbird (F), counter-balanced nest of sooty-capped hermit (G), and crotch-supported nest of white-eared hummingbird (H). *(After various sources)*

materials unequally on one side, achieving the same result (Ridgway, 1890). The nests of the latter species are otherwise unusual in being remarkably large and thick-walled, which increases the insulating value of the nest for these high Andean birds (Carpenter, 1976; Dorst, 1962). They nest in extremely well-protected and inaccessible locations, some-times in shallow caves of deep ravines, and as many as five active nests have been found within a radius of only 2 meters in such favored loca-tions (Smith, 1949)—an amazing concentration for any hummingbird, given the bleak environment. An equally remarkable breeding concen-tration of crimson topaz hummingbirds was reported by Ruschi (1979), who found 10 occupied nests of this species in an area of 100 square meters.

The length of time required to construct the nest probably varies greatly, but in a few observed cases the work has been virtually completed in a day or two (Bailey, 1974; Welter, 1935). More often it takes about a week, and sometimes the work may be spread out over two weeks (Legg and Pitelka, 1956). Frequently the female continues to add materials to the nest after she lays the eggs, and sometimes she continues this behavior well into incubation.

The eggs are pure white and almost invariably two in number. The tiny bee hummingbird of the West Indies probably lays the smallest egg of any species, but measurements are not available. However, the eggs of

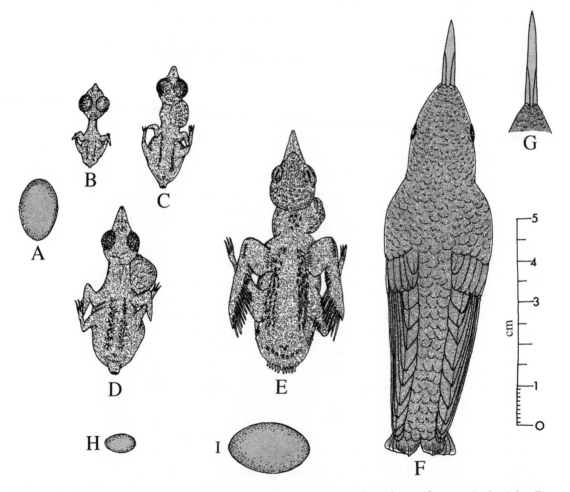

22. Eggs and nestlings of hummingbirds, including egg (A) and nestlings of blue-throated hummingbird at 1 day (B), 3 days (C), 6 days (D), 12 days (E), 24 days (F), and final bill length (G). Also shown are eggs of vervain hummingbird (H) and giant hummingbird (I). *(Mostly after Wagner, 1952)*

the slightly larger vervain hummingbird of the same area measure approximately 7.0 by 5.0 millimeters (Ridgway, 1890) and weigh only about 0.37 grams each. A normal clutch of two such eggs would thus be equal to about 34 percent of the weight of the adult female (Lack, 1976). The largest hummingbird eggs are those of the giant hummingbird, which average 20 by 12 millimeters (Figure 22) and probably weigh very close to 1.5 grams each; a clutch of two would thus represent about 15 percent of the weight of the adult bird. So, in common with other birds, the energy drain of laying eggs is probably less severe on females of larger species than of smaller ones.

In many species (or individuals) of hummingbirds, the female begins to incubate immediately after laying the first egg, and the eggs thus hatch in the same approximate time sequence with which they were laid. The eggs typically are laid in the morning and deposited about 48 hours apart. However, they are frequently laid on subsequent days and sometimes three days apart. When the eggs hatch synchronously or nearly so, incubation probably did not begin until the laying of the second egg.

Incubation periods of hummingbirds have commonly been seriously underestimated, perhaps because of their very small size; some published estimates of periods of as little as 9 to 12 days have appeared. In spite of the eggs' small size, incubation periods are actually long, perhaps because females usually have to leave the nest for extended periods of time to forage every day. This factor may cause a general prolongation of the minimum incubation period to 15 to 17 days (with a few reliable observations of 14-day periods). The longest-known incubation periods are those of the Andean hillstar, which average about 20 days, and sometimes require 22 to 23 days (Carpenter, 1976; Dorst, 1962).

Young hummingbirds are hatched in a nearly naked, blind, and totally helpless state. At the time of hatching they seem to be nearly all "head," but their eyes are tightly closed and the beak is barely indicated. Yet even newly hatched hummingbirds have a well-developed crop, and shortly after they hatch the female begins to "inject" extraordinary amounts of food into her young. She inserts her needlelike bill into the nestling's mouth and regurgitates food from her own crop to that of the young. Even nestlings but a few days old are fed large quantities of tiny insects and probably also nectar, which soon causes their crops to protrude from the sides of their necks like gigantic goiters (Figure 22C–E).

Hummingbird youngsters lack a distinct downy feather stage; instead, the definitive contour feathers emerge directly from the pinfeathers. Yet, in spite of the lack of downy insulation, the young birds are remarkably tolerant of short-term temperature fluctuations. Moreover, by the time they are about 12 days old, before they are well-feathered, they have often acquired a considerable degree of temperature control (Calder, 1971). Depending on the species, the female may continue to brood them until they are from 12- to about 18-days old. Species that rear their young under relatively cold conditions may have a prolonged fledging period; the Andean hillstar, for example, usually requires about 38 days, but favorable conditions may lessen it to as little as 22 days (Carpenter, 1976). Even after fledging the tail feathers and bill of young hummingbirds continue to grow for some time before they reach their adult length (Figure 22G), and maternal care and feeding of the young often continues for a while after the young leave the nest. Skutch (1973) has summarized information on the duration of parental care in various hummingbirds, and for five species the observed range of the last observed feeding was 40 to 65 days after hatching.

Sometimes adult hummingbirds attempt to feed young that are not their own. Thus Wagner (1959) observed wild adult white-eared hummingbirds feeding both nestling and fledgling birds that were not their own offspring. Under aviary conditions there have also been instances of adult birds "adopting" young not their own.

Even in tropical areas, most hummingbirds do not breed the year round, but rather exhibit seasonality in breeding that is probably associated with the relative intensity of blooming of preferred flower sources during wet or dry seasons. However, some equatorial species do breed throughout the year, as does the Andean hillstar in Ecuador (Smith, 1949) but not in southern Peru or northern Chile (Carpenter, 1976). Year-round breeding has also been reported for the Anna and Allen hummingbirds in southern California (Wells et al., 1978).

Although the incidence of multiple brooding still remains to be studied thoroughly, it is probably of relatively widespread occurrence in tropical hummingbirds. It has also been reported for several North American species, including the blue-throated hummingbird, where it seems to be fairly common. In other species such as the Anna, Allen, and black-chinned, it is less frequent, but probably all species attempt to nest a second time if their initial clutch or brood is lost prior to fledging.

In the species that sometimes exhibit multiple brooding, the female typically begins building a second nest while still feeding young from the first brood. Several instances of such concurrent care of two nests have been described for various North American species, including the white-eared (Skutch, 1973), ruby-throated (Nickell, 1948), and black-chinned hummingbirds (Cogswell, 1949).

In spite of the great perseverance and courage shown by female hummingbirds while defending their nests, the reproductive success of these birds in general is relatively poor. Such low rates for hatching and fledging young (Tables 8 and 9) are probably the result of high vulnerability of hummingbird nests to loss of eggs or young from accidents, weather-associated catastrophes, and predation. Indeed, one of the most successful species of nesting hummingbirds is the Andean hillstar,

Table 8

Various Nesting Success Rates Reported for Hummingbirds

Species	Authority	Total Nests	Eggs Laid	Young Hatched	Young Fledged	All Nests	Nests with Eggs
			Total Nests in Which			Percent Nesting Success	
Allen	Legg and Pitelka, 1956	18	16	—	4	22.2	25.0
Andean hillstar	Carpenter, 1976	19	28	—	16	—	88.9
Anna	Stiles, 1972b	85	68	42	23	31.3	44.2
Black-chinned	Stiles, 1972b	55	47	27	15	27.3	31.9
	Baltosser, 1986	157	—	—	—	34.4	—
Costa	Woods, 1927	—	29	—	12	—	41.4
Rufous-tailed	Skutch, 1931	22	17	10	6	45.4	58.8
White-eared	Wagner, 1959	39	—	—	12	30.7	—

Table 9

Various Hatching and Fledging Success Rates Reported for Hummingbirds

Species	Authority	Total Nests	Eggs Laid	Eggs Hatched	Hatching Success (Percentage of Eggs Laid)	Total Young Fledged	Fledging Success (Percentage of Eggs Laid)
Andean hillstar	Carpenter, 1976	19	37	—	—	22	59.4
Broad-tailed	Waser and Inouye, 1977	52	102	—	—	60	58.9
Costa	Woods, 1927	—	58	35	60.3	19	32.7
Green hermit	Snow, 1974	19	37	12	32.4	10	27.0
Rufous-tailed	Skutch, 1931	22	32	18	56.2	11	34.4
White-eared	Skutch, in Bent, 1940	9	18	9	50.0	3	16.7

which avoids high predation losses in its cold and unfavorable nesting environment. In other tropical species, as well as North American ones such as the Anna hummingbird (Stiles, 1972b), predation accounts for much of the nest mortality (Carpenter, 1976; Baltosser, 1986).

In addition to their persistent efforts at nesting, hummingbirds have long potential breeding spans. Very few have been banded in any number, but at least one banded female ruby-throated hummingbird survived at least 9 years (Baumgartner and Baumgartner, 1992). A female broad-tailed similarly survived to 12.1 years, and a male to 8.0 years (Calder and Calder, 1992). Two banded female calliope were recaptured six years after banding, and a male after five years (Calder and Calder, 1994).

Although reliable data on mortality rates in hummingbirds are not yet available, Baumgartner (1981) obtained some recapture data on ruby-throated hummingbirds. Of the 384 hummingbirds she captured between 1977 and 1979, she recaptured 88 birds (23 percent) the following year, 31 of 268 birds (11.5 percent) the second year after banding, and 10 of 110 birds (9.9 percent) the third year. These figures indicate a minimum annual survival rate of 23 to 46 percent, and, because undoubtedly some survive but are not recaptured, the actual rate must be considerably higher. Calder and Calder (1992) estimated a 50-percent annual survival rate for the broad-tailed hummingbird.

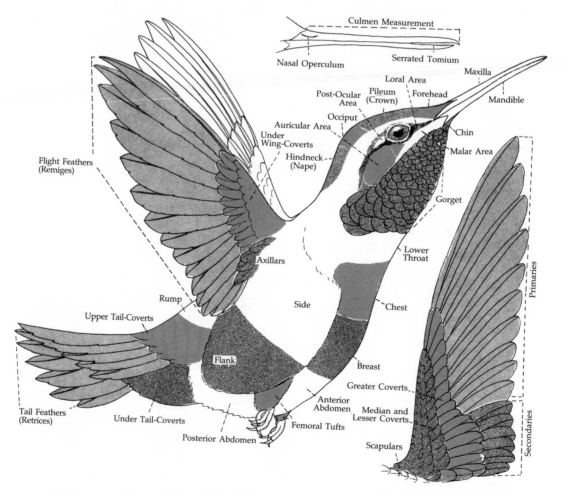

Body and feather areas of hummingbirds, showing features mentioned in the keys or text.

PART TWO

NATURAL HISTORIES OF NORTH AMERICAN HUMMINGBIRDS

This section includes all species of hummingbirds that have been reliably reported from anywhere north of the Mexican border, although several of them are not truly North American in the sense of having been proven to breed in the United States or Canada. West Indian species are also included, but only if they have been reported from the mainland of North America, as are Mexican species breeding north of Chiapas.

All descriptions are in typical field-guide terminology, which I have included in the Glossary and on the hummingbird figure facing this page. In addition, identifying characteristics of some species vary depending on the position of the bird and the angle at which the viewer sees it. Therefore, some descriptions are worded according to one or more of the following positions:

Position *a:* Viewer's eye between bird and light, bird's bill toward eye, bird nearly horizontal.

Position *b:* Viewer's eye directly above bird, bird's bill toward light, bird nearly horizontal.

Position *c:* Same as position *a,* but bird reversed (tail, instead of bill, toward eye).

LONG-TAILED HERMIT

Phaethornis superciliosus (Linnaeus)

Other Names
Buff-browned hermit, Guerrero hermit (*mexicanus*), Mexican hermit (*mexicanus*), Veracruz hermit (*veraecrusis*); Ermitánó colilargo, Ermitánó mexicano, Ermitánó rabudo (Spanish).

Range
Resident in Mexico from Nayarit in the west and Veracruz in the east, extending southeast to Guatemala, Belize, and Honduras. From there populations continue throughout Central America and into South America, reaching Peru, Bolivia, and Brazil.

North American Subspecies
P. s. mexicanus (Hartert). Resident from Nayarit to Oaxaca excepting Jalisco, where an apparent range gap occurs. Regarded as a possibly distinct species by Howell and Webb (1995).

P. s. veraecrusis (Ridgway). Resident from Veracruz to northeastern Chiapas.

P. s. longirostris (DeLattre). Resident from Chiapas southward to Honduras and Belize.

Measurements
Of *P. s. longirostris*: Wing, males 59–64.5 mm (ave. of 4, 62 mm), 2 females 60 mm. Culmen, males 39–44 mm (ave. of 4, 40.7 mm), 2 females 37.5, 40 mm (Ridgway, 1911). Eggs, 15.1–16.1 × 9.0–9.8 mm (Rowley, 1966).

Weights
The average of 18 males (mostly of South American taxa) in the Field Museum was 5.3 g (range 3.5–6.7 g); that of 24 females was 5.1 g (range 4.1–6.6 g). The average of 17 males from Costa Rica was 6.19 g (*SD* 0.3 g); that of 12 females was 5.82 g (*SD* 0.19 g) (Stiles, 1995). The average of males from

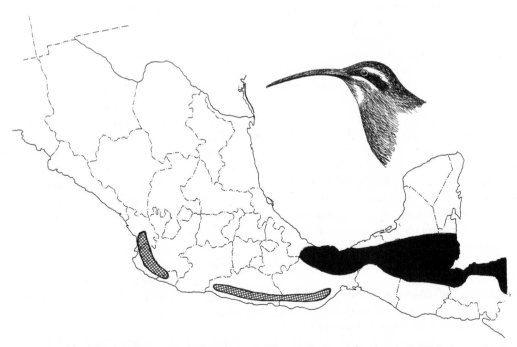

Residential range of the long-tailed hermit, including the typical form (inked) and the "long-tailed" Mexican endemic taxon (cross-hatching). (*Adapted from Howell and Webb, 1995*)

Costa Rica was 6.13 g (*SD* 0.25); that of 29 females was 5.87 g (*SD* 0.23) (Stiles and Wolf, 1979).

Description (After Ridgway, 1911)

Adult male and female. Both sexes identical in plumage, minor differences in linear body measurements and mass may distinguish the smaller females; crown dusky and slightly glossed with bronze-green; upperparts dull metallic bronze-green; feathers sometimes edged with buffy; rump and upper tail-coverts more barred with dusky; tail basally pale and white-tipped, but grading into dull black subterminally; and middle pair of rectrices narrowed terminally, elongated and mostly white; each side of the head with a dull black to dusky streak from the lores to the auriculars, above and below whitish stripes separate this area from the crown and throat; iris brown; feet brownish; and bill mostly blackish above and yellowish below.

Immature. Similar to adult but with broader brownish-buff margins on back and crown feathers. Like other hummingbirds so far known (Ortiz-Crespo, 1974), juveniles are likely easily distinguished by their anterior–posterior bill corrugations that are rather deeply incised along the sides of the bill. These ridges typically extend from the dorsal edge of the maxilla, proceed forward and downward at an angle of about 30°, and usually end near the lower edge of the maxilla. Birds lacking such corrugations are likely to be at least a year old. "Subadults" (birds under a year old but in adult plumage) often can be distinguished by this method, although the corrugations are more shallow, less obvious, and may not extend to the anterior parts of the bill. Juvenile hummingbirds typically have grayish or buffy feather tips on their back and nape feathers, and the rectrices of the juvenal plumage are retained until the first adult plumage is attained, probably at about 10 to 11 months of age (Baltosser, 1987).

Identification

In the hand. The combination of a very long, decurved bill (culmen 37–44 mm), and the long, highly graduated tail (central feathers 63–68 mm) that pales from brown to nearly white toward the tip immediately identifies this otherwise quite uniformly bronzy-brown species.

In the field. Rather easily recognized by its brownish overall color, elongated tail, and long, decurved bill. No other similar-sized hermit occurs in the region. Calls include a series of nasal *chirr* and *chip* notes, squeals, chattering notes, explosive *sweik* calls, and a song that is a series of single buzzy notes (variously transliterated as *chink, churr, shree,* etc.) uttered at a rate of about 60–70 phrases per minute, uttered from song perches over periods of up to 30 minutes.

Habitats

Associated with humid lowland forests and forest edges or old second-growth forests, often near water, and especially near growths of *Heliconia.* Stiles (1980) observed the birds most frequently (53 percent of 125 observations) in forest understory habitats, but they also were fairly commonly observed in forest edge and light gap sites (34 percent of observations) and rather rarely in forest canopy or nonforested sites.

Movements

No long-distance movements have been documented. This species is evidently residential at the lowland rainforest locality La Selva, Costa Rica, where it is "extremely abundant" nearly every month of the year (Stiles, 1980). In a study of color-marked males, Stiles and Wolf (1979) found that sexually active birds tended to forage within 100 to 400 meters of their display grounds, at least during the greatly prolonged display period (about 10 months). A maximum foraging distance of about 2700 meters from an established lek territory was observed in a sample of 146 color-marked resident males, and a similar maximum distance was observed among adult females moving between foraging areas. Young males were much more inclined to move from one lek to another than were resident males; the maximum recorded movement of any single bird moving between leks was about 1500 meters. Females exhibited overall movement tendencies similar to those typical of young males, whereas established adult males exhibited little tendency for performing any inter-lek movements.

Foraging Behavior and Floral Ecology

Stiles (1975) undertook a comprehensive study of the use of the available *Heliconia* species at La

Selva, Costa Rica, where nine species of this important food-plant group occur. Long-tailed hermits are "traplining" species, undertaking rather long flights from one flowering cluster to the next, and not trying to maintain territorial control over any. Species of *Heliconia* (Musaceae) pollinated by this and other hermits exhibit nonoverlapping peaks of flowering, presumably to avoid cross-pollination and competition for pollinators. One species most commonly used by the long-tailed hermit is *H. pogonantha* (= *"rostrata"*), which flowers over a long period, but with a peak during the early dry season. Another heavily used species, *H. tortosa* (or *"H-18"*), flowers during the wet season, which is when most of the *Heliconia* species bloom most profusely. *H. pogonantha* had one of the greatest amounts of nectar production, and highest concentrations and caloric contents of all the *Heliconia*

species, whereas *H. tortosa* had a much lower estimated overall caloric value to hummingbirds. All five of the local *Heliconia* species regularly visited by hermits have long, curved corollas that variously tend to correspond to the curved-bill shape of the hermits, whereas the three *Heliconia* species with essentially straight corollas were mostly visited by nonhermit species. In *H. pogonantha* the overall inflorescences are pendant and the individual corollas are decurved, forcing the bird to bend its head upwards while hovering, and causing the pollen to be attached to the top of the bill or the head. In *H. tortosa* the pollen is also attached to the upper side of the bill or head, but in this case it is a result of the quite different flower shape (upcurved corolla) and the erect position of the inflorescence, so that the head may be held horizontally or tilted downward while feeding (Figure 23).

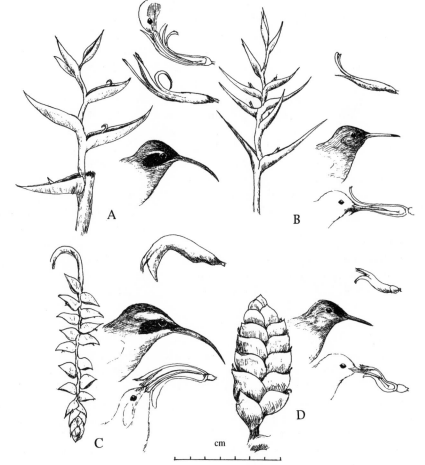

23. Comparisons of *Heliconia* species used by hummingbirds in Costa Rican rainforest, showing bill and flower structures of various resident hummingbirds and the commonly exploited *heliconia* species: bronzy hermit and *H. tortosa* (A), rufous-tailed hummingbird and *H. lathispatha* (B), long-tailed hermit and *H. pogonantha* (C), and crowned woodnymph and *H. imbricata* (D). Associated bill penetration and typical pollen deposition patterns are also shown.
(Partly adapted from Stiles 1975, 1979)

A

B

C

D

cm

In other *Heliconia* species used less frequently by the long-billed hermit, such as *H. lathispatha*, the inflorescence is erect and the corolla appears long but is actually is about the same effective length as in *H. pogonantha*. In this species of *Heliconia*, relatively short-billed hummingbirds have no difficulty entering the flower, and pollen then is deposited on the mandible and chin of these birds. Another species of *Heliconia* with an even shorter corolla is *H. imbricata*, which is used significantly by coexisting nonhermit species such as the fork-tailed emerald, the white-naped jacobin, and the rufous-tailed hummingbird. Pollen from this *Heliconia* tends to be deposited toward the tip of the bird's mandible (Stiles, 1975).

Stiles and Wolf (1979) observed long-tailed hermits feeding on a total of 21 species of plants at their La Selva study site, but 27 percent of all their nearly 1000 foraging observations on the long-tailed hermit involved its use of *H. pogonantha*. Of the remaining observations, 9 percent involved use of *H. imbricata*, 9 percent involved *H. acuminata* (or "H-3"), and 7 percent involved *H. wagneriana*. Collectively, *Heliconia* species accounted for two-thirds of all foraging observations at La Selva. Two species of wild gingers (*Costus*) accounted for another 14 percent, and two genera of acanthaceids (*Jacobinia* and *Aphelandra*) likewise accounted for a joint total of 8 percent, so that few other plant taxa were significant sources of nectar in this locality. Another study (Fraga, 1989) has indicated that the long-tailed hermit is the principal pollinator of *Aphelandra sinclariana*, although it is also visited by bees and by the nectar-stealing little hermit.

Breeding Biology

A very large amount of information on the social behavior of this species is now available through the work of Stiles and Wolf (1979). Like several other hermit hummingbirds, lekking behavior is well developed in this species (Davis, 1958). Leks of long-tailed hermit hummingbirds are persistent, with individual males gaining initial access to a lek (typically on its periphery) as immatures, and surviving as lek residents for periods as long as four years (estimate based on a sample of 37 known-aged males). Survivorship from year to year of the lekking males averaged about 50 percent for all age classes of males. However, the older males tended not only to dominate younger ones during territory contents, but also attained the majority of copulations, a pattern typical of arena-forming birds generally (Johnsgard, 1994).

The social displays of lekking males include several vocalizations, of which singing is the most characteristic. It consists of a single phrase that is repeated monotonously, at a rate of 60 to 70 times per minute, for extended periods of up to 30 minutes, and may occupy at least half of the male's daylight hours on the lek. Individual variations in songs exist, as well as dialect variations between leks. Male visual displays (see Figure 20) include a "float" display, in which a male slowly shuttles back and forth in front of another bird, which often responds by exposing its orange throat toward it. Often when hovering in front of a rival (or a female), the resident male will hover in front of it, open and snap shut its bill in a "gape and bill-pop" display, named for the distinctive associated dry and snapping vocalizations associated with it. Another common display is the "perch exchange," which one bird begins by hovering in front of a perched rival (or female), then circles around behind it and supplants the previously perched individual. This exchange continues for as many as ten rapid sequences. Two males may also perch "side by side" and perform mutual gaping and vigorous tail-wagging movements. Similar side-by-side postures and movements occur in a sexual or precopulatory context between perched males and females, although copulations also can be preceded by the float display or by perch-exchange displays. Homosexual as well as heterosexual mountings have been observed (Stiles and Wolf, 1979), which is not surprising in view of the species' lack of sexual dimorphism or obvious sex differences in social behavior.

At La Selva, this species displays actively until August, but peak breeding occurs during the major blooming period of *H. pogonantha* in February or March. Nesting drops off rapidly after August, and very few flowers are available during October and November. Skutch (1964) found eight nests, spanning the overall period of January to August. All were in heavy forest habitat, were elevated about 1.5 to 2.5 meters above ground, and were fastened to the tips of small palms of a species bearing needle-like spines on the fronds.

The nests were all cone-like, broadest at the top, tapered to fit the inside surface of the supporting palm frond, and with a thin vegetative "tail" that dangled below the end of the frond. They were made of rather stiff and wiry plant fibers, apparently held together with spider or insect silk. Like swifts, hummingbirds also incorporate saliva into the nests while constructing them, producing an amalgamation of dried saliva, silk, and plant materials. The distinctive tails on the long nests of *Phaethornis* hummingbirds perhaps help serve as counterweights, and the total weight of the nest and its contents also causes the supporting leaf to fold over, thus hiding and helping to protect the nest from above (Sick, 1993). Among six nests found by Davis (1958), most attached to palm leaves (but not of thorny species), all had vegetational "tails," and all had two eggs.

Skutch (1964) noted that all the nests he found produced clutches of two eggs or nestlings; the eggs were laid at intervals of two days in each of two instances. Five incubation sessions observed by Skutch averaged 44 minutes, and five "recess" periods averaged 25.8 minutes. Addition of silk to the nest and supporting frond occurred throughout the incubation period, which in one case lasted 17 or 18 days, and in another at least 17 days. During the entire incubation period and while brooding, the female invariably sat with her head oriented inward (directly toward the supporting frond). Because of her long bill, she had to hold her head back beyond the vertical, and her tail was similarly forced upward in order to clear the edge of the nest.

The two nestlings observed by Skutch hatched on the same day, and very soon after hatching each chick adopted the same position in the nest as that of the parent, namely facing the inner surface of the supporting leaf. The two chicks thus maintained parallel positions in the nest throughout the relatively long (22–23 days) nestling period. When feeding her young, the female never landed on the nest, but instead hovered above the chicks, inserting her bill in the throat of each in turn. Typically each chick was fed two or three times in alternate sequence. The female never cleaned the nest; instead the nestlings voided in the usual hummingbird method of expelling their fluid wastes out over the sides. Only one of the nests observed by Skutch was successful in fledging young; the others disappeared or were destroyed by unknown agents before the young fledged. In two nests observed by Davis (1958), the fledging period was at least 18 days in one case and about 25 to 27 days in the other.

Evolutionary and Ecological Relationships
Stiles (1975, 1980) has discussed the complex ecological relationships between this species of hummingbirds (and others) and the role of *Heliconia* in influencing the birds' breeding seasons and general foraging needs. Similarly, the plants must rely on the birds for their own pollination, and several interacting adaptations among the *Heliconia* species have evolved. These include spacial and temporal partitioning of blossoming by the plants, ethological mechanisms influencing flower choice by different hummingbird species, and major differences in flower structure and bill length that help determine where pollen is deposited on the birds' bills or faces. Considerable insect–hummingbird competition for nectar occurs in several species of hummingbird-adapted flowers; thus, for example, in some *Heliconia* species, basal nectaries are present that long-billed hummingbirds can easily reach but bees cannot. Nevertheless, aggressive behavior by bees may be effective in keeping hermit hummingbirds such as the long-tailed hermit from gaining ready access to preferred nectar-rich flower resources (Gill et al., 1982).

LITTLE HERMIT

Phaethornis longuemareus (Lesson)

Other Names
Boucard's hermit, Longuemare hermit; Ermitano chico, Ermitaño chileanchito, Ermitaño enano (Spanish).

Range
Resident in eastern Mexico from Veracruz south to the Guatemala-Belize borders; thence south throughout Central America and extending into tropical South America to the Guianas, Ecuador, Peru, and Brazil. Also resident on Trinidad.

North American Subspecies
A. l. adolphi (Gould). Resident in Mexico south to Chiapas.

Measurements
Of *A. l. adolphi:* Wing, males 38.2–42 mm (ave. of 5, 39.5 mm), one female 37 mm. Culmen, males 21–22.5 mm (ave. of 5, 21.7 mm), 2 females 21 and 22 mm (Ridgway, 1911). Seven Trinidad males had wings averaging 42 mm (range 41–43 mm); 5 females averaged 42.4 mm (range 42–44 mm) (ffrench, 1991). Eggs, 11.7–11.9 × 7.1–7.9 mm (Skutch, 1951).

Weights
An unsexed sample of 58 birds averaged 3.0 g (*SD* 0.38 g) (Brown and Bowers, 1985). Nine Costa Rican males averaged 2.49 g (*SD* 0.17 g), 7 females 2.62 g (*SD* 0.17 g) (Stiles, 1995). Six Trinidad males averaged 3.2 g (range 3–3.5 g), 3 females weighed 3.0 g each (ffrench, 1991). Haverschmidt (1968) reported Surinam birds as weighing 2.8–3.3 g.

Description (After Ridgway, 1911; Wetmore, 1968)
Adult male and female. The sexes very similar as adults, but the female paler cinnamon below, especially on the chin and throat; crown and upperparts dull metallic bronze to greenish bronze, the

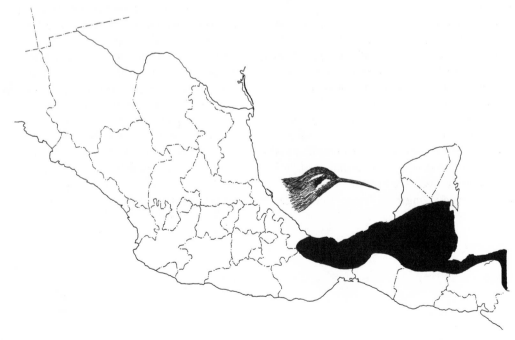

Residential range of the little hermit. *(Adapted from Howell and Webb, 1995)*

lower rump becoming chestnut; tail bronze, with paler tips, the middle pair of rectrices nearly white at their tips; sides of the face crossed by a dusky streak from lores to auriculars, above and below which are cinnamon-toned malar and supra-ocular streaks; iris dark brown; feet dull yellowish; bill dull black above and yellow below.

Immature. Not yet specifically described, but the age criteria for the long-tailed hermit probably apply equally to this species, especially the ridged condition of the upper mandible.

Identification
In the hand. This is a very small brownish-bronze hummingbird with a graduated tail that is about as long as the wing (central rectrices 33–38 mm) and a relatively long, decurved bill (culmen 21–23 mm).

In the field. This tiny hermit is unique in its insect-like size, long and decurved bill, and dark auricular patch. It is likely to be seen foraging close to the ground in forested areas, and it often utters squeaky and jerky notes. The male's "song" consists of an extended series of *chip* or *sik* notes, uttered as the perched birds elevate their bills and vibrate their tails rapidly.

Habitats
Associated mainly with lowland forests, forest edges, brushy undergrowth, and areas of second growth. One of the important ecological studies of this species is that of Stiles (1980), who studied this and 21 other hummingbird species at La Selva, a lowland rainforest locality in eastern Costa Rica. There the little hermit is an abundant permanent resident throughout the year, where it mainly resides in forest understory locations (60 percent of 88 site observations by Stiles), but it also has been observed to some extent along the forest edge or in forest gaps and, least frequently, in non-forested habitats. Skutch (1951) observed that it prefers the shade of tall but dense second-growth thickets and light woodlands, and when present in primary forest it is usually to be found near the edges rather than in the forest interior.

Movements
No long-distance movements have been documented or seem likely to occur, as this species is seemingly sedentary, and established traditional lek sites have persisted in particular areas for at least 18 years (Skutch, 1964).

Foraging Behavior and Floral Ecology
In spite of its very small size, the long, curved bill of this species allows it access to nectar sources in flowers that are otherwise available only to hummingbirds of substantially larger size. Skutch (1951) observed that in central Costa Rica these birds seem to favor plants such as *Stachytarpheta* (Verbenaceae), which has small, violet flowers that are not obviously adapted to hummingbirds. They also visit the spectacular blossoms of the hummingbird-adapted scarlet passion-flower *Passiflora vitifolia* (Passifloraceae), which has a corolla so long that only the long-tailed hermit can easily reach its nectar. Since little hermits cannot reach the nectar reserves at the base of this species' flowers, they content themselves with taking small insects attracted to the blossoms and getting nectar from accessory nectaries located on the flower bracts. Snow and Snow (1972) estimated that only 8 percent of their observations of foraging little hermits involved insect hunting, and all of this was spent in surface searches rather than in aerial hawking, which is probably a relatively difficult activity for the long- and curve-billed hermits.

The long-tubed blossoms of the skull-cap *Scutellaria costaricana* (Labiatae) are actually pierced by the bird's bill while it hovers above the flower, a process requiring great skill on the part of the bird, since it must penetrate only one side of the flower to reach the nectar and not pass its bill all the way through the corolla tube. At La Montura, Costa Rica, it has been seen foraging in this manner on *Cousarea* species (Rubiaceae), a species with flowers having a corolla length of about 20 millimeters. Thus, because of their flexible foraging abilities the species is not dependent on any single flower type, which probably helps facilitate the very extended nesting season that has been documented for Costa Rica (Stiles, 1985).

In Trinidad this species' most favored herbaceous plants include several *Heliconia* species (Muscaceae), *Cephaelis mucosa* (Rubiaceae), *Pachystachys coccinea* (Acanthaceae), *Justica* species (Acanthaceae), and *Costus spiralis* (Zingiberaceae)—all plants with red, orange, or pink blossoms. Perhaps the most heavily used of the shrub group is *Palicourea crocea* (Rubiaceae), which also has red blossoms. Among 280 feeding records of little hermits, a total of 40 plant species were visited. Of these, 62 percent were herbaceous plants up to 3 meters high; 27 percent were shrubs; and the remaining 11 percent included a few trees, bromeliads, and vines. Plants bearing red flowers were visited 75 percent of the time (270 observations), and 88 percent (257 observations) of the flower visits involved plants with corolla tubes ranging from 10 to 39 millimeters in length (Snow and Snow, 1972).

Breeding Biology
Stiles (1980) observed that at the lowland rainforest site La Selva, Costa Rica, this species and the long-billed hermit have an extended breeding season. Nests are being found nearly throughout the year, suggesting that there is no strong dependence on a single plant species for the foods needed for breeding. Molting occurs over much of the year in a rather ill-defined pattern, but with a peak in June and July, so that a substantial overlap between molting cycles and breeding periods occurs in both of these hermits.

Skutch (1951, 1964) has described the social behavior of this species, and especially has reported on the lekking behavior exhibited by the birds. He reported that the leks are active throughout most of the year and may be spread out over a distance of 15 meters or more. The numbers of males occupying a lek is still unstudied, but probably the overall pattern is much like that of the long-tailed hermit. Males sit on singing perches and vocalize persistently while pointing the bill upward, swelling the throat, and vigorously wagging the tail back and forth or pumping it up and down. The song is repeated endlessly, at about 2-second intervals throughout the day during the peak of the display period. Perch-exchange behavior similar to that described for the long-tailed hermit has been seen. There also are "floating" displays, in which one bird tilts both its tail and its bill upward, forming a curious boat-like or crescent-shaped profile, or at times bends it downward while hovering in front of another perching individual. In once case, Skutch (1961) observed an apparent male floating above a presumed female for approximately 10 minutes, hov-

ering slightly above the female's head and oscillating laterally back and forth over a distance of a few inches, simultaneously moving slowly up and down. Every few seconds the male would do an aerial about-face, perform a complete circle or a turn and a half, or would wildly shuttle back and forth while producing louder wing noises. In one observation of this type, the apparent male landed briefly on the female's back, who then moved aside to avoid him. Gaping toward another individual is also common and probably serves in both courtship and aggression (ffrench, 1991).

Snow (1968) reported that singing assemblies of this species exhibit local variations (dialects) in their vocalizations, and that birds occupying nearby perches tend to produce similar songs. Subsequently, additional cases of song-sharing by male singing assemblages have been reported for hummingbirds such as the green hermit, Anna, and two species of violet-ears (Gaunt et al., 1994).

Nesting in this species is quite prolonged seasonally. Stiles (1980) reported that nine nests at La Selva, Costa Rica, were found between February and December—six of them between April and July. Skutch (1951) mentioned finding 13 occupied nests in south-central Costa Rica, of which all were active from April to December, and 7 were found during June and July. Two general breeding periods (April to August; November to January) seem to be present there, with little or no nesting during the driest (February, March) and wettest (September, October) months. In Trinidad the total reported breeding span extends from December to June, with 11 of 15 nesting records for January and February (ffrench, 1991).

Like the long-tailed and other hermits, the nest of the little hermit is cone-like and is attached near the drooping tip of a leaf having a more or less concave lower surface and a pointed tip. Twelve of 13 nests observed by Skutch (1951) were attached to palm fronds; the other was bound to a coffee leaf. All of the palm nests were associated with a species having very sharp spines on the stem and midrib. The nests ranged in height from less than a meter to about 2 meters above the substrate, and in each a vegetational "tail" of varied length formed its lower tip. The plant materials comprising the bulk of the nests were quite diverse. However, spider or insect silk was an integral part of all nests, the strands binding the nest together and also holding it to the supporting frond. This webbing was constantly renewed and supplemented throughout the 16-day incubation period. One nest observed from its initiation required 4 days of building, and all of the construction was done by the female alone.

In nearly all of the nests observed by Skutch (1951) there were two eggs or nestlings; however, in two nests there was only a single nestling that probably represented a survivor of an earlier brood or clutch of two. In three cases the egg-laying interval was established as two days, but in one case the interval was a single day. During incubation one female spent an average of 60.7 minutes on the nest per session during nine sessions, with a mean recess period of 15.1 minutes during 15 breaks in incubation. Another female averaged 27.2 minutes during nine incubation sessions, and her intervening recesses averaged 10.4 minutes. A third averaged 21.4 minutes on the nest (seven sessions) and 13.3 minutes off the nest. These three birds collectively devoted 70 percent of Skutch's total observation time to incubation, and presumably were foraging the rest of the time. Among the nests where such information could be determined, the incubation period was 16 days (Skutch, 1951, 1964).

Almost as soon as the chicks hatched, they oriented their bodies in the nest cup so that their heads faced the leaf that supported the nest, and they rarely altered this position during the entire 21- to 23-day fledging period. While feeding the young, the female typically hovered in the air above them, but when they were very young she was observed to grasp the rim of the nest while regurgitating food into the gullets of the chicks. Two very young chicks (1–3 days old) were fed four times in about 2 hours, whereas at 12 to 14 days of age they were fed twice as often (four times per hour). The eyes of the chicks were fully open by about 10 days of age. By about 15 days the chicks were well feathered, but they remained crouched in their nests until they were ready to fly about 6 days later (Skutch, 1964).

Evolutionary and Ecological Relationships
The nearest relatives of the little hermit include the reddish hermit, an even smaller (1.8–2.2 g), quite widespread species of the eastern South American lowlands. Another close relative is the minute hermit, a similarly tiny species of southeastern Brazil that is sometimes regarded as a geographic replacement form of the little hermit.

WEDGE-TAILED SABREWING

Campylopterus curvipennis (Lichtenstein)

Other Names
Curve-winged sabrewing (*curvipennis*), Long-tailed sabrewing (*excellens*), Nightingale hummingbird, Singing hummingbird, Tuxtla Sabrewing (*pampa*); Chupaflor gritón, Fandangero colicuña, fandangero colilargo (Spanish).

Range
Resident from San Luis Potosi and Tamaulipas south through eastern Mexico to the Yucatan Peninsula, Guatemala, and Belize.

North American Subspecies
C. c. curvipennis. Resident from Tamaulipas to Oaxaca, north of the Isthmus of Tehuantepec.

C. c. pampa (Lesson). Resident south of the Isthmus from Campeche and Yucatan to Guatemala and Belize.

C. (c.) excellens (Wetmore). Resident of southern Veracruz and western Chiapas in the area of the Isthmus of Tehuantepec. Often recognized as a distinct allospecies (e.g., Howell and Webb, 1995).

Measurements
Of *curvipennis:* Wing, males 65–68.5 mm (ave. of 9, 66.8 mm), females 60–68 mm (ave. of 4, 63.6 mm). Culmen, males 26–29.5 mm (ave. of 9, 26.6 mm), females 26–27.5 mm (ave. of 4, 26.6 mm) (Ridgway, 1911).

Of *excellens:* Mean measurements are greater; 22 males had wing chords averaging 71.1 mm, (SD 2.21 mm), 18 females had wing chords averaging 65.6 mm (SD 0.98 mm) (Winker et al., 1992).

Of *pampa:* Measurements (especially culmen length) of average less than those of *curvipennis.* Eggs, no information.

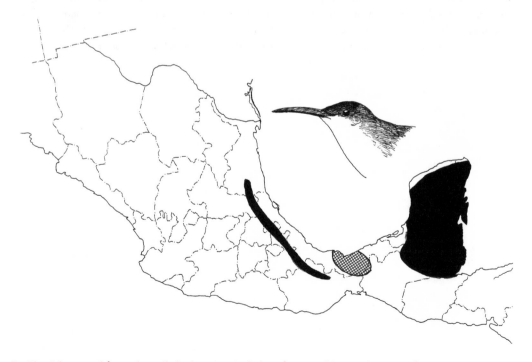

Residential range of the wedge-tailed sabrewing, including the typical form (inked) and the "long-tailed" Mexican endemic taxon (cross-hatching). *(Adapted from Howell and Webb, 1995)*

Weights

An unsexed sample of 20 birds from Mexico and Belize averaged 6.2 g (Feinsinger, 1976). Twenty-two males of nominate *excellens* had an average mass of 10.8 g (*SD* 1.05 g), and 18 females had an average mass of 6.8 g (*SD* 0.52 g) (Winker et al., 1992).

Description (After Ridgway, 1911)

Adult male. Crown bright metallic violet to purple, and the rest of the upperparts metallic green to bronze-green, tail-coverts and tail more bluish green; rectrices graduated in length, the middle pair distinctly the longest; remiges purplish dusky, and the underparts variably gray (darker in *pampa*); gray extends on the sides of the head to the auriculars; a small post-ocular white spot present; three outer primaries with shafts greatly thickened and flattened basally, these feathers strongly curved; Iris dark brown; feet brown; bill black above and somewhat lighter below, except in *pampa,* where entirely blackish.

Adult female. Virtually identical to the male, but with shorter central rectrices and smaller wing measurements (the mass differences cited above suggest an even greater degree of sexual dimorphism as compared with other North American hummingbirds); lateral rectrices tipped with brownish gray, the crown color slightly duller violet; shaft-thickening of the outer primaries more poorly developed than in males.

Immature. Similar to adults but with brownish buffy tones on the underparts, and the crown less glossy; juveniles of both sexes have buffy feather tips, less blue on the crown, and a duller green back color than adults (Winker et al., 1992).

Identification

In the hand. The enlarged shafts of the three outer primaries identify this as a sabrewing. The wedge-shaped tail, with the central feathers noticeably the longest and with no white spots on the outer rectrices, easily separates it from the violet sabrewing as well as from all other Mexican hummingbirds.

In the field. Easily recognized by its large size, a glittering violet-purple crown, green upperparts, white underparts, and wedge-shaped green tail. A small, inconspicuous white spot is present behind each eye. Calls include various squeals and squeaks, often uttered from hidden perches in vine tangles. The male's advertising song is a long series of hesitant squeaking notes that begin as a sequence of insect-like chips or trilled notes, then proceed into an extended gurgling warble or an excited series of jumbled notes. Thus, it is one of the most melodius singers of all North American hummingbirds, as implied by its vernacular name, nightingale hummingbird.

Habitats

Associated with open woodlands and humid, densely shaded forests, forest edges, and second growth, but sometimes also using more open habitats such as garden-edge locations. The species usually occurs at elevations of under 300 meters, but rarely extends as high as 1200 meters. Generally it is both solitary and inconspicuous, except perhaps when males are singing.

Movements

No long-distance movements have been documented, nor are they likely to occur in this apparently sedentary species.

Foraging Behavior and Floral Ecology

Foraging is done from low to high levels in the ecological strata of the forest, but no specific information is available on this species' preferred foods or its foraging behavior.

Breeding Biology

Singing by males occurs among dense tangles of vegetation, at heights varying from near ground level to high in the canopy. Birds may sing as solitary individuals or in groups of up to three birds. One singing male was observed to cock his wings, wave them back and forth with slow flapping movements, and simultaneously tilt his head back and turn it from side to side, while also spreading his tail and cocking it up and down. Singing is sometimes also done in flight, as the bird goes from perch to perch, and is often preceded by a short, trilled call. The song itself is an extended mixture of whistles and gurgles. The birds are smaller than violet sabrewings but have similar songs; interspecific aggression has been observed that was evidently the result of interspecific territoriality, in which male song similarities elicited aggressive responses from the larger violet sabrewing (Winker et al., 1992).

Nests of sabrewings are typically well-camouflaged and cup-like, and are attached saddle-like by spider webbing to horizontal branches. The nests are usually fairly substantial structures, with the usual hummingbird clutch of two eggs present. Nest records exist for the period March to July (Howell and Webb, 1995), but male singing in Veracruz has been observed from October through March. Males in breeding condition have been reported over a slightly broader period, between September and May (Winker et al., 1992).

Evolutionary and Ecological Relationships
Apart from an obvious relationship to the other sabrewings, no nearest relative is clearly apparent from plumage traits. The unusual wedge-like tail probably serves as much for regulating flight control as for display, and this species reportedly is able to shift from the typical hummingbird flight mode, with very rapid wingbeats, to a more typically avian pattern involving slower wing movements. No function for the greatly flattened and enlarged shafts of the outermost primaries has been proved, although because this trait is best developed in males it might be assumed that it has some social signalling function, rather than being a generalized aerodynamic adaptation. Gould (1849–61) suggested that the strengthened primaries might improve aerial dominance behavior during aggressive interactions. It also has been suggested that these specialized feathers somehow function as a sound-production mechanism during aerial displays (see next species account: violet sabrewings). Yet, if anything, the enlarged shafts should reduce the degree of primary vibration during high-speed flight, rather than increase it. Certainly little wing noise is made by violet sabrewings during normal flight. In addition, Greenewalt (1960a) stated that in the sabrewings the primaries "straighten out" during flight, so perhaps the modified shafts simply improve aerodynamic efficiency in some manner that might relate to their reputed flexibility in flying methods.

VIOLET SABREWING

Campylopterus hemileucurus (Lichtenstein)

Other Names
DeLattre sabrewing; Chupaflor, Fandangero morado (Spanish).

Range
Resident in southern Mexico from Veracruz and northern Oaxaca (with possible local breeding in Guerrero) south through Chiapas to Guatemala, Belize, and the rest of Central America south to Panama.

North American Subspecies
C. h. hemileucurus (Lichtenstein). Resident from Mexico south to Nicaragua.

Measurements
Wing, males 76.5–82.5 mm (ave. of 38, 79.5 mm), females 68.5–78 mm (ave. of 30, 73.9 mm). Culmen, males 26–31 mm (ave. of 38, 28.9 mm), females 30–35.5 mm (ave. of 30, 32.1 mm) (Ridgway, 1911). Eggs, 16.8–17.3 × 11.6–11.8 (Wetmore, 1968).

Weights
An unsexed sample of 56 birds averaged 10.5 g (*SD* 0.5 g) (Brown and Bowers, 1985). The average of 4 males was 6.4 g (range 6.2–6.6 g; that of 4 females was 6.4 g (5.0–7.1 g) (Smithe and Paynter 1963).

Description (After Ridgway, 1911)
The sexes differ as adults, but in both sexes the shafts of the outer three primaries are variably enlarged and strongly curved, more strongly so in males.

Adult male. A dusky forehead and crown, but the rest of the head (except for a small white post-ocular spot) bright metallic blue-violet, this color

Residential range of the violet sabrewing. Arrowhead indicates an extralimital breeding location. *(Adapted from Howell and Webb, 1995)*

extending down the back and underparts, becoming more greenish on the upper tail-coverts and upper wing-coverts; remiges dull blackish or faintly glossed with purple; middle pair of rectrices bluish black, with the lateral three pairs broadly tipped with white; iris dark brown; feet brown; bill black.

Adult female. Generally duller and more metallic green where the male is violet, but often some blue or violet on the throat (as spots or a solid patch of color); underparts gray.

Immature. Young males resemble adult females, but are glossed with bluish green below, and increasingly slow metallic violet above as they mature. Immature females have more bronzy green upperparts than adults, and have little or no blue present on the throat.

Identification
In the hand. Like the violet sabrewing, the shafts of the three outer primaries are greatly enlarged over much of their length, but in this species the tail is not wedge-shaped (central rectrices 50–60 mm), and the outer rectrices are extensively white.

In the field. Among the easiest of Mexican hummingbirds to identify because of its large size and overall violet (males) or green (females) color, except for the white patches on the outer tail feathers, which seemingly are constantly flashed in flight as it hovers above flowers or threatens other birds. When landed on a food source, it often holds its wings open vertically, while waving them rapidly, as if to keep other birds away. Its vocalizations include squeaky notes that range from weak calls to loud rattles or chips and a song that includes sharp chipping notes and warbles.

Habitats
Associated with dense montane forests and cloud forests, including forest edges and openings. It usually resides above 500 meters' elevation, and breeds mainly above 1000 meters and extends locally to 2500 meters. It also at times descends almost to sea level, at least as a transient. In western Mexico, Hutto (1985) found it to be mostly associated with pine-oak-fir forests (89 percent of observations), and secondarily with tropical

deciduous forest (11 percent). In Costa Rica it favors wet mountain forests and their edges (Stiles and Skutch, 1989). It is not present in the low-altitude forest of La Selva, and it is an uncommon nonbreeder at 1000 meters at La Montura, a premontane forest on the Atlantic slope. At Monteverde (1300–1500 meters) it is a common breeding resident of the cloud forest. In Panama it occurs as low as 400 meters outside the breeding season, but mainly breeds between 1500 and 2400 meters.

Movements
No movements have been documented, but there may be some seasonal southward movements along the Caribbean coast of Mexico during winter (Howell and Webb, 1995).

Foraging Behavior and Floral Ecology
These large hummingbirds tend to forage in undergrowth vegetation, including understory shrubs such as *Cephaelis* (Rubiaceae). Wagner (1946a) observed that when *Marcgravia* (Marcgraviaceae) is in bloom the birds drink its abundant nectar and eat insects that are caught in it. Insects that are attracted to the small, green blossoms of the planted *Inga* (Leguminaceae) trees are also greatly favored. At times they also hawk small insects in flight, such as gnats.

Skutch (1967) observed that the birds forage on a variety of ornithophilous plants including *Heliconia* species and also sip nectar from the flowers of cultivated bananas. He once observed a female or young male take possession of a scarlet passion flower (*Passiflora vitifolia*), a species having a deeply flower with a sleeve of thick tissue that prevents short-billed hummingbirds from reaching the nectar. This bird was able to drive its bill past the protective sleeve and reach the nectary in the same manner and with the same effectiveness as did the long-billed and green hermit hummingbirds.

When they encounter smaller species of hummingbirds, they easily dominate them and readily threaten or chase away any intruders. They nevertheless often operate as a trapliner species, sometimes feeding on flowers of the shrub *Malvaviscus arboreus* (Malvaceae) as well as on the previously mentioned *Inga* trees (Feinsinger, 1976).

Breeding Biology

The only significant available information on the breeding behavior and biology of this species was provided by Skutch (1967). He observed a singing assembly of four males, strung out over a forest edge at intervals of about 15 meters, along the shrubby margins of a heavy montane forest that filled a deep ravine. The males sat on perches that were generally about 5 meters high and well hidden, but were sometimes visible from the adjoining pasture. Their songs were extended vocalizations made up of a train of single notes, with a slow tempo and frequent squeaky notes. Tail-spreading, or tail-spreading in combination with a rapid vibration of the feathers, exhibited the white-tipped outer feathers in a stunning visual display. Singing began fairly early in the morning and continued until late afternoon. The singing assemblies (and associated breeding behavior) evidently are active nearly throughout the year in that region of Costa Rica, heard as late as mid-September by Skutch.

Surprisingly little information exists on aerial displays, and possible associated social signalling functions of the three specialized outer primaries in adult males. Sick (1993) stated that during "sky-diving" before the female, the male produces a wing-generated sound similar to the sounds made (by tail vibrations) in male common snipes during their somewhat similar aerial displays. However, no firsthand account seems to be available to confirm this possibility.

Skutch (1967) reported finding only a single nest of this species. It was in a montane forest at about 1150 meters' elevation, on the horizontal branch of a small tree leaning out above a narrow and steep ravine. It was about 4 meters above the bottom of the ravine, and was a deep, open cup, with a covering of green moss draped all around it. It was so large and bulky that Skutch initially thought it was the nest of a flycatcher. The nest contained two eggs when discovered in late May. The incubating female usually sat with the axis of her body at right angles to the branch, but sometimes was parallel to it, with her head, tail, and wings all projected above the rim of the nest. The female attended her nest an estimated 64 percent of the time during one morning, with incubation sessions lasting from 11 to 85 minutes (5 sessions averaging 43 minutes each), and recesses of from 10 to 40 minutes (5 recesses averaging 24 minutes). The nest disappeared prior to hatching, perhaps having been swept away by floodwaters. There is also a Costa Rican nesting record (eggs collected) for early July.

Evolutionary and Ecological Relationships

This is a very distinctively plumaged species, with no obvious plumage similarities associating it with either of the other two Central American sabrewings. Probably its nearest living relative is the white-tailed sabrewing of Venezuela, Trinidad, and Tobago, which is quite similar in its plumages and has even larger white tail markings than are present in this one.

WHITE-NECKED JACOBIN

Florisuga mellivora (Linnaeus)

Other Names
Jacobin; Chupaflor mielero, Chupaflor nuqui-blanco, Colibri nuquiblanco, Jacobino nuquiblanco (Spanish).

Range
Resident in southern Mexico, from Veracruz and northern Oaxaca eastward through Guatemala and Belize, and continuing south throughout Central and tropical South America to Peru, Bolivia, and Brazil.

North American Subspecies
F. m. mellivora. Resident from Veracruz and Oaxaca south to northern South America.

Measurements
Wing, males 62–71.5 mm (ave. of 46, 68.4 mm), females 62–70 mm (ave. of 16, 65.2 mm). Culmen,

males 16.5–22 mm (ave. of 46, 19.3 mm), females 17.5–22 mm (ave. of 16, 19.0 mm) (Ridgway, 1911). Eggs, 15 × 10 mm (Wetmore, 1968).

Weights
An unsexed sample of 28 birds averaged 7.4 g (*SD* 0.63 g) (Brown and Bowers, 1985). Eight Field Museum male specimens averaged 6.8 g (range 6–8 g), 9 females averaged 6.6 g (range 5.5–8.4 g). Five males from Costa Rica averaged 7.02 g (*SD* 0.38 g), 5 females averaged 6.48 g (*SD* 0.26 g) (Stiles, 1995). Five unsexed birds averaged 6.8 g (range 6–8 g), 3 males averaged 6.6 (range 6.5–7 g) (ffrench, 1991).

Description (After Ridgway, 1911)
Adult male. Head, neck, and chest metallic blue to violet-blue; behind this the sides of the neck, upper breast, and entire upperparts metallic green, except

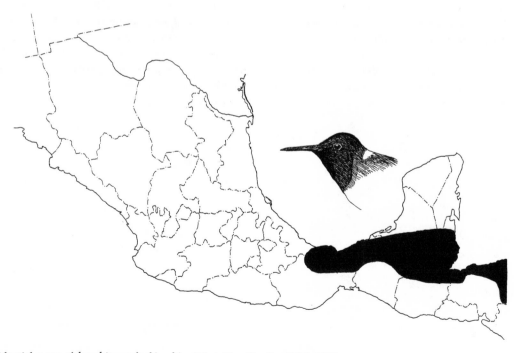

Residential range of the white-necked jacobin. *(Adapted from Howell and Webb, 1995)*

for a white crescent-shaped mark around the lower hindneck; tail white, the rectrices narrowly edged with black; remiges slate-black, and underparts forward to the breast pure white; iris dark brown; feet dusky; bill dull black.

Adult female. Upperparts metallic bronze-green or greenish bronze; tail more bluish green, with rectrices becoming blackish toward their tips and variously tipped with white or grayish; chin, throat, and chest white with dusky to metallic green spots that produce a scale-like pattern, and the sides of the breast also mottled or scaled like the underparts; abdomen and flanks white; the rear underparts becoming more dusky; sometimes quite like males, with blue heads, white nape-bands, and white tails.

Immature male. Similar to adult males but duller, with the central rectrices mostly dusky and broadly black-tipped; rictal and malar areas tinted with light brown or cinnamon. There is probably a complete postjuvenal molt (Stiles, 1980).

Identification
In the hand. In adults of both sexes the metallic-colored, central upper tail-coverts are so elongated that they appear to be rectrices, hiding the actual central pair of rectrices, which are almost entirely white, as are the other eight true rectrices.

In the field. Males are easily identified by their rather large size and entirely white tail, as well as their distinctive violet-blue head. Females are strongly patterned with scaly or scalloped spots over their entire underparts, and the tail is white only at the tips of the outer feathers. Birds intermediate in plumage between these types are not rare; they include young males and male-like females. The calls include squeaky notes, trilling, and various "rolled" vocalizations. During aerial foraging or feeding, the tail may be somewhat cocked and flashed open occasionally.

Habitats
Associated with lowland forests, plantations, forest edges, second growth with scattered tall trees present, and forest streams, from sea level to about 300 meters' elevation. At La Selva, Costa Rica, 32 percent of 49 sightings of this species were

in forest understory habitats; 24 percent were in forest canopy sites; and the rest were in forest-edge, light gaps, or nonforest habitats. It is likely that males mainly occur in the forest understory and females in the canopy, with both sexes using light gaps, but this sexual difference is not yet fully established (Stiles, 1980).

Movements
No movements have been documented by banding. The species disappears from and evidently moves out of La Selva between August and November, and at the same time becomes abundant at Tortugero and other coastal sites in eastern Costa Rica, suggesting some seasonal migrations (Stiles, 1980). The species is also known to be a long-distance migrant in South America (Snow and Snow, 1972).

Foraging Behavior and Floral Ecology
In Trinidad, 57 percent of 34 foraging observations occurred among vines, and the remainder among large trees. Nearly all (90 percent of 34 records) of the flowers visited were red; the majority (64 percent) of 28 observations involved flowers with corolla tubes 20 to 29 millimeters; and most feedings (86 percent of 27 records) of the flowers had corolla tubes 4 to 5 millimeters in diameter (Snow and Snow, 1972).

Foraging occurs at a wide variety of flowering leguminous trees (*Inga, Bauhinia, Erythrina*), other tropical flowering trees (*Vochysia*, Vochysiaceae; *Symphonia*, Guttifereae), woody vines (*Norantea*, Marcgraviaceae), and epiphytes (*Columnea*, Gesneraceae), as well as various *Heliconia* species (Stiles and Skutch, 1989; ffrench, 1991). The vine *Norantea*, which may reach the tops of the tallest trees and whose red flowers have a corolla length of 23 millimeters, is especially attractive to this long-winged and strong-flying hummingbird.

Only a relatively small percentage of the species' foraging time is spent chasing insects, and this is done by hawking rather than surface-searching. Hawking for aerial insects over extended periods and at heights of about 10 to 15 meters is usually performed within the confines of the forest, but these birds also have been seen chasing insects as high as 50 meters above jungle rivers (Snow and Snow, 1972).

Breeding Biology

Males are reported to be usually solitary, but small parties have been seen, calling and chasing one another, suggesting that lek-like aggregations may be part of the species' social behavior. The male-like plumages of some females are of interest, and deserve attention as to their possible significance.

Nesting activity in Costa Rica has been reported to occur between January and June, and in Panama active nests have been found in January and February. In Mexico the breeding period evidently extends from April to August. The nest is a cup-like structure, placed at heights of 1 to 3 meters on the upper surface of a leaf of understory palms, in locations protected from above by another leaf (Stiles and Skutch, 1989).

Evolutionary and Ecological Relationships

The nearest probable relative of this species is the black jacobin of eastern Brazil, a species in which the male is nearly all blackish violet except for the contrasting white tail.

GREEN VIOLET-EAR

Colibri thalassinus (Swainson)

Other Names
None in general English use; Verdemar, Colibre orejiviolaceo verde, Chupaflor pavito (Spanish).

Range
Breeds in the upper tropical and temperate zones of Mexico from Jalisco, San Luis Potosi, and Hidalgo south through Middle America to Peru and Bolivia. Accidental in North America (Texas).

North American Subspecies (After Friedmann et al., 1950)
C. t. thalassinus (Swainson). Breeds commonly in the temperate zone (up to 2850 meters in Michoacan) from Jalisco and San Luis Potosi south to Guatemala. Probably winters at lower elevations.

Measurements
Wing, males 63–70 mm (ave. of 6, 66.9 mm), females 60–63 mm (ave. of 16, 61 mm). Culmen, males 18–22 mm (ave. of 6, 20.2 mm), females 19–22 mm (ave. of 16, 20.3 mm) (Ridgway, 1911). Eggs, ave. 13.6 × 8.8 mm (extremes 13.1–13.9 × 8.7–9.1 mm).

Weights
The average of 24 individuals of both sexes was 5.24 (Feinsinger and Colwell, 1978); averages of 15 males and 12 females were 5.7 and 5.0 g, respectively (Wolf et al., 1976). An unsexed sample of 84 birds had a mean mass of 5.9 (*SD* 0.34 g) (Brown and Bowers, 1985).

Description (After Ridgway, 1911)
Adult male. Above metallic green or bronze-green, darker or duller on pileum; tail metallic bluish green or greenish blue (middle rectrices sometimes green or even bronze-green) crossed by a broad subterminal band of blue-black; remiges dark brownish slate or dusky, very faintly glossed with

Residential range of the green violet-ear. (*Adapted from Howell and Webb, 1995*)

bronzy purplish or violet, the secondaries more or less glossed at tip and on edges with metallic green or bronze-green; loral, suborbital, and auricular regions rich metallic dark violet-blue, sometimes invading (somewhat) sides of neck; malar region, chin, and throat bright metallic green—varying from slightly bluish to yellowish emerald green—each feather with a darker mesial streak or spot; center of chest dark metallic blue or violet-blue, passing laterally into metallic-green; breast, sides, flanks, and abdomen metallic green or bronze-green, duller posteriorly, where sometimes slightly broken by faint pale grayish brown or pale grayish buff tips or margins to the feathers; under tail-coverts metallic green or bronze-green, margined or edged with pale buffy brownish or grayish buffy; bill dull black; iris dark brown; feet dusky grayish brown.

Adult female. Similar to the adult male, but smaller and slightly duller in coloration.

Immature. Similar to the adult female, but much duller in coloration, the upper tail-coverts and feathers of rump (sometimes pileum also) narrowly and indistinctly margined with pale grayish buffy; green of underparts much duller and suffused with grayish, the feathers indistinctly margined terminally with pale grayish; blue of chest absent or only slightly indicated.

Identification

In the hand. The only member of the violet-ear group likely to be found in North America; the nearest other species is *C. delphinae* (of Guatemala southward), which is generally brownish rather than metallic green. The presence of a subterminal band of bluish black on the tail, the ear-patch, and the large size (wing 60–70 mm) all serve to identify the species.

In the field. Occurs in oak woods and clearings, and appears almost uniformly green; utters a loud double note, *chip-tsirr*, often while high in trees, repeated endlessly during the breeding season; also utters snapping little rips (Slud, 1964).

Habitats

In Mexico, this species breeds in high mountain forests. In the Valley of Mexico these include not only the original forests of mixed oak, cypress, and pines, but also cutover areas with overgrown gullies, shrubbery, and remaining high trees at the edges of fields. Between breeding seasons the males sometimes occur in forests of firs (*Abies religiosa*) at 2640 to 3450 meters of elevation in the Valley of Mexico, whereas in Chiapas they have been found at corresponding periods in open areas or clearings of forests between 990 and 1980 meters of elevation (Wagner, 1945).

Farther south, the species is also associated with the highlands; in Guatemala it breeds from 1500 to 3300 meters of elevation, in Costa Rica from 1500 to 3000 meters, and in Venezuela from 900 to 3000 meters. Probably in all these areas the higher altitudes are associated with the breeding season, and lower altitudes are used outside the breeding season. On the Sierra de Tecpam, Guatemala, the birds breed mainly in forests of cypress (*Cupressus bethamii*) above 2700 meters, where forest openings or edges provide abundant flowering plants, particularly certain species of *Salvia* (Skutch, 1967). Generally in Costa Rica the species seems to thrive in parklike pastures and "elfin woodland" that succeed the partial removal of mountain forests. Its total vegetational range extends from the middle of the subtropical belt upward to timberline on the high volcanoes (Slud, 1964).

In El Salvador, the birds have been recorded from 1900 meters in the pine zone to 2400 meters in the cloud forests, and are found mainly in sunny areas such as along trails or clearings (Dickey and van Rossem, 1938). In Honduras the species is common in cloud forests, and less frequently occurs in pine-oak forests, from 990 to at least 2370 meters' elevation (Monroe, 1968).

In Panama, the birds occupy open mountain slopes, where shrub and tree growth is scattered or open. Their altitudinal range there is from 1335 to 2070 meters, or (rarely) to 2625 meters (Wetmore, 1968).

In South America the species is found in the subtropical and temperate zones, extending in Venezuela through cloud forest and open woodland to the edge of the páramo, between 890 and 2970 meters' elevation (de Schauensee and Phelps, 1978).

Movements

This is a distinctly mobile species, which may perform some real migrations in Mexico. Wagner (1945) believed that at the northern end of the species' range in the Valley of Mexico all the females, the young, and a variable percentage of the adult males fly south in October or early November and return to their breeding grounds in July, possibly migrating as far as Chiapas, Guatemala, or even farther south. Males, however, remain in the area in varying numbers, moving to fir forests where flowering plants are available in all but the driest winters, when they too may have to migrate.

Skutch (1967) found no evidence of violet-eared hummingbirds crossing the Isthmus of Tehuantepec. Instead, he believed that probably the birds of central Mexico simply move to lower altitudes during the nonbreeding season, as is the case elsewhere in Central America. He noted that, although the birds breed in Costa Rica at altitudes above 1650 meters, during the dry season they can be found on the plains at the base of the mountains at about 630 meters. There are differences in the timing of these seasonal migrations, which are associated with regional differences in the length and intensity of the dry season. Thus, in the western highlands of Guatemala the dry season lasts from mid-October to mid-May, whereas in the higher parts of the Costa Rican mountains it extends only from about January to early April. Likewise, in El Salvador, there are apparently seasonal shifts of one or both sexes. Dickey and van Rossem (1938) found only females and young of the year in the vicinity of Los Esemiles, the males apparently having moved to higher or lower altitudes.

Probably the few occurrences of this species in the United States can be attributed to their mobile tendencies. As summarized by Oberholser (1974), there were three or four sightings of the birds in Texas during the 1960s and early 1970s. The first was seen in Hidalgo County, in heavy brush of the Santa Ana National Wildlife Refuge, during July 1961. Another was seen in Cameron County in April 1964, feeding on shrubs with ruby-throated hummingbirds. A third was seen and photographed at various times between August 25 and September 18, 1969, in Travis County, between Austin and Oak Hill; during the same period there was a possible sighting in Austin at a *Leucophyllum*

bush. A definite fourth record occurred in 1976, when a bird was seen at a feeder in the Wimberly area of Hays County between July 3 and August 13 (*American Birds* 30:96). A year later the species was seen at the same site between May 21 and July 6 (*American Birds* 31:199). There are now about 16 records for Texas (*American Birds* 48:963). There is also a recent sight record for southern Ontario (*American Birds* 45:1113).

Foraging Behavior and Floral Ecology

In Mexico, the distribution of montane flowers, especially several species of *Salvia,* seems to determine the local distribution of this species (Wagner, 1945). Likewise, in Guatemala several mint species, including both red-flowered types such as *S. cinnabarina* and the blue-flowered *S. cacaliaefolia,* are favorite foraging plants. There the birds also feed on the scarlet blossoms of a bean (*Phaseolus*) that climbs over old stumps in cornfields, the purple-white flowers of an alpine thistle (*Cirsium consociatum*), and the red-flowered cupheas (*Cuphea infundibulum*) found in pastureland groves. They have also been seen, along with several other hummingbird species, feeding on a planted shrub (*Stachytarpheta*) sometimes used in hedgerows through the pastures (Skutch, 1967).

Limited studies involving colored sugar solutions (Lyerly et al., 1950) with a single female of this species indicated that, among the four colors used—red, green, blue, yellow—she preferred yellow-colored solutions significantly less than the others. There were no significant differences among the other colors.

Observations by Lyon (1976) in the montane pine habitat of Oaxaca indicated that males of this species controlled relatively small feeding territories (average of four, about 230 square meters). They dominated the smaller white-eared and bumblebee hummingbirds, but were in turn dominated by the blue-throated and magnificent hummingbirds. Some violet-ears established and maintained their territories for short periods, but eventually these birds were displaced by blue-throats or magnificent hummingbirds.

A study by Wolf et al. (1976) in Costa Rica indicated that this species concentrated most of its foraging and all of its territorial behavior on only two species of flowers (*Centropogon valerii* and

C. brumalis). The birds also often obtain insects by hawking them in the air (Skutch, 1967).

Breeding Biology
In some parts of Mexico males are evidently sexually active almost the entire year, being silent only during the molt in April and during days of unfavorable weather. They typically choose exposed perches, usually high in a tree, and continuously utter their repeated calls. Often several males gather in a small area, but this is probably because of the distribution of food plants rather than any social tendencies (Wagner, 1945).

The "song" of the male is a metallic *k'chink chink k'chinky chink,* endlessly repeated from morning to evening, often uttered from an exposed dead twig between 4.5 and 12 meters above ground, rarely much higher or lower. Singing is not restricted to a single perch, but rather the bird may use several twigs separated by 6 to 9 meters, frequently moving from one to another. The birds begin to sing at dawn, as soon as the light is strong enough to see form and color, and they pause during the day only to forage, court females, and chase intruders from their territories. During an 83-minute period of observations, one male foraged on *Salvia* blossoms for about 7 minutes and spent more than 70 minutes singing from his perches. The periods of singing ranged from less than a minute to 11 minutes of nearly continuous vocalizations (Skutch, 1967).

Males usually sing in sight and hearing range of several other males, but they leave one another alone as long as their territorial limits are respected. Depending on the type of terrain and the locations of good perching sites, these territories may be from 45 to 90 meters apart. In both Costa Rica and Mexico a particular male having a distinctively recognizable voice may defend the same territory in subsequent seasons; in Mexico one such bird occupied the same tree for four consecutive years (Wagner, 1945).

Wagner described three different forms of calls, based on three rising degrees of sexual excitement. The usual one, apparently comparable to that described by Skutch as repeated metallic notes, was described by Wagner as variants of *huit ti titatia, huit tita.* During the peak of sexual excitement the male continuously repeats a *huitta*

huitta phrase, or a softer variant during flight. Often at this time he will fly from tree to tree in an undulating flight, keeping his wings spread and quivering for several seconds on alighting. This level of display occurs when the females have completed their nests and are looking for mates. Evidently the far-reaching call and display flights of the males help the females to locate them.

As soon as a female comes into view, the male follows her through the length of his territory. The two birds fly side by side in a fluctuating wavy path through a stretch that probably includes the nesting area of the female. In the later stages of the flight they may clap their wings in a pigeon-like manner. The female may then descend from the crown of a tree in a wavering flight with wing-clapping, skirt the ground, and then fly up and perch on a small twig. She repeats this movement until the male leads her on a final wild flight preceding copulation. Actual copulation has not been described, but it evidently does not occur in the vicinity of the nest (Wagner, 1945). However, in the related species *C. coruscans,* copulation occurs immediately after a hovering display flight by the male directly in front of the female (R. J. Elgar, *Avic. Mag.* 88:26–33, 1982).

In central Mexico, females arrive at the breeding grounds at the end of July, and immediately begin to build a nest. There they produce only one brood per season, and replace lost clutches only if the nest is destroyed during the first half of the breeding cycle (Wagner, 1945). In Costa Rica nests have been found from October to March, and male singing extends over a somewhat longer period (September to March). The season in Costa Rica is long enough to allow for two or perhaps three broods (Skutch, 1967). In El Salvador, young birds are common during February and March, suggesting a similar breeding period there (Dickey and van Rossem, 1938).

Nests in Mexico have been found mostly in fairly open areas; five of eight found by Wagner were in densely overgrown barrancas with steep sides covered by small oaks. The nests themselves were on oak branches less than 2 meters above the ground. Another nest was in the branches of a small oak that had grown up from an old root, and the two remaining nests were placed on the forks of a stem of *Salvia polystachya* in dense brush growth.

Skutch found four nests in Guatemala among the horizontal lower branches of cypress saplings, from 1 to 2 meters above ground level. He found four more in Costa Rica, all of which were attached to downward-drooping stems or dangling roots or vines. The latter nests were made by the race *C. cabanidis,* and Skutch suggested that there may be some racial differences in nest construction and nest-site preferences. Although both races apparently prefer to build their nests close to the ground, *C. thalassinus* seems more regularly to construct its nest of mosses and to attach many shriveled leaves and grass blades to the outside. The many long, hanging grass leaves typical of at least the northern race easily distinguish the nests from those of other Mexican hummingbirds. The inner lining may be either of moss, or less frequently of plant down or feathers. The nests are from 50 to 75 millimeters in diameter, and are often slightly less deep, but grass strands may hang down from the sides by nearly as much as 250 millimeters.

Wagner's (1945) studies found that each female had a nesting territory of some 600 to 1000 square meters, which was surrounded by a neutral zone separating it from adjacent territories. Nests of the species were from 52 to 95 meters apart, but sometimes females of broad-tailed or white-eared hummingbirds would nest within 15 meters of an occupied violet-ear nest.

The exact length of time for nest construction has not been reported, although Wagner (1945) stated that from 55 to 65 days are needed to complete a nesting cycle—from the start of the building until the complete independence of the young. This includes a 16- to 17-day incubation period, a normal fledging period of 23 to 25 days, and a postfledging dependency period of 5 to 7 days; thus, building probably requires 10 to 20 days. Incubation must begin with the laying of the first egg, since the interval between the hatching of the two young may be as much as 24 hours. Moore (1947) reported a male of a related violet-ear species (*C. coruscans*) assisting in incubation, and Schäfer (1952) also observed one incubating and feeding young at a nest; Schmidt-Marloh and Schuchmann (1980), however, did not find positive evidence of this for the green violet-ear.

The fledging period varies considerably in this species, with extremes of 19 to 28 days depending on weather, food availability, and the number of young in the nest. In one case, by the 8th day after hatching the pinfeathers were beginning to break and the eyes were starting to open. By the 11th day the eyes were open, the back feathers were breaking free of their sheaths, and the pinfeathers of the tail were emerging. By the 15th day the body feathering was nearly complete, the primaries were 14 to 15 millimeters long, and the tail feathers were emerging from their sheaths. Feathering was completed by the 18th day, and by the 22nd day, at fledging, all the feathers were completely unsheathed. The bill was more than half the adult length and had become entirely black, having gradually changed from yellowish at hatching (Wagner, 1945).

Skutch (1967) has contributed some observations on the care of the young in this species. In a nest that contained two young that were 5 and 7 days of age, the female brooded them eight times over a 4-hour period, for intervals of from a few seconds to 24 minutes, but collectively for only 48 minutes (20 percent of the total period). During the afternoon hours the total brooding time was likewise only 22 percent of the total observation period. The young birds thus endured several hours of very cold, nearly freezing air, and also some periods of intense radiation during the middle part of the day. During an 8-hour period they were fed 21 times, apparently mostly *Salvia* nectar as well as minute insects caught by the mother in the air above the nest.

Evolutionary and Ecological Relationships

The green violet-ear is clearly part of the closely related *Colibri* group, and on the basis of plumage would seem to be the nearest relative to *C. coruscans*, which coexists with it in South America from Venezuela southward.

Not much is yet known of bird–plant relationships for this species, but Colwell et al. (1974) reported that the Costa Rican shrub *Centropogon valerii* is primarily pollinated by the green violet-ear. In the southern Costa Rican highlands the birds' territories often center on clumps of this plant, which they defend from their own and other species. Observations by both Skutch and Wagner suggest that various species of *Salvia* are also locally important plants for the green violet-ear.

GREEN-BREASTED MANGO

Anthracothorax prevostii (Lesson)

Other Names
Prevost's mango; Chupaflor pechiverde, Mango pechiverde (Spanish).

Range
Resident in Mexico from southern Tamaulipas on the east slope and Oaxaca on the west slope, southeast across the Isthmus of Tehuantepec to the Yucatan Peninsula, and along both slopes southward throughout Central America to Panama, where replaced by the recently recognized species *verguensis* (AOU, 1995).

North American Subspecies
A. p. prevostii. Resident from Tamaulipas to Costa Rica.

Measurements
Of *A. p. prevostii:* Wing, males 63.5–68.5 mm (ave. of 24, 66 mm), females 62–69 mm (ave. of 15, 64.7 mm). Culmen, males 24–29 mm (ave. of 24, 26.2 mm), females 25–31 mm (ave of 15, 28.1 mm) (Ridgway, 1911).

Weights
A sample of 14 birds of both sexes averaged 6.4 g (*SD* 0.44 g) (Russell, 1964; Dunning, 1993). Five females averaged 6.1 g (range 5.7–6.5 g); two males each weighed 5.8 g (Russell et al., 1979).

Description (After Ridgway, 1911)
Adult male. Metallic bronze-green or greenish bronze above, the middle pair of rectrices similar, but the other rectrices deep purplish maroon to violet-purple, and edged with black; remiges brownish slate, chin black, bordered by metallic blue-green on chest, sides of neck and head, breast, and flanks; under tail-coverts and feet dusky; iris dark brown; bill black.

Breeding (hatched), residential (inked), and non-breeding (stippled) ranges of the green-breasted mango. *(Adapted from Howell and Webb, 1995)*

Adult female. Less bronze above than the male; often pure metallic green; lateral rectrices with brownish gray replacing the purplish tones, and the feather tips grayish white; middle of the chin and throat uniform or patterned black; below this the median chest, breast, and abdomen metallic blue-green, bounded on each side with a narrow white stripe below the green flanks and sides of the head and neck.

Immature. Young of both sexes resemble adult females, but have white on the chin and upper throat; chestnut spotting is present on the chin and throat, sometimes also on the sides of the chest; there may be rufous borders or subterminal bands on many of the underpart feathers of immatures, and the median line is blacker in young males than in females.

Identification

In the hand. This fairly large hummingbird has a black, distinctly decurved bill (culmen 25–30 mm), velvety black median throat markings, and a medium-length tail (32–38 mm) that is either bright purple (males) or has conspicuous white feather tips (females).

In the field. These hummingbirds are relatively large, and the male has a distinctive purple tail. The male is otherwise almost entirely green except for a velvety blackish throat. Females are green above, with a white-tipped dusky tail and a distinctive green stripe extending down the mid-ventral area from the blackish upper throat to the abdomen, which is separated from the green sides and flanks by narrow white borders on either side. The vocalizations include repeated chipping notes, high shrill notes, and a song that consists of a series of high-pitched and thin *see* notes, often uttered in flight for from a high song perch.

Habitats

Associated with semi-open sites in relatively dry and open lowland forests, forest clearings, and forest edges. The species is especially typical of savannas, pasturelands, streamside borders, coffee plantations, roadsides, and brushy fields having scattered tall trees. It ranges from sea level to about 1500 meters' elevation.

Movements

No movements have been documented. In Brazil, banding studies have shown that the related black-throated mango may migrate over 1500 kilometers (Ruschi, 1967). There is also a recent photographically documented record of the green-breasted mango from Corpus Christi, Texas (*American Birds* 46:288).

Foraging Behavior and Floral Ecology

Flowering leguminous trees (*Inga, Caesalpina, Erythrina*) are favored food plants, as are various flowers of lianas, either for the plants' nectar or for the insects that they attract. The birds typically feed on flowers at fairly low heights, but also hawk insects well above land and may sing or quietly sit on fairly high perches. They also have been seen foraging on open fruit husks, evidently finding small invertebrates there (Wagner, 1946a).

Breeding Biology

Little information is available. In Mexico the breeding season reportedly extends from March through June (Howell and Webb, 1995). In El Salvador breeding may occur somewhat later, from October to February; and in Costa Rica the nesting season extends from December to April or May, the relatively dry season, when most other hummingbirds are also breeding.

The nest is built on a bare branch that may be situated as high as 30 meters but sometimes is located fairly low in vegetation. Other closely related mangos typically build small, cup-like nests, saddled on horizontal branches. The usual hummingbird clutch of two white eggs is laid, and in the case of the closely related black-throated mango the incubation period has been established at 16 to 17 days. The fledging period for that species is usually 24 days, and at least on Trinidad there may be two or three broods per season. The interval between nestings ranges from 9 to 26 days when the same nest is used for succeeding broods, and up to 37 days when a new nest is constructed (ffrench, 1991).

Evolutionary and Ecological Relationships

This is only one of many species of rather robust hummingbirds known as mangos, a name based on the species' well-known attraction to flowers of the mango tree (*Mangiphera,* Anacardiaceae) in the West Indies. The Veraguan mango of Panama is certainly the nearest relative of the green-breasted mango (Wetmore, 1968), but as of 1995 was regarded by the American Ornithologists' Union (AOU) as a distinct species. The black-throated mango, ranging from Panama southward well into South America, is also an obvious, very close relative, the adults differing mainly in the underpart markings of both sexes.

EMERALD-CHINNED HUMMINGBIRD

Abeillia abeillei (Lesson and DeLattre)

Other Names
Abeille hummingbird; Chupaflor de barba esmeralda, Colibri barbiesmeralda (Spanish).

Range
Resident in Mexico from Puebla and Veracruz south to Oaxaca and Chiapas, and in Central America south to Nicaragua.

North American Subspecies
A. a. abeillei. Resident from eastern Puebla and western Veracruz south to Honduras.

Measurements
Of *A. a. abeillei:* Wing, males 44–50.5 mm (ave. of 16, 47.8), females 41.5–46 mm (ave. of 3, 43.8 mm). Culmen, 10–11.5 mm (ave. of 16, 10.6 mm), females 11–12 mm (ave. of 3, 11.5 mm) (Ridgway, 1911). Eggs, no information.

Weights
Four birds of unstated sex averaged 2.7 g (Dunning, 1993).

Description (After Ridgway, 1911)
Adult male. Metallic bronze-green or greenish bronze on the crown and entire upperparts, including the middle pair of rectrices; other rectrices black, the feathers tipped with brownish gray; a small postocular white spot, and the sides of the head and lower throat blackish (these areas surround a gorget of metallic emerald green); remaining underparts mostly brownish gray; iris brown; feet brownish; bill black.

Adult female. Similar to the adult male, but the lateral rectrices tipped with pale brownish gray, and the underparts much paler grayish white, with no gorget or black present.

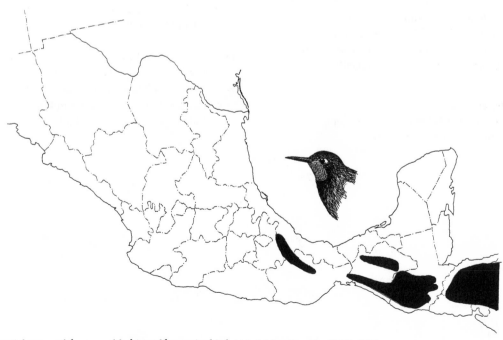

Residential range of the emerald-chinned hummingbird. *(Adapted from Howell and Webb, 1995)*

Immature male. Similar to the adult female, but with some scattered emerald green feathers in the throat (Howell and Webb, 1995).

Identification
In the hand. The very short, straight bill (culmen, 10–12 mm) separates this species from all other Mexican hummingbirds except for (a) the two coquettes, which have white rump bands that are lacking in the emerald-chinned, and (b) the similarly small sparkling-tailed hummingbird, which also has distinct white rump patches.

In the field. This is a very small hummingbird with a very short, straight bill, and white post-ocular spots in both sexes. The male has a glittering green gorget surrounded by velvety black, and dark gray underparts; the female is pale grayish white in these same areas. The vocalizations include a liquid, rattling trill, and the male's "song" is a series of high, thin, and slightly squeaky *chips* uttered singly or in clusters of notes (Howell and Webb, 1995).

Habitats
Associated primarily with cloud forests, but extending to humid evergreen or mixed pine-evergreen forests and forest edges, usually between 1000 and 2000 meters' elevation. It usually forages inconspicuously at low heights and rarely extends into open habitats.

Movements
No movements have been documented in this essentially unstudied species.

Foraging Behavior and Floral Ecology
No specific information is available on this short-billed, tiny hummingbird, which is inconspicuous and elusive. Only flowers with very short corollas can be exploited by this species (unless the hummingbird can pierce the corolla of long-tubed species and steal the plant's otherwise unavailable nectar); therefore, the species is likely to be rather flexible and opportunistic in its food sources. Like other very short-billed hummingbirds, it is likely to pierce when trying to obtain nectar from blossoms with deep corollas. Wagner (1946a) observed the birds feeding on insects attracted to arums (Araceae) and also stated that they hovered up and down the trunk of a tree, occasionally stopping to obtain an insect from its surface. He thought that dipterans constituted the major insect prey of Mexican hummingbirds, although spiders were captured and fed to the young in large numbers during their nestling periods.

Breeding Biology
Little information is available. The males are reported to sing from low- to medium-height perches in the forest understudy; and the nest is a deep cup, attached to a vertical shoot, placed fairly low in the understory vegetation. Nesting in Mexico probably occurs during February and March (Howell and Webb, 1995).

Evolutionary and Ecological Relationships
This species is usually placed in a monotypic genus, but is probably a fairly close relative of the violet-headed hummingbird, a similarly small and short-billed species ranging from Honduras to Bolivia.

RUFOUS-CRESTED COQUETTE

Lophornis delattrei (Lesson)

Other Names
DeLattre coquette, Short-crested coquette (*brachy-lopha*); Coqueta corona leonada, Coqueta cresti-corta (*brachylopha*) (Spanish).

Range
Endemic resident in the mountains of Guerrero as an isolated (possibly specifically distinct) popula-tion; also occurs disjunctively (in a long-crested form) in Costa Rica, Panama, and western South America from Colombia to Peru and Bolivia.

North American Subspecies
L. d. brachylopha (Moore). Endemic to Guerrero. Considered a distinct species by Banks (1990) and Howell and Webb (1995), but not by Ornelas (1987) nor apparently by Sibley and Monroe (1990).

Measurements
Wing, males 39.9–41.9 mm (ave. of 3, 41.0 mm), females 42.4–45.2 (ave. of 2, 43.8). Culmen, 3 males 12.7 mm, females 12.8–13.0 mm (ave. of 2, 12.9 mm). Crest of males, 10.5–11.7 mm (vs. 17.9–21.5 mm in males of the southern population) (Ornelas, 1987). Measurements of the few-known *brachylopha* specimens (as here reported) are similar to those of the population from Panama and Colombia, but are slightly larger; male crest lengths are much shorter. Eggs, no information.

Weights
A mean mass of 2.8 g has been reported for *delattrei* (Stiles and Skutch, 1989). No weights from birds representing the endemic Mexican taxon are available.

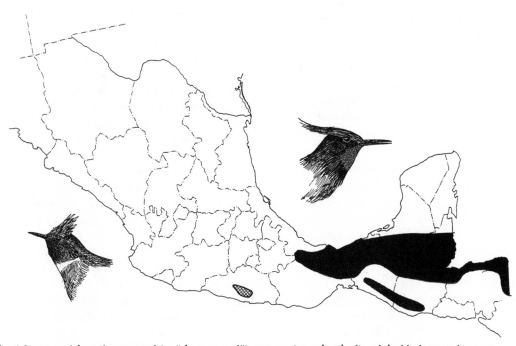

Residential ranges of the rufous-crested (or "short-crested") coquette (cross-hatched) and the black-crested coquette (inked). *(Adapted from Howell and Webb, 1995)*

Description (Partly after Ridgway, 1911, and Based on the Nominate Taxon)

Adult male. Crown and elongated crest of narrow, rather rigid feathers cinnamon rufous; the nape, hindneck, and most upperparts bronze-green, but a band of white to cinnamon-blue crosses the upper rump; central rectrices dark bronzy, but the others cinnamon-rufous, with blackish tips; chin and throat metallic yellowish emerald green, the feathers with cinnamon-buff bases (this color slightly exposed on the sides of the throat); underparts dull metallic bronze-green except for a white throat-band below the gorget; iris brown; feet grayish brown; bill red to reddish brown, with a grayish tip (in the nominate populations), or blackish (in the Mexican taxon). The male specimen shown in the painting by Mark Marcuson (Plate 3) is of the long-crested, reddish-billed Central and South American form, whereas those shown in the painting by Sophie Webb (Plate 24) are of the endemic Mexican taxon, the adult male having a short, bushy crest, and a black bill.

Adult female. Differs from the male in lacking a crest; forehead tinted with rufous, the throat also cinnamon-rufous to cinnamon-white (in the nominate form) or white (in the Mexican taxon); dusky spots on the lower throat sometimes forming a transverse patch, the lateral rectrices tipped with rufous, and with broad black subterminal bands; central rectrices metallic green, like the back; abdomen grayish brown, the flanks unspotted brownish gray, and the under tail-coverts cinnamon to pale cinnamon; bill more dusky toward the tip than in the male.

Immature male. Apparently much like the adult female, but the forehead, lores, chin, and upper throat all light cinnamon-rufous to neutral gray; older immature males with various amounts of green or dull black in the throat region, and the cinnamon-colored crest variably developed.

Identification

In the hand. The very short, straight bill (culmen 9–11.5 mm) and white bands across the rump identify this as a coquette, and the whitish upper breast and unspotted flanks separate it from the black-crested coquette.

In the field. This tiny hummingbird is readily identified by its whitish rump band and rufous forehead or crest, as well as its short, straight bill. The only other Mexican hummingbird with a continuous white rump band is the black-crested coquette, which is geographically separated from this species. Usually seen perched on twigs, it typically wags or pumps its tail while hovering or foraging. Its calls are still undescribed.

Habitats

Associated with humid evergreen or semi-humid forests, forest edges, or forest openings, including pine-evergreen forests and semi-deciduous forests, mostly at about 1000 to 1500 meters' elevation, possibly drifting lower when not breeding.

Movements

No movements have been documented. Coquette hummingbirds probably wander rather widely in their search for foods associated with small and abundant flowers, and they are too small to compete successfully for food with other hummingbirds in the vicinity. They thus "trapline" within broad home ranges, rather than try to establish localized feeding territories. However, in homing experiments performed by the late Brazilian hummingbird expert A. Ruschi, marked frilled coquettes did not return when displaced more than 15 kilometers (cited in Sick, 1993).

Foraging Behavior and Floral Ecology

No detailed information exists on this tiny species. It is known to forage around trees such as the leguminous *Inga* species, which is often planted in Central America to provide shade for coffee bushes. Trees of this genus have white flowers with a relatively short corolla length (approximately 8–14 mm) and apparently produce large amounts of nectar (Stiles, 1985). Coquette hummingbirds of closely related species in Costa Rica also are attracted to hedges of *Stachytarpheta* (Verbenaceae), which produces many small purple flowers (Skutch, 1961), and to the ubiquitous second-growth trees of the genus *Cecropia* (Cecropiaceae), whose numerous but minute flowers may offer abundant pollen but probably provide no nectar. Coquette hummingbirds often perch on exposed twigs, and while in flight pump their tails up and down, in the same distinctive manner as woodstars. Because of their rump-band, the birds often appear rather bee-like (Ridgley and Gwynne, 1989), and they are easily confused with similar-sized sphinx moths (Sick, 1993). Sick has suggested that the several similarities in appearance and flight behavior between moths and coquette hummingbirds, even including tail-pumping by sphinx moths of the genus *Aellopus*, may represent a case of adaptive Batesian mimicry on the part of these moths, which may gain some protection from aerial predators by resembling these relatively elusive birds.

Breeding Biology

No information exists for the Mexican population of rufous-crested coquettes, nor indeed for the somewhat better studied southern population. Courtship in the coquette hummingbirds apparently includes lateral oscillating shuttle flights of the male in front of the female (Skutch, 1961). The nest and eggs of both of the rufous-crested taxa are apparently still undescribed. In other members of this little-studied genus the nest is known to be a simple cup-like structure attached by spider webbing to a small, often relatively horizontal, branch or twig. For example, the nests of the white-crested coquette have been found attached to slender twigs as high as about 25 meters above ground, but most are evidently situated within about 6 meters of the substrate (Skutch, 1961). The incubation of the remarkably tiny (1.5–2.8 g) frilled coquette of South America requires only 12 to 13 days (Sick, 1993), which appears to be the shortest incubation period so far reported for any hummingbird species.

Evolutionary and Ecological Relationships

The nearest relative of this species is probably the tufted coquette, which occurs in northeastern South America and Trinidad. In adult males of this species, the brown feathers of the lower gorget are extended to an extreme length, and each is tipped with a glittering green spangle.

BLACK-CRESTED COQUETTE

Lophornis helenae (DeLattre)

Other Names
Princess Helena coquette; Coqueta crestinegra, Penachudo (Spanish).

Range
Resident in Mexico from southern Veracruz and northern Oaxaca eastward through Chiapas (including both the Pacific and Caribbean slopes) to Guatemala, thence south through Central America along the Caribbean slope to Costa Rica.

North American Subspecies
None recognized.

Measurements
Wing, males 38–42.5 mm (ave. of 21, 39.8 mm), females 37.5–41 mm (ave. of 12, 38.9 mm). Culmen, males 10–12.5 mm (ave. of 21, 11.3 mm), females 11–12.5 mm (ave. of 12, 11.3 mm (Ridgway, 1911).

Weights
An unsexed sample of 4 birds had a mean weight of 2.6 g (Brown and Bowers, 1985); 2.8 g has been reported as a mean mass (Stiles and Skutch, 1989), as has 2.9 g (Stiles, 1985).

Description (After Ridgway, 1911)
Adult male. Crown dark metallic green, with filamentous feathers of the occipital crest greenish black; nape, hindneck, and posterior upperparts metallic brown-green except for a band of white or buffy crossing the rump; central rectrices bronzy basally, the others cinnamon-rufous, with bronzy blackish on the outer webs; chin and throat feathers a brilliant metallic yellowish green, bounded posteriorly by a patch of velvety black feathers on the lower throat, flanked by longer feathers of buff, streaked on their upper webs with black; breast, underparts, sides, and flanks white, spotted with

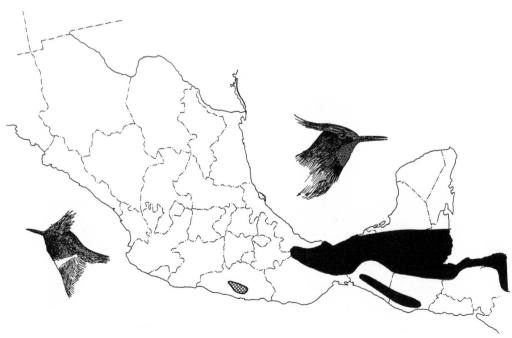

Residential ranges of the rufous-crested (or "short-crested") coquette (cross-hatched) and the black-crested coquette (inked). *(Adapted from Howell and Webb, 1995)*

metallic bronze; Iris dark brown; feet brownish; bill red, with a darker tip.

Adult female. Metallic green above, with a narrow white band across the rump, as in males; the central rectrices also metallic-colored, with blackish tips, and the lateral rectrices cinnamon-rufous, with a subterminal black band; the sides of the face blackish, and the throat brownish buff to cinnamon-buff with many dusky or bronzy flecks; the breast more uniformly metallic bronze, and the other underparts white, spotted with metallic brown; bill probably brown to blackish above, and reddish below.

Immature male. Resembles the adult male, but the occipital crest less developed, and the gorget area whitish, with dark flecking.

Identification

In the hand. The very short, straight bill (culmen 10–12.5 mm) and the white rump-band identify this species as a coquette, and it is further separated from the rufous-crested coquette by a black facial pattern and brown-spotted underparts.

In the field. This tiny hummingbird is easily recognized by the combination of its blackish face patch and a complete white rump band. The red bill is short and straight. Males also have a distinctive filamentous crest and a black and buff necklace below their glittering green gorget. The birds produce an insect-like humming in flight, but otherwise are mostly silent. Their probable song has been described as a repeated, upslurred *tsuwee* (Howell and Webb, 1995).

Habitats

Associated with humid forests, forest edges, forest openings, and clearings or coffee plantations, usually between 500 and 1500 meters' elevation. In Honduras the birds are associated with rainforests below 750 meters' elevation (Monroe, 1968).

Movements

No movements have been documented, but the coquettes are believed to be fairly mobile hummingbirds, in spite of their small size.

Foraging Behavior and Floral Ecology

Little specific information exists. As a food plant Stiles (1985) listed only *Hampea appendiculata* (Tiliaceae), a tree with moderately abundant whitish yellow flowers that have an effective corolla length of 8 millimeters and apparently high rates of nectar production. This observation was made by Stiles at a premontane rainforest site (La Montura, Costa Rica), at an elevation of about 1000 meters, and where this coquette is rare, and apparently only a non-breeding visitor. According to Monroe (1968), poinsettias are a popular food plant in Honduras.

Breeding Biology

The biology of this species is essentially unknown. Birds in breeding condition have been documented during January in Guatemala, but nests have not been described. Other species of *Lophornis* are known to have the typical hummingbird open-cup style nest, placed on top of a more or less horizontal branch or sometimes a fork, and attached by spider webbing to the supporting structure (Sick, 1993). The nestling period of the related white-crested coquette has been established at 21 to 22 days (Skutch, 1976), among the shortest-known fledging periods reported for hummingbirds generally (Skutch, 1973).

Evolutionary and Ecological Relationships

The unique male plumage of this species offers no obvious clues as to its nearest living relative. The dark plumage of both sexes suggest affinity with species such as the festive coquette, of northeastern South America.

FORK-TAILED EMERALD COMPLEX
(GOLDEN-CROWNED, COZUMEL, AND CANIVET'S EMERALDS)

Chlorostilbon canivetii (Lesson)

Other Names
Canivet's emerald (*canivetii*), Salvin's emerald (*osberti*), golden-crowned emerald (*auriceps*), Cozumel emerald (*forficatus*); Chupaflor esmeralda, Esmeralda de Canivet (*canivetii*), Esmeralda de Cozumel (*forficatus*), Esmeralda Mexicana (*auriceps*) (Spanish).

Range
Three well-defined populations exist in North America. One resides in Mexico from Tamaulipas south through the Caribbean slope to the Yucatan Peninsula, northern Guatemala, and Belize. The second occurs along the Pacific slope from Sinaloa and Durango south to Oaxaca. The third

is an insular population on Cozumel Island. Other populations of this species (depending on how it is delineated taxonomically) extend south in Central America, through Panama, to Venezuela and Colombia. It also occurs on the Caribbean islands of Curacao, Aruba, Bonaire, and Trinidad.

North American Subspecies
C. (c.) canivetii (Canivet's emerald). Resident from Tamaulipas to Honduras.

C. (c.) auriceps (Gould). Golden-crowned emerald. Resident from Sinaloa south to Oaxaca. Considered a distinct species by Howell (1993a) and

Residential range of the fork-tailed emeralds, including the taxa *canivetti* (inked), *osberti* (cross-hatching), *auriceps* (diagonal hatching), and *forficatus* (horizontal hatching). Arrowheads indicate additional insular breeding locations. (*Adapted from Howell and Webb, 1995*)

Howell and Webb (1995); recognized by the American Ornithologist Union (AOU, 1995).

C. (c.) forficatus (Cozumel emerald). Resident on Cozumel and adjoining islands. Considered a distinct species by Howell (1993a) and recognized by the AOU (1995).

Measurements
Of *C. (c.) auriceps:* Wing, males 43–44 mm (ave. of 4, 43.4 mm), 2 females, 43 and 44 mm. Culmen, males 13–13.5 mm (ave. of 4, 13.2 mm), 2 females, 13 and 15.5 mm (Ridgway, 1911). Egg (of *canivetii*), 11.9 × 7.2 mm (Rowley, 1966).

Weights
Seven males of *assimilis* (in Costa Rica) averaged 2.62 g; 12 females averaged 2.48 g (Feinsinger, 1976). Six individuals (sex unspecified) of *assimilis* averaged 2.34 g (*SD* 0.21 g) (Tiebout, 1993). An unspecified sample of *C. (c.) auriceps* averaged 2.1 g (*SD* 0.3 g) (Arizmendi and Ornelas, 1990).

Description (After Ridgway, 1911)
The various taxa occurring in Mexico differ mostly in tail-length and degree of tail-forking; the tail is longest and most deeply forked in *C. (c.) auriceps* (males 38–45 mm, females 30–34 mm) and *C. (c.) forficatus* (males 39–44 mm, females 31–35 mm), and is shortest and least deeply forked in *C. (c.) canivetii* (males 30–38 mm, females 27–32 mm).

Adult male. Crest and rest of head golden green to golden, and the remaining upperparts ranging from golden green to golden bronze; tail glossy blue-blue to blackish, with the middle rectrices tipped with brownish gray; underparts also bright metallic golden green, except for white femoral tufts; Iris brown; feet dusky; bill carmine, with a dusky tip.

Adult female. Similar to the adult male on the upperparts, but somewhat duller; the tail less strongly forked and mostly metallic green basally, with darker subterminal areas and grayish tips, especially laterally; a dusky auricular patch,

margined above with a white postocular stripe and below with brownish gray, which extends back from the throat to the under tail-coverts; bill mostly bright red, as in male.

Immature. Male similar to the adult female, but with varying amounts of metallic green spotting on the throat and underparts, the wing-coverts more greenish blue, a more strongly forked tail; later like the adult male, but lacking the metallic crown patch, and the abdomen dusky gray, spotted with metallic green; bill probably also more blackish than in adults. Young females probably also have less colorful bills that show striations on the upper mandible.

Identification
In the hand. The combination of a distinctly forked tail, with the outer tail feathers 30 to 48 millimeters long and the middle ones 14 to 22 millimeters; and a rather short, straight, and mostly bright red bill (culmen 13–16 mm) identify this species.

In the field. Males of this species group have a distinctively forked tail and a bright red bill, but otherwise are nearly uniformly glittering green. Females have a slightly forked tail. They are green above and rather uniformly grayish below, with distinct white stripe behind and above the eye, and a rather dusky mask from the eye backwards. Their vocalizations include repeated wiry *tseee-tseeree* notes and various other dry-sounding rattles and chatters.

Habitats
Associated with diverse habitats, including open woodlands, second growth, shrubby forest edges, brushy habitats, and similar rather open environments from sea level to nearly 2000 meters. Hutto (1985) reported that in a western Mexico survey, 83 percent of this species' sightings were in tropical deciduous forest and 17 percent were in cloud forest.

In a sample of 57 sighting locations at La Selva, Costa Rica, Stiles (1980) observed males more or less equally divided among forest edges or light-gaps (31.5 percent), forest canopies (28 percent, but especially favored during the dry season), understory sites (17.5 percent), and nonforested sites (21 percent). By comparison, among 52 sightings, females were most often seen in forest understory sites (40 percent), and secondly were observed in forest edge or light-gap sites (31 percent), with the remaining sightings mainly occurring in nonforested sites (17 percent). In Jalisco, Mexico, most observations of foraging were observed at subcanopy and middle understory levels (Arizmendi and Ornelas, 1990).

Movements

No specific information on movements is available, but this species is apparently something of a wanderer, moving from place to place as food supplies permit.

Foraging Behavior and Floral Ecology

This species generally favors small, short-tubed, and generally insect-pollinated plants, including a variety of shrubs, herbs, and vines not exploited by other hummingbirds (Howell and Webb, 1995). In a study of a Mexican hummingbird guild, Feinsinger (1976) observed that the fork-tailed emerald exploited a wide range of resource species and flower densities, and acted as a non-territorial trapliner species mainly in competition with the green violet-ear, which it appears was able to outcompete. Similarly, studies by Arizmendi and Ornelas (1990) indicate that, because of its small size, it is excluded from favored food plants by larger hummingbirds and even large bees. Rather, it feeds mostly from insect-adapted and short-tubed plants such as the arborescent *Vitis mollis*, which alone accounted for 66 percent of more than 2800 feeding observations.

Breeding Biology

Breeding in this species group (at least *C. (c.) auriceps*) occurs during spring, between February and July, in Mexico (Howell and Webb, 1995).

The breeding season is later in Costa Rica, occurring after the rainy season, from November to March or April.

Wolf (1964) described a single nest he found in Oaxaca during March. The nest was located neara road, and in dense, low, second-growth, about75 centimeters above ground. It was a pendant structure, attached on one side to a broad leaf, and on the other side to two grass stems. The body of the nest mainly comprised bark strips, lined internally with a downy material. A similar pendant nest, found by Rowley (1962), was constructed with lateral supports in a manner resembling that of a red-eyed vireo. Nests from Costa Rica have been described as having distinctive strips and chips of bark, and placed as high as 3 meters above ground in herbaceous plants. Rowley (1966) also mentioned seeing a nest that had been collected by a native boy. That nest had been placed in a low bush about a meter above ground and was close to a running creek.

In other species of this genus, the nest is typically a suspended structure, attached to a hanging wire, root, or grass stalk, and sometimes situated beneath an overhanging bank. In the related glittering-bellied emerald, the female may spend ten days completing her nest; in that same species, a nest may be superimposed on a previous season's nest (Sick, 1993).

The nest observed by Wolf already contained two eggs at the time of discovery, and one of the eggs hatched that same day. The second egg hatched two days later, but this youngster died soon after hatching, and its body was left in the bottom of the nest. Feeding of the chick averaged about twice per hour for the first 15 days of observation, and between three and four feedings per hour during the last 10 days. By the time the surviving young was 10 to 12 days old, its eyes were opening, and wing-fluttering was first seen at 18 days. By 25 days the chick could flutter for more than a meter, and fledging had occurred by the time the nest was visited the next day.

Evolutionary and Ecological Relationships
Howell's (1993a) analysis of morphology and
color resulted in his recommendation for splitting
the previously recognized single species ("fork-
tailed emerald") into the three listed here. Howell
also suggested merging the Central American
forms *osberti* and *salvini* into a separate species
(*C. salvini*, or "Salvin's emerald"). This form
presumably is the nearest relative of the *canivetti*
species group.

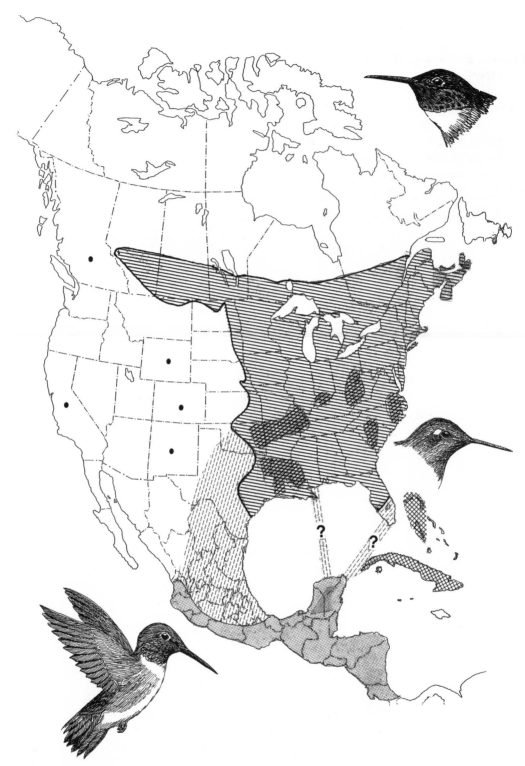

Breeding (hatched) and wintering (stippled) ranges and migration routes (broken hatching) of the ruby-throated hummingbird. Areas of greater breeding densities are shown by cross-hatching. Dots represent states or provinces with extralimital sightings. The Florida–Yucatan migration is speculative; occasional overwintering occurs on Cuba. The residential range of the Cuban emerald is also shown (cross-hatching).

CUBAN EMERALD

Chlorostilbon ricordii (Gervais)

Other Names
God bird; Zunzún, Zumbador, Zumbete, etc. (Spanish).

Range
Resident on Cuba, the Isle of Pines, and the northern Bahama Islands (Grand Bahama, Abaco, and Andros; rare and local on New Providence). Straggles occasionally to southern Florida.

North American Subspecies
? (actual subspecies unknown.)

C. r. bracei (Lawrence). Resident on the Bahama Islands.

C. r. ricordii (Gervais). Resident on Cuba and the Isle of Pines.

Measurements
Of *C. r. ricordii*: Wing, males 50–55 mm (ave. of 13, 52.3 mm), females 48–52.5 mm (ave. of 12, 50.7 mm). Culmen, males 14.5–18.5 mm (ave. of 13, 17.1 mm), females 17.5–19 mm (ave. of 12, 18.1 mm) (Ridgway, 1911). Eggs, ave. 12.62 × 8.2 mm (extremes of 5, 12.5–13.0 mm × 8.0–8.3 mm) (U.S. National Museum specimens).

Weights
The average weight of 4 males from Cuba was 3.35 g (range 3.25–3.75 g); that of 3 females was 3.54 g (range 3.25–3.87 g) (George Watson, personal communication).

Description (After Ridgway, 1911)
Adult male. Above dark metallic bronze-green, darker and decidedly duller on pileum; four middle rectrices dark metallic bronze or greenish bronze, the next pair similar but with inner webs greenish black; two outer pairs of rectrices greenish black or black faintly glossed with bluish green or greenish blue, the outer web of next-to-outer pair slightly bronzed; remiges dark brownish slate or dusky, faintly glossed with violaceous; underparts brilliant metallic green (yellowish emerald green); femoral tufts and under tail-coverts white, the latter sometimes with a few small spots or streaks of grayish on lateral feathers; maxilla dull black; mandible pinkish with tip dusky; iris dark brown; feet dusky.

Adult female. Above similar in color to adult male; beneath brownish gray (between drab gray and smoke gray); the sides, from neck to flanks inclusive, metallic green, with feathers gray beneath surface; anal tufts white; a grayish white postocular spot; bill, etc., as in adult male.

Identification
In the hand. The only large hummingbird likely to appear in the southeastern United States (peninsular Florida) that has a strongly forked blackish or violet tail, a partially exposed nasal operculum, a wing length of at least 48 mm, and a small white spot behind the eye.

In the field. Appreciably larger than the ruby-throated hummingbird and extensively and uniformly green above and below. The tail is blackish and deeply forked in males and less so in females, and the under tail-coverts are white or mostly white. Both sexes have a small white spot behind the eye, and females have whitish underparts, becoming greenish on the sides. It is normally found in wooded areas and copses, and in Florida has been observed mostly during fall and winter months in southern areas.

Habitats
This species is widely distributed on Cuba, apparently occupying a variety of lowland habitats, but probably primarily open forests of both humid and arid types. It may also occur in the mountains of Cuba. On Great Bahama the birds frequent brushy undergrowth of open pine woods, and generally occur where there is any considerable growth of bushes on both of the larger and smaller islands. Barbour (1943) stated that they occurred in parks, gardens, and wild, open country; and George

Watson (personal communication) found them widely distributed in mixed forests in the Sierra Crystal in Oriente, coastal scrub forest at Playa Giron, and garden parkland in Soledad.

Movements

This species is resident in Cuba and the northern Bahama Islands. However, it is an occasional vagrant to New Providence Island and has appeared in southern Florida on several occasions. The first of these was in 1943, when an individual was seen by numerous observers from October 20 to November 8. It was next seen on June 12 and 15, 1953, about 16 kilometers south of Cocoa (Sprunt, 1954).

Individuals were observed at Naranja in January 1961 and at Stock Island on March 27 of that year (*Audubon Field Notes* 15:323). In 1964 the species was again seen at Cocoa Beach between October 10 and 19 (*Audubon Field Notes* 18:26), and a female was observed at Hypoluxo on August 10, 1977 (*American Birds* 31:991).

Foraging Behavior and Floral Ecology

No specific information is available on these subjects. The species seems to be a generalist forager, "occurring wherever there are flowers" (Barbour, 1943).

Breeding Biology

Chapman (1902) found a nest on the branch of a coffee bush in southern Cuba in March. It was composed of green mosses, bound about with strips of bark, which hung in flowing streamers 12 to 13 centimeters below. Two eggs were present. According to Barbour (1943), nests could be found in any month. Renesting or multiple nesting was evidently frequent, since one bird that lost her initial nest in a hurricane built a new one on a chandelier. Four broods were raised from this single nest, and never once was her mate observed during this period. The incubation and fledging periods are unreported, but studies of a related species—the fork-tailed emerald—indicated a fledging period of 24 to 25 days (Wolf, 1964). Courtship and mating displays are also unknown, but have been described for the related blue-tailed emerald (Elgar, 1980).

Evolutionary and Ecological Relationships

This species' nearest relative is the Hispaniolan emerald, which Ridgway (1911) included with it in the genus *Riccordia,* mainly because their outer tail feathers are relatively wide as compared with those of more typical species of *Chlorostilbon.*

There have been no descriptions of any hybrids involving the Cuban emerald, or of its ecological relationships with other hummingbirds and plants.

DUSKY HUMMINGBIRD

Cynanthus sordidus (Gould)

Other Names
None in general English use; Colibri prieto, Chupamirto prieto (Spanish).

Range
Endemic resident in Mexico from Michoacan southeast to Oaxaca.

North American Subspecies
None recognized.

Measurements
Wing, males 56.5–57 mm (ave of 3, 56.8 mm), female 55 mm. Culmen, males 27.5–29 mm (ave. of 3, 28.3 mm), female 29 mm (Ridgway, 1911). Eggs, 12.5–12.7 mm × 8.3–8.5 mm (Rowley, 1984).

Weights
Three birds of unspecified sex had a mean mass of 4.4 g (range 4.3–4.7 g) (Dunning, 1993).

Description (After Ridgway, 1911)
Adult male. Upperparts generally dull metallic bronze-green to greenish bronze, the tail duller and more grayish, with dusky bases, especially laterally; a small postocular steak of whitish extends back above the dusky auriculars; underparts deep sooty gray, darkest on the throat, but with white femoral tufts and a tuft of white feathers on each side of the rump; iris brown; feet dusky; bill red, with a dusky tip.

Adult female. Very similar to the adult male, but with a broad subterminal area of gray or brown on the lateral rectrices.

Immature. Similar to adults, but the outer rectrices narrowly edged with buffy (Howell and Webb, 1995); the bill is likely variably blackish, with parallel oblique striations on the upper bill surface, as in apparently all young hummingbirds.

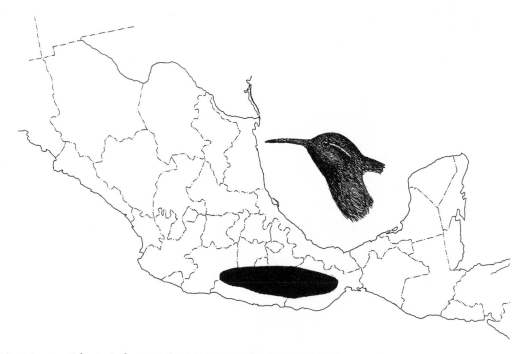

Residential range of the dusky hummingbird. *(Adapted from Howell and Webb, 1995)*

Identification

In the hand. This hummingbird can be identified by the combination of its dusky grayish underparts and a nearly straight red bill that is broadened at the base as in the broad-billed hummingbird, its likely nearest relative. However, the broad-billed hummingbird is greener above, a clearer gray below, and has more conspicuous white tail spotting.

In the field. This medium-sized hummingbird is one of the dullest in color of all North American species, with dusky gray underparts, dull bronzy-green upperparts, and a dark-tipped red bill. The whitish postocular stripe, above a dusky auricular area, is one of the few useful fieldmarks. Vocalizations include chattering calls, buzzy chipping notes, and a series quiet warbles.

Habitats

Associated with open habitats, especially arid subtropical scrub habitats such as roadsides, gardens, and similar brushy sites. Sites with available elevated perches such as agaves are favored. The species' overall range is about 900 to 2200 meters, but most nesting records are from about 1600 meters' elevation (Binford, 1989). Hutto (1985) reported that 66 percent of his habitat sightings were in thorn forest; 26 percent were in disturbed deciduous forest; and 7 percent were in tropical deciduous forest.

Movements

No movements of any kind have been documented.

Foraging Behavior and Floral Ecology

This is a little-studied species that is said to forage in the middle to upper levels in trees (Howell and Webb, 1995), but no specific information on favored food plants is available.

Breeding Biology

Birds in nesting or breeding condition have been reported from March to May, in August, and in November and December (Howell and Webb, 1995). It appears that the species is flexible and adapts its nesting to periods of available moisture in this semi-arid region. One of the few accounts of breeding is that of Rowley (1984), who found two nests with eggs and an active nest without eggs during the months of November (the nest without eggs), December, and March. Like those of broad-billed hummingbird, the nests were small, cup-like structures, constructed in apparently similar manner. Nothing else of significance is known about their breeding biology, but it probably closely resembles that of the broad-billed hummingbird.

Evolutionary and Ecological Relationships

This species is presumably most closely related to the broad-billed hummingbird, which has a slightly overlapping range and a plumage pattern that is extremely similar but only slightly more colorful and contrasting.

BROAD-BILLED HUMMINGBIRD

Cynanthus latirostris (Swainson)

Other Names
None in general English use; Chuparrosa matraquita, Chupaflor piquiancho (Spanish).

Range
Breeds from southern Arizona, southwestern New Mexico, and southwestern Texas south to Chiapas on the west coast of Mexico, including Tres Marias Islands, and to Tamaulipas and northern Veracruz on the east coast (Friedmann et al., 1950). Northern populations winter in Mexico.

North American Subspecies
C. l. magicus (Mulsant and Verreaux). Breeds in southern Arizona, southwestern New Mexico, and (rarely) southwestern Texas south on both slopes of the Sierra Madre Occidental to Colima and Aquascalientes, Mexico.

C. l. propinguus (Moore). Resident of central Mexico from Guanajauto to Michoacan.

C. l. toroi (Berlioz). Resident of Michoacan and northern Guerrero.

C. l. doubledayi (Bourcier). Resident on Pacific slope of western Guerrero and western Chiapas. Includes *nitida* (Salvin and Godman). Sometimes regarded as a distinct species, but the race *toroi* is somewhat transitional (Binford, 1989).

Measurements
Wing, males 49–57 mm (ave. of 31, 51.6 mm), females 49.5–54 mm (ave. of 19, 51.6 mm). Culmen, males 18.5–22 mm (ave. of 31, 20.4 mm), females 19.5–23.5 mm (ave. of 19, 21.4 mm) (Ridgway, 1911). Eggs, ave. 12.6 × 8.5 mm (extremes 11.5–13.5 × 7.5–9.8 mm).

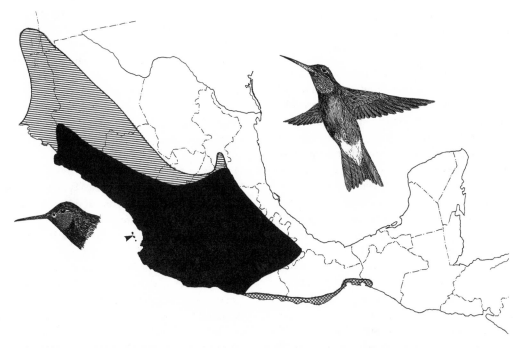

Breeding (hatched) and residential (inked) ranges of the broad-billed hummingbird. The residential range of the parapatric taxon *doubledayi* is shown (cross-hatching); breeding sites of the insular taxon *lawrencei* are also indicated (arrowhead). (*Adapted from Howell and Webb, 1995*)

Weights

The average of 7 males of *C. l. magicus* was 3.72 g (range 3.2–4.4 g); a single female weighed 3.4 g (specimens in Museum of Vertebrate Zoology, University of California). A mixed-sex sample of 19 birds had a mean mass of 3.1 g (*SD* 2.5–4.0 g) (Binford, 1989).

Description (After Ridgway, 1911)

Adult male. Above metallic bronze-green, usually duller on pileum, where sometimes passing into dull grayish brown on forehead; tail glossy blue-black or dark steel blue, the four middle rectrices tipped (more or less broadly) with deep brownish gray, the remaining rectrices sometimes narrowly margined with the same; remiges dusky brownish gray or dull slate color, faintly glossed with purplish, the outermost primary narrowly edged with pale gray or grayish white; chin and throat bright metallic greenish blue or bluish green (the color more blue anteriorly, more green posteriorly), passing into metallic bronze-green on breast, sides, flanks, and abdomen; under tail-coverts dull white, usually more or less distinctly grayish centrally, the shorter ones sometimes with dusky, slightly metallic, spots; and tufts and tuft on each side of rump white; bill purplish red or carmine, dusky terminally; iris dark brown; feet dusky.

Adult female. Above similar in color to adult male but duller, especially on pileum, which is usually dull grayish brown or brownish gray anteriorly; remiges paler grayish brown; middle pair of rectrices bronze-green passing into blue-black or greenish black terminally (the extreme tip sometimes green or bronzy); other rectrices with basal half (more or less) bronze-green, the remaining portion blue-black tipped with brownish gray (most broadly on lateral pair); underparts sooty gray or dark drab-gray (browner than mouse gray); the sides of chest glossed (more or less) with metallic green or bronze-green; under tail-coverts mostly dull white (grayish centrally); anal tufts and tuft on each side of rump white; a white or grayish white postocular spot, and below this a dusky area extending to beneath the eye; maxilla dull black, sometimes brownish basally; mandible dusky for terminal half (more or less), reddish basally; iris and feet as in adult male.

Immature male. Similar to the adult female but feathers of pileum, hindneck, back, scapulars, rump, etc., tipped or terminally margined with pale buffy brown or grayish buff; rectrices as in adult male; chin and throat (in older specimens) intermixed with metallic bluish green or greenish blue feathers, these margined terminally with pale grayish or buffy brown.

Immature female. Similar to the adult female but feathers of upperparts margined terminally or tipped with pale buffy brown (sometimes more cinnamomeous on pileum and rump).

Identification

In the hand. Likely to be confused only with the white-eared hummingbird, which also has a reddish bill widened near the base and a white stripe behind the eye. However, in this species the white eye-stripe is barely indicated, the black ear-patch is less definite, and the tail is bluish black and slightly notched in both sexes.

In the field. Found in arid country often dominated by mesquite or agaves, sometimes also in mountain canyons. Both sexes have reddish bills, distinguishing them from magnificent hummingbirds, and the male's lack of a long white eye-stripe distinguishes him from the male white-eared. Females also have less white behind the eye than do female white-eared, and they have more grayish underparts and a bluish black tail rather than a greenish one. They utter chattering kinglet-like calls, and the male produces a high-pitched humming sound during aerial display.

Habitats

In the United States this species is mostly limited in summer to rocky canyons in desert-like mountain habitats, where streams or springs provide growths of sycamores or mesquites. In Arizona it is fairly common in mesquite-sycamore vegetation from the Guadalupe Mountains westward to the Baboquivaris, and northward at least to the Santa Catalinas (Phillips et al., 1964). Nests in Mexico have been found from slightly above sea level in Sinaloa to 435 meters in Sonora, and most specimens from northern Mexico have been collected between 150 and 3000 meters' elevation (Moore, 1939b). Farther south, the species is found in nearly all habitats in Colima, from sea level to about 2100 meters, the upper limits in pine-oak forest. It is especially common there in thorn forest and on high grassy slopes in oak woodland (Schaldach, 1963).

Movements

In the United States, this species is clearly migratory. In the Rio Grande valley of Texas it has been reported from May 17 to October 20 (Oberholser, 1974). In Arizona, it is usually present from mid-March until mid-September, with occasional arrivals in early March and sometimes lingering until October 1 (Phillips et al., 1964). In Sonora, it is a permanent resident in the tropical zone northward to the vicinity of Guaymas along the coast and interiorly to the vicinity of Moctezuma (van Rossem, 1945). There is no good evidence of migratory movements farther south in Mexico.

Probable post-breeding wandering has resulted in some extralimital sightings in the United States. Thus, there is a single record of the species from Utah (*American Birds* 33:201), and at least seven records from California (*American Birds* 31:374). Most of the California records are for San Diego county from September to April, but there is one record as far north as Pacific Grove, Monterey County (*Audubon Field Notes* 23:622).

Foraging Behavior and Floral Ecology

The observations by Moore (1939b) are almost the only ones available on the foraging of this species. He observed the birds at flowers of the ocotillo, a yellow-flowered *Opuntia,* and paintbrush, but noted that the favorite foraging plant in Sonora was a red-flowered shrub called the "tavachin" (probably of the genus *Caesalpinia* or *Poinciana*), which was growing in sandy arroyos. A nesting female visited this species approximately every 15 minutes during the late afternoon.

In the United States, these birds have been observed foraging on ocotillos, agaves, pentstemons, and other blooming plants (Oberholser, 1974). Samples from the stomachs of four birds collected in Arizona provide some idea of their insect consumption, which included leafhoppers, jumping plant lice, root gnats, flower flies, spiders, daddy-longlegs, and miscellaneous bugs and hymenopterans (Cottam and Knappen, 1939).

Breeding Biology

Nesting in Sonora and Sinaloa apparently occurs over a wide span of time, probably from January to August, with nests being found in January, March, and May (Moore, 1939b). A series of 16 egg dates from Mexico extend from January 16 to May 21, and half of these fall between March 25 and May 11, indicating the peak of the season (Bent, 1940). Birds in breeding condition have been collected in Queretaro in late November and December, and nesting has been noted in December (Friedmann et al., 1950). Records for Arizona are rather limited, but Bent (1940) indicated that five egg records extended from April 14 to July 15. In Texas breeding apparently extends from early May to early August (Oberholser, 1974). There are no specific nest records for New Mexico.

Practically nothing has been written on display behavior, but notes by F. C. Willard (in Bailey, 1928) indicated that the male performs a "pendulum swing back and forth in front of the female," accompanied by a sound much like the *zing* of a rifle bullet that is higher in pitch than that of any of the other small hummingbirds.

Four nests found by Moore (1939b) in northern Mexico were all within 2 meters of the ground. All had some grass stalks in the body of the nest, were lined with white plant down, and were adorned with bits of leaves and bark on the outside surface; but none had any lichens. They all were very small as well, having an internal diameter of only about 19 millimeters. One was placed on a small tree overhanging the bank of an arroyo; another was attached to the stalk of a vine; a third was in a bush covered by dry vines; and the fourth was in an "espino" tree. Three of the nests contained two eggs each, which were laid two days apart in at least one case.

A nest found in Texas was situated on a triple fork of a small willow about 3 to 4 meters above the ground, almost overhanging the Rio Grande. One Arizona nest was also found over water, about 1.5 meters up in a small willow, and another was found in a hackberry bush 1.2 meters above a creek. Of three nests from Sonora, two were in mesquites and the other was in an apricot tree (Bent, 1940).

Brandt (1951) described some Arizona nests found in Sabino Canyon. One was on a small willow twig about a meter above ground, close to a stream channel. Nearby was an old nest on a dead, drooping willow twig, also only about a meter above water. Close to this old nest was a newly built nest on a drooping sycamore limb, a little higher above a dry stream bed. All three nests were apparently situated in such a way as to mimic the compact little balls of leaves and vegetation formed during periods of high water, and were largely composed of such vegetational debris, which made them extremely inconspicuous.

According to Oberholser (1974), the outside of the nest may be variably decorated with bits of plant stems, leaves, or even white cotton thread or the blooms of plants; only rarely does it include lichens.

No information is available on incubation and brooding behavior or on incubation and fledging periods.

Nesting success rates for 20 nests, as estimated by Baltosser (1986), included an incubation success rate of 75 percent, a nestling success rate of 88.9 percent, and an overall *nesting success* (i.e., percentage of nests producing at least one fledged young) of 66.7 percent. Predation by snakes was believed to be a significant source of nest losses. The estimated seasonal productivity per female, assuming that double brooding occurred during the long breeding season, was 2.32 fledged young per season. Baltosser attributed the relatively high nesting success of this species

to its choice of well-protected nest sites, which invariably were beside rock outcrops or in extremely dense thickets.

Evolutionary and Ecological Relationships
Short and Phillips (1966) described a hybrid obtained in the Huachuca Mountains involving this species and the magnificent hummingbird, and they commented on the similarities of the genera *Cynanthus* and *Eugenes*. They did not suggest that the genera should be merged, but believed that the two are not as distantly related as is implied in present classifications such as those of Peters (1945) and the American Ornithologists' Union (AOU, 1957).

Mayr and Short (1970) have concluded that the broad-billed hummingbird is a close relative of the dusky hummingbird of western and central Mexico, and hybridization between these two species has been reported (Friedmann et al., 1950). They did not consider these to represent a superspecies, but regarded them as closely related to *Amazilia*. No other definite hybrids are known, although—as noted in the account of the violet-crowned hummingbird—the type specimen of *Amazilia "salvini"* has at times been considered a hybrid between the violet-crowned and broad-billed hummingbirds.

Too little is known of the species' foraging ecology to comment on possible interspecific ecological relationships.

CROWNED WOODNYMPH

Thalurania (furcata) ridgwayi (Gmelin)

Other Names
Blue-crowned woodnymph (*townsendi*), Common woodnymph, Forked-tail woodnymph (*furcata*), Honduran woodnymph (*townsendi*), Mexican woodnymph (*ridgwayi*), Ridgway's woodnymph (*ridgwayi*); Ninfa Mexicana, Ninfa coronada, Ninfa azulada (Spanish).

Range
Resident in Mexico from Nayarit to Colima (*ridgwayi*); an apparent range gap exists from there to eastern Guatemala and Belize, where a similar taxon (*townsendi*) continues southeastward through Central America to Honduras. Other woodnymph populations extend through Costa Rica and Panama (*venusta*) to South America, where (depending on accepted taxonomic limits) the species reaches as far as Colombia and western Venezuela.

North American Subspecies
T. (f.) ridgwayi (Nelson). Resident from Nayarit to Colima. Considered a distinct species from the more forked-tailed Central American taxa (*townsendi* and *venusta*) by Escalante-Pliego and Peterson (1992) and by Townsend and Webb (1995). Judged conspecific with these and also with *colombica* of Panama and Colombia, as well as with the *furcata* complex of South America by the American Ornithologists' Union (AOU, 1983) and by Wetmore (1968).

Measurements
Of *T. (f.) ridgwayi:* Wing, adults ave. 56.27 mm (*SD* 0.36 mm). Exposed culmen, adults ave. 17.11 mm (*SD* 0.18 mm) (Escalante-Pliego and Peterson, 1992).

Of *townsendi:* Wing, males 53.5–55.5 mm (ave. of 3, 54.3 mm), females 48–51.5 mm (ave. of 4, 49.7 mm).

Residential ranges of the crowned woodnymph, including the Mexican endemic *ridgwayi* (cross-hatching) and the Central American taxon *townsendi* (inked). *(Adapted from Howell and Webb, 1995)*

Culmen, males 19.5–20 mm (ave. of 3, 19.7 mm), females 20–20.5 mm (ave. of 4, 20.1 mm) (Ridgway, 1911).

Weights

Five males of *ridgwayi* in the Field Museum had a mean mass of 3.8 g (range 3.5–4.2 g); 2 females each weighed 3.5 g. Thirteen males of *townsendi* had a mean mass of 4.59 g (*SD* 0.22 g); 9 females had a mean mass of 4.04 g (*SD* 0.15 g) (Stiles, 1995). Field Museum specimens representing various South American woodnymph taxa include 51 males with a mean mass of 4.4 g (range 2.5–5.2 g), and 24 females with a mean mass of 4.1 g (range 3.5–5.2 g).

Description *T. (f.) ridgwayi* (After Ridgway, 1911)

Adult male. Forehead and crown metallic violet-blue; otherwise the head, upperparts and anterior underparts dark metallic blue-green or bronze-green; tail black, with a faint bluish gloss, and the remaining underparts below the chest blackish (in most individuals) or greenish (in about a third of the population); iris brown; feet blackish; bill black.

Adult female. Similar to the male, but lacking the metallic crown patch, and instead bronzy green, like the other upperparts; the rectrices bluish black, the two outer pairs broadly tipped with grayish white; the sides of the head, throat, and underparts including the sides pale brownish gray, with dull grayish white under tail-coverts and femoral tufts.

Immature male. Resembles the female, but the underparts dark gray, the under tail-coverts bluish black, and some emerald green often visible on the throat; the tail less deeply forked than in adults (Howell and Webb, 1995). There is a complete postjuvenal molt in this species (Stiles, 1980).

Identification

In the hand. This is a rather small hummingbird, with a straight, black bill (culmen 19.5–20.5 mm) and a very slightly forked tail (the difference between the shorter central rectrices and the long outer rectrices averaging about 5.4 mm, according to Escalante-Pliego and Peterson, 1992) that is glossy bluish black in males and blackish but white-tipped in females.

In the field. The violet-blue crown of males, together with an otherwise almost uniformly glittering green body, dark grayish black underparts, and a glossy blue-black tail, should easily identify this species. Females are glittering green above and rather uniformly grayish below, with the corners of the tail tipped with white. Females of the stripe-tailed hummingbird are quite similar in size and appearance, but have white-striped tail feathers.

Habitats

Associated with humid forests, forest edges, clearings, streamsides, and coffee plantations, from sea level to about 1000 meters (locally to 1250 meters). Breeding occurs at fairly low altitudes, but outside of the breeding season the altitudinal range is greater.

Movements

No major movements have been documented. Stiles (1980) found that the birds moved from forest habitats to forest edges and second growth, as favored heliconias came into flower. One marked male moved more than a kilometer from a *Hamelia* tree to an *H. imbricata* patch, and a female moved more than half a kilometer from an *H. imbricata* site to an apparent nesting area. At La Selva, Costa Rica, much of the population seems to depart the entire area from October through December. At this time the birds become much more common along the coastal lowlands, but there may also be a movement up mountain slopes after breeding to non-breeding habitats as high as 1200 meters, where the birds evidently concentrate on *Heliconia* species.

Foraging Behavior and Floral Ecology

Stiles (1985) determined at La Montura, Costa Rica that crowned woodnymphs foraged on *Renealmia cernua*, a herbaceous member of the ginger family (Zingeribaceae) with orange flowers having a short corolla length (14 mm) and relatively low nectar production. Feeding also was seen on *Cephaelis elata* (Rubiaceae), a shrub or small tree bearing white and red flowers. It has a short corolla length (11 mm) and low nectar production but produces an abundance of flowers when in maximum blossom.

Breeding Biology
At La Selva, Costa Rica the breeding season is quite extended. Fifteen egg records were spread from December to June, with an apparent peak in April (27 percent) (Stiles, 1980). More generally in Costa Rica the breeding season extends from February to May or June (Stiles and Skutch, 1989). In Mexico, birds in breeding condition have been obtained during February and March (*ridgwayi*); in Central America, specimens of *townsendi* in breeding condition have been obtained during January and February (Howell and Web, 1995).

The nest of *Thalurania* species in general is a cup placed on a more or less horizontal branch or fork, but no specific information on the nesting of this widespread and often fairly common species is apparently available.

Evolutionary and Ecological Relationships
Other members of this genus occur in South America, such as the long-tailed woodnymph and violet-capped woodnymph, which differ from the present species by rather minor and mostly sex-limited traits such as the male's tail-length in the long-tailed species and the color of the male's underparts in the violet-capped species.

XANTUS HUMMINGBIRD

Hylocharis xantusii (Lawrence)

Other Names

Black-fronted hummingbird; Colibri de Xantus, Chuparrosa de Xantus (Spanish).

Range

Endemic resident of the southern part of Baja California, including islands of the Gulf of California north to Isla San José.

North American Subspecies

None recognized.

Measurements

Wing, males 50–54 mm (ave. of 10, 52.3 mm), females 43.5–52.5 mm (ave. of 10, 48.9 mm). Culmen, males 17–19 mm (ave. of 10, 17.7 mm), females 17–19 mm (ave. of 10, 17.8 mm) (Ridgway, 1911). Eggs, 11.4–13.5 × 7.7–9.9 (ave. of 20, 12.3 × 8.4) (Bent, 1940).

Weights

An unsexed sample of 11 birds had a mean mass of 3.5 g (Brown and Bowers, 1985).

Description (After Ridgway, 1911)

Adult male. Forehead, lores, auricular region, and chin all black, but the rest of the upperparts metallic bronze-green; rump and upper tail-coverts margined with rusty, and the middle pair of rectrices chestnut, with bronze-green edges; the other rectrices chestnut, sometimes with bronzy edgings; a conspicuous white superciliary stripe separates the black cheeks from the metallic green crown; the throat also metallic green, but becoming more yellowish on the breast and cinnamon-rufous on the posterior underparts and sides, with some metallic green spotting present on the sides; white femoral tufts present; iris dark brown; feet dusky; bill red, with a dusky tip.

Residential ranges of the Xantus (cross-hatched) and white-eared (inked) hummingbirds. The arrowhead represents an extralimital breeding location of the latter. *(Adapted from Howell and Webb, 1995)*

Adult female. Similar to the male but with a brownish gray forehead, and the rectrices with a variably evident blackish subterminal band; lower cheeks, chin, and entire underparts pale cinnamon-buff except for white femoral tufts; mandible duller red than in the male, and dusky for about the terminal half.

Immature. Males similar to the adult female, but the crown tinged with rusty brown, and the throat spotted with metallic or yellowish green. Females similar to adult females, but the bill more blackish, and in both sexes with parallel oblique ridges on the upper mandible.

Identification
In the hand. The nearest relative of this species, the white-eared hummingbird, has overlapping wing, culmen, and tail measurements; therefore, *xantusi* is best recognized by its cinnamon underparts and the chestnut to cinnamon (rather than blackish) tones of the lateral rectrices.

In the field. Easily identified in its native Baja habitat, where no other similar species occurs. The very similar white-eared hummingbird of mainland Mexico lacks the cinnamon tones of the underparts present in this species. Vocalizations include buzzy and chipping notes, and the apparent male song is a series of soft and buzzy notes that may approach a twittering or warble.

Habitats
Associated with arid scrub, brushy woodlands, and gardens, from sea level to 1500 meters. Arriaga et al. (1990) suggested that this species is probably mutually dependent on the endemic species of madrone (*Arbutus peninsularis*, Ericaceae). The hummingbird feeds almost exclusively on this tree's nectar during late-winter months, when few if any other plants are in bloom, and the tree evidently needs the hummingbird for achieving its pollination.

Movements
There are two recent records of this species straying to southern California (San Diego, 1986; Ventura, 1987), including an unsuccessful nesting in Ventura (*American Birds* 41:330, 1986; 42:321, 1987). This displacement represents an extralimital record

of roughly 600 kilometers from the species' known native breeding range. Otherwise there is no indication that these birds move any significant amount in the course of a year.

Foraging Behavior and Floral Ecology
Little information is available on this species. Arriaga et al. (1990) studied the interactions of these hummingbirds and the endemic madrone of the Baja peninsula. This medium-sized tree, associated with hillside and canyon woodland habitats, produces abundant flowers with pale pinkish corollas. During a total of 475 hummingbird visits to madrones, the birds spent 361 minutes foraging and 317 minutes in territorial activities. Most of the visits were to the upper part of the crown and to the external section of the crown. The birds preferred trees with extensive crown cover; such trees supported the greater number of inflorescences. During the summer, just after the rainy season, the birds also have been observed foraging on the blossoms of such diverse plants as *Castilleja bryantii* (Scrophulariaceae), *Lepechinia hastata* (Labiateae), *Behria tenuiflora* (Amaryllidaceae), *Lobelia laxiflora* (Campanulaceae), *Calliandra peninsularis* (Mimosoidaceae), and *Mirabilis jalapa* (Nyctaginaceae). Several species of small arthropods also have been found in stomachs, and the birds reportedly spend much time hunting for minute insects around oak and pine trees (Lamb, 1925; Bent, 1940).

Breeding Biology
The best available sources of information on this little-studied species are still that of Lamb (1925) and the later summary by Bent (1940). Bent reported that 12 breeding records for a variety lowland locations ranged from April 5 to May 17, and 14 records for upland localities ranged from July 19 to August 5. Lamb observed evidence of even earlier breeding activity at a lowland site of about 150 meters' elevation; all but 3 nests of the 12 he found in late March and early April already had eggs about to hatch or contained nestlings. He observed no aerial dive displays during courtship, but did see a considerable amount of chasing behavior. A wide variety of trees were used for nest sites in Lamb's lowland study site. There the birds typically nested near water, usually

suspending their nests on twigs of various trees, but building saddle-like nests in two cases. However, at an upland site (situated at about 1500 meters' elevation) live oaks were always chosen for nesting. At this site their nests were nearly all hung from the tips of oak twigs and situated within 2 meters of the ground. The nests placed in oaks were all highly decorated with lichens, whereas the lowland nests lacked such lichen decorations but sometimes had small

fragments of other vegetable matter. All were open-cup nests with very bulky sides. The incubation and fledging periods, still unreported, are probably much like those of the white-eared hummingbird.

Evolutionary and Ecological Relationships
This species is obviously a vicariad of the white-eared hummingbird and would probably hybridize with it if given the opportunity.

WHITE-EARED HUMMINGBIRD

Cynanthus leucotis (Vieillot)
(*Hylocharis leucotis* of AOU, 1933)

Other Names
None in general English use; Orejas blancas, Chupaflor orejiblanco (Spanish).

Range
Breeds casually in the mountains of southeastern Arizona, and south through the forests of both eastern and western Sierras from Sonora, San Luis Potosi, and Tamaulipas, Mexico, south to Nicaragua (Friedmann et al., 1950). Northern races winter in Mexico and Guatemala.

North American Subspecies
C. l. borealis (Griscom). Breeds casually in southeastern Arizona (Huachuca Mountains) and in the temperate zone of northern Mexico from 1170 to 3000 meters on both slopes of the Sierra Madre Occidental and the Sierra Madre Oriental, winter-

ing up to 1920 meters from Sonora and Chihuahua to northern Sinaloa and northern Durango. Also reported from Tamaulipas, but birds from this area may be closer to *C. l. leucotis* (Friedmann et al., 1950).

C. l. leucotis. Resident of southern Mexico from Sinaloa, Durango and San Luis Potosi south to Chiapas.

C. l. borealis (Griscom). Resident of northern Mexico; casual in southern Arizona.

Measurements
Wing, males 52.5–59.5 mm (ave. of 17, 55.3 mm), females 49–55 mm (ave. of 15, 51.9 mm). Culmen, males 14.5–18.5 mm (ave. of 17, 16.6 mm), females 16–18.5 mm (ave. of 15, 17.4 mm) (Ridgway, 1911). Eggs, ave. 12.5 × 8.0 mm (11.9–12.7 × 7.9–8.3 mm).

Residential ranges of the Xantus (cross-hatched) and white-eared (inked) hummingbirds. The arrowhead represents an extralimital breeding location of the latter. *(Adapted from Howell and Webb, 1995)*

Weights

Breeding females average 3.25 g (Wagner, 1959). The average of 158 males was 3.6 g (SD 0.3 g); that of 51 females was 3.2 g (SD 0.2 g) (Lyon, 1976). A sample of 158 males had a mean mass of 3.6 g (SD 0.30 g); 51 females had mean mass of 3.2 g (SD 0.2 g) (Dunning, 1993).

Description (After Ridgway, 1911)

Adult male. Forehead, loral and malar regions, chin, and upper throat rich metallic violet or violet-blue, passing into velvety black on suborbital and auricular regions and into duller black, faintly glossed with bluish or greenish, on crown; occiput and hindneck dark metallic bronze or bronze-green; back, scapulars, wing-coverts, and rump varying from bright metallic green to bronze-green or golden green; upper tail-coverts similar but usually more bronzy (sometimes golden bronze), and, together with feathers of rump, more or less distinctly margined with rusty; middle pair of rectrices bright bronze-green, bronze, or golden bronze; the next pair similar but darker; the remaining rectrices bronzy black, tipped (more or less distinctly) with bright bronze or bronze-green; remiges purplish dusky, the inner secondaries glossed with bronze-green; a broad white post-ocular stripe, extending backward and downward above and behind upper margin of auricular region to side of neck; middle and lower throat brilliant metallic emerald green (more yellowish green posteriorly), abruptly defined against the dark violet or violet-blue of upper throat and chin; chest, breast (except medially), sides, and flanks metallic bronze or bronze-green, interrupted by grayish margins to the feathers, the basal grayish also showing where feathers are disarranged; median line of breast and abdomen dull grayish white, sometimes tinged with brownish buffy; femoral tufts dull white; under tail-coverts grayish brown, faintly glossed with bronze, centrally, broadly margined with dull whitish; basal half (more or less) of bill coral red, terminal portion dull blackish; iris dark brown; feet dusky.

Adult female. Above similar to the adult male, but pileum dusky brown, the feathers (especially on forehead) sometimes margined with pale rusty brown, and lateral pair of rectrices broadly tipped with brownish gray; a broad black suborbital and auricular patch and white postocular stripe, as in adult male; underparts pale brownish gray or dull grayish white, spotted with metallic bronze-green, this predominating laterally; median line of breast and abdomen plain dull grayish white or pale brownish gray; under tail-coverts grayish centrally (the shorter ones bronzy or bronze-green) broadly margined with dull grayish white; maxilla dull black, mandible reddish, with terminal portion dusky; iris and feet as in adult male.

Immature male. Pattern of coloration as in adult, but no blue on head, and brilliant emerald-green of throat merely indicated; pileum dull dusky green-ish, the feathers margined with dull tawny, this prevailing on occiput; chin and upper throat dull grayish white spotted with dusky; lower throat metallic emerald-green, the feathers distinctly margined with grayish white; prevailing color of rump, superficially, dull tawny, the upper tail-coverts distinctly margined with the same; outer pair of rectrices broadly tipped with light brown-ish gray, the second pair more narrowly tipped with the same.

Immature female. Like adults, but duller, the feathers of pileum margined with rusty, and spotting of underparts much duller and less metallic.

Identification

In the hand. Most likely to be confused with the broad-billed hummingbird, since it too has a red-dish base to the bill, which is somewhat broader than deep basally. A long white eye-stripe is present in both sexes, below which a black ear-patch extends forward to the lores. In females this is more dusky, and their throat and under-parts are spotted with greenish, rather than uniformly grayish as in females of the broad-billed hummingbird.

In the field. The same criteria as noted above (red-dish bill, white eye-stripe, and dark black ear-patch) serve as good fieldmarks, and the females' greenish tail and greenish flanks and throat are useful in separating this species from females of the otherwise similar broad-billed hummingbird. This species is found in pine-oak woods near streams, and especially in oak woodlands of mountains. Males utter a clear, repeated *tink* call that sounds like a small bell, delivered constantly from perches.

Habitats

Little is known of this species' habitats in the United States, since it is seen only rarely north of the Mexican border. U.S. habitats include woodlands in Cave Creek Canyon of the Chiricahua Mountains and Ramsey Canyon of the Huachuca Mountains in Arizona, the Chisos Mountains of Big Bend National Park in Texas, and the Animas Mountains of New Mexico.

In Mexico the species generally occurs in the temperate zone between 1170 and 3000 meters, nesting at least between 2250 and 3300 meters and wintering up to at least 1900 meters. Schaldach (1963) reported it to be the most abundant breeding bird in open, grassy fields within the arid and humid zones of the pine-oak forests of Colima; and Rowley (1966) also noted that it was the most abundant hummingbird in the upper levels (cloud forest and boreal forest) of his study area in Oaxaca. Moore (1939a) stated that it was the most common hummingbird in the mountains of northwestern Mexico above 1500 meters. It was the dominant species in relation to others of its family there, maintaining control of its preferred food flowers even against larger species such as magnificent and blue-throated hummingbirds. According to Wagner (1959), the species' altitudinal range is from 1200 to 3900 meters, whereas at the southern end of its range in El Salvador it occurs between 1000 and 2400 meters (Dickey and van Rossem, 1938). In the Soloma region of Guatemala it occurs commonly between 1750 and 2850 meters in scrub oak thickets, pine forests, and along the edges of oak and cloud forests, nesting in adjacent cornfields (Baepler, 1962). More generally in Guatemala it occurs from 1200 to 3300 meters above sea level, preferring the more open woods of oaks, pines, and alders and the clearings and brushy mountainsides (Skutch, in Bent, 1940).

All sorts of scrubby growth, but especially the undergrowth of oak forests, seem to represent optimum habitat for this species. The diverse habitats also include pine woods, rather dense pine-oak forests, high mountain fir forests, partially open mountain country with scattered trees and shrubs, suburban gardens, and even vacant lots with scattered shrubs and flowers.

Movements

Texas records of this species extend from April 27 to August 13. Arizona records generally fall between June 9 and August 14 in the Huachuca Mountains, although two specimens were supposedly taken in Arizona on October 1 (Phillips et al., 1964) and one sighting occurred on September 13 (*American Birds* 25:88). Other than Arizona and Texas, the only other records for the United States are from New Mexico, where the species has been seen a few times in the middle elevations of the Animas Mountains during June and July (Hubbard, 1978).

In the Valley of Mexico, considerable seasonable population variations occur, depending on temperature and rainfall. There, maximum breeding occurs during the summer months from June through September, when both rainfall and insect life also peak. However, elsewhere in Mexico it occurs at other times of the year, such as during spring months in the Colima area and during winter months in the high mountains of Guatemala. Thus, the timing and extent of seasonal movements obviously varies greatly throughout the species' range (Wagner, 1959).

In the vicinity of Mexico City a proportion of the high montane population migrates, with the weather determining the degree of migration. Although the species is capable of nesting there at any time of the year, during the dry season there is an absence of flexible plant materials needed for nest-building. Thus, there are variations between permanent residents and relatively migratory populations. If there is unusually cold and wet weather between the end of October and the beginning of December, the birds overwinter in the Mexico City area and do not leave for the higher mountain areas for breeding until May or June (Wagner, 1959).

Foraging Behavior and Floral Ecology

According to Wagner (1959), this species shows no special preference for red flowers. It extracts insects from a variety of flowers, and during the winter also captures insects in flight. In Guatemala,

one of the principal food plants at the beginning
of the nesting season is a burmarigold (*Bidens
refracta*), which has yellow flowers and produces
relatively little nectar. Later, the birds specialize
on various species of mints, especially the red-
flowered *Salvia cinnabarina,* but sometimes also
the blue-flowered *S. cacalioefolia.* In Mexico they
use the blue-flowered *S. mexicana* blossoms but
avoid the larger and more elongated red flowers
of *S. cardinalis,* which are regularly visited by
the larger blue-throated hummingbird (Wagner,
1959).

In an Oaxaca study area, Lyon (1976) found that
this species had the most flexible foraging behavior
of the six major species of hummingbirds present.
It visited a wide range of blossom sizes and exhibited an unusual raiding pattern as well, involving
secretive low approaches to stands of *Penstemon*
within the territories of larger and more dominant
blue-throated and magnificent hummingbirds. As
the average territory sizes of the blue-throats
decreased during the summer months, their
efficiency of territorial defense increased, thus
excluding the white-ears from the *Penstemon*
stands and forcing them to become more dependent on scattered or smaller-flowered species,
especially *Cuphea jorullensis.* The size of male territories in this species averaged 430 square meters,
as compared with 720 square meters in the magnificent and 780 square meters in the blue-throated.

Breeding Biology
In Guatemala, male white-eared hummingbirds
become sexually active near the end of the rainy
season, just as plants that blossom during the dry
season are starting to open. Males gather into
"singing assemblies" of as many as seven birds.
These groups are well spread out, but probably
within hearing distance of one another. The actual
distance between individual birds may be from
18 to 30 meters, and the total assemblage may
spread out over 180 meters. Some males also
display solitarily, well beyond the hearing of all
others. Typically the birds sit on exposed perches
from less than 1 meter above ground to as much as
12 meters in the air. Each male utters a low, clear
tink note that is usually bell-like and repeated endlessly, especially in early morning hours. The seasonal singing activity corresponds to the nesting

period of the females; certainly its major purpose is
to attract females, but it probably also deters other
males from intruding in the territory. Frequently
the nearest flowers are at some distance from the
singing-perch, and thus the territory serves mainly
as a mating station rather than as a means of estab-
lishing dominance over a local food resource. In
Guatemala these territories are defended for three
or four months, from early September until the end
of the year (Skutch, in Bent, 1940).

Around Mexico City, the birds sing to some
degree during the winter months, but song intensity increases during spring, and by midsummer
the males begin to congregate on their common
singing grounds. There they call from morning to
night in groups of as many as seven individuals,
according to Wagner (1959). Only during a short
period between late March and late May are the
birds sexually inactive in the Mexico City area, but
in Colima they apparently breed during that
period, which suggests year-round potential for
breeding, depending on local conditions.

Courtship involves several stages, beginning
with a female's selection of one individual from
the group of males, which she lures to her nesting
area. She lands on a tree branch and is immediately courted by the male, which whirs around,
above, in front of, and beside her. Sometimes the
female moves to a new location, only to be followed by the male; this procedure is repeated
several times, with the sitting intervals becoming
shorter and the intervening flights longer. While
in flight the partners sometimes hover, facing
each other, and whenever the female lands, the
male repeatedly "invites" her to rejoin him in
the air. The ensuing nuptial flight is wild and darting, during which the birds alternate looping
maneuvers and short periods of facing each other
while hovering. Its final phase is a rapid curving
flight during which the birds presumably fly to a
mating site.

Nest sites are nearly always in shrubs or fairly
low trees. In Guatemala, the usual site is a composite shrub (*Baccharis vacinoides*) that grows abundantly on the mountainsides. All but 1 of 17 nests
found by Skutch (in Bent, 1940) were in bushes of
this plant, at heights of 1.5 to 6 meters above
ground. However, in Mexico a favored location is
on oaks; all 28 nests found by Wagner near Santa

Rosa were on oaks (*Quercus nitens*, less often *Q. reticulata*), as were 2 of 4 nests described by Moore (1939a) from northwestern Mexico.

Oaks make an attractive nest site probably because the undersides of oak leaves have a downy hairy covering that can be removed, especially where leaf-miner larvae have been active or where leaf galls are present. The nests are constructed mainly out of such materials, which are held down with spider webbing and covered exteriorly with greenish mosses and grayish lichens for camouflage. The nests are usually almost 50 millimeters in diameter, with the rim of the cavity noticeably incurved and about 25 millimeters in diameter; they range from 25 to nearly 75 millimeters in height (Skutch, in Bent, 1940).

From 15 to 20 days are required to build the nest, and 60 to 70 days may elapse from the onset of nest-building to fledging of the young. In the Santa Rosa area of Mexico, resident birds may raise three broods per year (Wagner, 1959). Skutch (in Bent, 1940) suggested that two broods may be raised in a season, and described a female that built a second nest 12 meters away from the first, only a week after fledging the single offspring of her first nesting effort.

Nests of several females are sometimes located fairly close to one another, at times only 35 to 50 meters apart. Further, new nests are often constructed on the foundations of old ones. Incubation lasts from 14 to 16 days, with the longer periods typical of winter months. Likewise, in Mexico the fledging period varies from 23 to 28 days, with the growth of the young dependent on weather and seasonal variations in day-length (Wagner, 1959).

In his report on incubation behavior and nesting development, Skutch (in Bent, 1940) noted that one female devoted nearly eight hours of one day to incubation, and about four hours were spent off the nest, presumably for foraging. By the time the young birds were 7 or 8 days old, their pinfeathers began to sprout, and their eyelids began to part at 9 or 10 days after hatching. Four days later the green portions of their contour feathers appeared, and at 16 days the flight feathers began to emerge from their sheaths. Two days later the tail feathers did the same. The young were brooded nightly until they were 17 or 18 days old, by which time they were well covered with feathers, and in Skutch's observations fledging required 23 to 26 days. One youngster, 40 days old and two weeks out of the nest, was still fed occasionally by its mother, even though she was now incubating her second clutch of eggs. Wagner (1959) noted that by the time the young are 20 days old they may weigh about 25 percent more than the average adult weight. However, in spite of such maternal care, there is a high incidence of nest and nestling mortality. Skutch (in Bent, 1940) reported that of 18 eggs in 9 nests, only 3 youngsters survived to fledging; Wagner (1959) found that among a total of 39 nests only the nestlings of 12 fledged.

Evolutionary and Ecological Relationships

In the judgment of Mayr and Short (1970), this species and the xantus hummingbird of Baja California comprise a superspecies, with the latter a relict form and the most strongly differentiated avian form of that area. I agree with the close relationships of these two species, and I further believe that "*Basilinna*" and "*Cynanthus*" should both be considered congeneric with the typical "sapphires" of the genus Hylocharis.

In spite of the wide ecological and geographic range of the species, no hybrids involving the white-eared hummingbird are known. Probably a good part of its success can be attributed to is likely being a generalist forager, using a wide array of blossom sizes and flower heights (Lyon, 1976).

BLUE-THROATED GOLDENTAIL

Hylocharis eliciae (Bulcier and Mulsant)

Other Names

Blue-throated sapphire, Elicia goldentail; Chupaflor colidorado, Zafiro gorjiazul (Spanish).

Range

Resident of Mexico from Veracruz and eastern Oaxaca eastward across southern Chiapas to the Guatemala border, and with additional populations continuing along both slopes of Central America to Panama.

North American Subspecies

None recognized.

Measurements

Wing, males 47.5–52 mm (ave. of 20, 49.6 mm), females 44.5–48.5 mm (ave. of 20, 46.8 mm). Culmen, males 17–18.5 mm (ave. of 20, 17.6 mm), females 17–19 mm (ave. of 20, 17.9 mm) (Ridgway, 1911).

Weights

An unsexed sample of 13 birds had a mean mass of 3.6 g (Brown and Bowers, 1985). A mixed-sex sample of 6 birds had a mean mass of 3.39 g (Feinsinger, 1976). Stiles and Skutch (1989) reported a mean mass of 3.7 g.

Description (After Ridgway, 1911)

Adult male. Crown, sides of head, and anterior upperparts all metallic green, the hues becoming more golden on the rump, and the tail brilliant golden bronze; chin and upper throat whitish, this color merging with metallic violet-blue on the lower throat and breast, forming a brilliant gorget surrounded by metallic green except along the median breast, which is grayish brown; the other underparts also grayish brown except for white femoral tufts; iris dark brown; feet grayish brown; bill deep reddish, with a dusky tip.

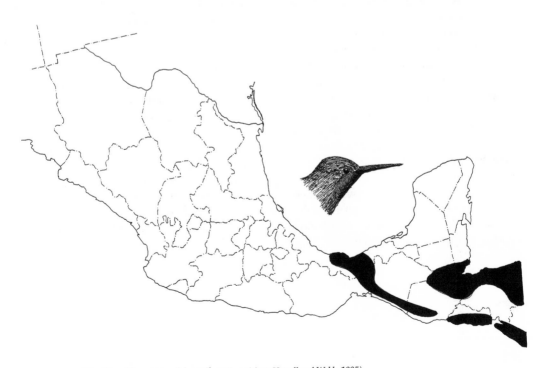

Residential range of the blue-throated goldentail. (*Adapted from Howell and Webb, 1995*)

Adult female. Similar to the adult male, but with paler underparts, and with the violet-blue gorget more restricted and broken up by dull whitish feather margins.

Immature. Apparently undescribed, but probably recognizable by the lack of any metallic gorget feathers when still quite young, a more blackish bill, and pale or buffy edgings to the outer rectrices; the upper bill surface probably also has the parallel striations or ridges apparently typical of all young hummingbirds.

Identification
In the hand. The violet-colored gorget (poorly developed in females) and tail of this humming-bird, which together with the rump is a brilliant golden bronze, sets this species apart from all other Mexican hummingbirds.

In the field. Both sexes of this small humming-bird have a glittering golden tail and rump, contrasting with an otherwise more greenish body and, in males, a well-developed gorget of blue-violet. Females have a much-less-developed gorget, and both sexes have a short, black-tipped reddish bill. The vocalizations include lisping notes, a chatter, squeaky notes, and a song by males that includes a series of about four *see-bit* notes or a more extended series of *chi-chi* notes.

Habitats
Associated with the edges of humid to semihumid forests, woodlands, forest openings or edges, and plantations, from near sea level to 1000 meters. Skutch (1972) reported that in Costa Rica the birds are not inhabitants so much of old-growth forests as they are of light, moderately tall second-growth woodland and groves. There the species is most common at an elevation of about 750 meters. Skutch observed that the species became more abundant in his locality as forests were gradually replaced by pastures, coffee plantations, tobacco fields, and other kinds of plantations. In Panama the species is reportedly mostly a forest inhabitant, occurring among mangrove woodlands, in swampy woodlands near the mouths of rivers, and at elevations generally below 1000 meters (Wetmore, 1968).

Movements
No movements have been documented.

Foraging Behavior and Floral Ecology
Wetmore (1968) described this bird as being attracted to flowering shrubbery, and others have seen it feeding at mid-level to upper levels of forests. The species' short bill would suggest it is a generalist in its foraging adaptations. Feinsinger (1976) described it as basically territorial, but probably a facultative trapliner. In his study area, the birds defended flowering *Lobelia laxiflora* (Campanulaceae) from February through April, and the leguminous tree *Inga brenesii* when it was in flower during October and November. They often lost encounters for possession of such trees with larger hummingbird species such as the rufous-tailed. Both of these species have corolla lengths of about 16 to 23 millimeters. The *Inga* trees were partly attractive to hummingbirds because of their effectiveness in attracting insects in spite of their low nectar production, whereas the *Lobelia* produced a high quantity of nectar daily.

Breeding Biology
Skutch (1972) reported on singing assemblies in this species that he observed in Costa Rica. One such assembly was in an open stand of "burio" (*Heliocarpus,* Tiliaceae) and other rapidly growing trees and included three participating males. They perched at heights of about 6 to 15 meters, singing on leafless twigs near the edge of the grove. Another was situated among a similar stand of second-growth trees including those of the genera *Inga* (Leguminaceae), *Cecropia* (Cecropiaceae), and *Croton* (Euphorbiaceae), where there were also three participating males. They likewise sang at perches on exposed twigs, the two most distant perches being about 30 meters from each other. This assembly (perhaps involving multiple genera-tions of males) persisted for at least 10 years, by which time the woods were becoming heavier and denser than the birds seemed to prefer. The males sang for about 8 or 9 months of the year; singing was reduced during very rainy periods or very dry periods.

The song types of the two groups were quite different from each other, but members of the same assembly sang distinctly similar phrases, suggesting that dialects develop and are transmitted between generations through learning by the participating members. Skutch (1972) noted that in 8 of the 13 species of hummingbirds that he personally knew to have singing assemblies, the sexes were very similar as adults, whereas those species of hummingbirds in the United States that were sexually dimorphic and produced elaborate aerial displays appeared to lack singing assemblies.

Skutch (1972) reported finding only two nests of this species. One was saddled over the horizontal branch of a small tree, with a single leaf adding lateral support. It was about 6 meters above ground and was cup-like, with some lichens added to the outside surface. The nest and eggs vanished about ten days after it had been found in late December. The other was attached to a spray of bamboo and was situated about 2 meters above ground. It was on a thin, descending branch and had lateral support from a small twig. It also had lichens on the outside, and some green moss on the underside. Small nestlings were present in the nest when it was found in mid-April, and these birds fledged at about the end of that month.

Evolutionary and Ecological Relationships
Probably the nearest relative of this species is the gilded hummingbird of Brazil, Bolivia, and Uruguay, which also has an extremely bright, golden-green tail.

WHITE-BELLIED EMERALD

Amazilia candida (Bulcier and Mulsant)

Other Names
None in general English use; Chupaflor candido, Esmeralda vientre-blanco (Spanish).

Range
Resident in Mexico from southern San Luis Potosi southeastward to the Isthmus of Tehuantepec, and extending along both slopes to the Yucatan Peninsula, the Guatemala border, and thence southward in Central America to Nicaragua.

North American Subspecies
A. c. candida. Resident from Chiapas south to Nicaragua.

A. c. genini (Meise). Resident in Veracruz, Puebla, and Oaxaca; presumably north to San Luis Potosi.

Measurements
Wing, males 49–52.5 mm (ave. of 23, 51.1 mm), females 48.5–52 mm (ave. of 12, 49.9 mm). Culmen,

males 15–17.5 mm (ave. of 23, 17.3 mm), females 16–19 mm (ave. of 12, 17.3 mm) (Ridgway, 1911).

Weights
A sample of 13 birds of both sexes had a mean mass of 3.8 g (*SD* 0.36 g) (Dunning, 1993). Stiles and Skutch (1989) reported 4 g as a mean mass for the species.

Description (After Ridgway, 1911)
Adult male. Upperparts metallic bronze to bronze-green, the back becoming more greenish and the tail metallic bronze with a subterminal blackish band (except on the central rectrices); the two outermost pairs of rectrices crossed with brownish gray bands near their tips; the underparts up to the sides of the head white; Approaching the metallic feathers, the white becomes spotted or overlaid with metallic green from the malar region to the flanks; a small white postocular spot is often present.

Residential range of the white-bellied emerald. (*Adapted from Howell and Webb, 1995*)

Adult female. Not always distinguishable from the male, but the terminal parts of the outer rectrices may be paler grayish.

Immature. Very similar to the adults; very young birds may have buff- or pale-edged feathers, especially on the outer rectrices, and probably have a duller, predominantly blackish bill, with parallel striations on the upper (maxillary) surface.

Identification
In the hand. Like several other medium-sized Mexican hummingbirds, this species is immaculate white on the underparts and has a straight, mostly red bill. However, this species has a green forehead no different in color from the rest of its upperparts, and its outer tail feathers are inconspicuously banded with brownish gray near their tips (separating it from the very similar but somewhat larger green-fronted hummingbird, which has an unpatterned tail).

In the field. This relatively plain hummingbird is white below and glittering green above in both sexes. Like the rather similar and widely sympatric but slightly larger azure-crowned, its bill is mostly red, but its forehead is green, not blue. Calls include shrill chipping notes, often rolled or trilled, and the male's song is series of often double or treble phrases of chips or squeaks.

Habitats
Associated with humid evergreen forests, semideciduous forests, and forest edges, as well as coffee plantations and forest clearings, from sea level to 1500 meters. In Oaxaca the species breeds commonly at low elevations in tropical evergreen forest, and it is uncommon at higher elevations, up to the lower edge of cloud forest (Binford, 1989).

Movements
No specific movements have been documented, but strong local movements probably occur, when wintering birds move to the Pacific slope of Oaxaca and Chiapas (Binford, 1989; Howell and Webb, 1995).

Foraging Behavior and Floral Ecology
Little studied, this species is said to forage at much the same kinds of plants as the rufous-tailed, often at fairly low heights. Hummingbirds of this general taxonomic group, such as the rufous-tailed, are territorial species and can dominate many other smaller species of hummingbirds; but in turn they often are dominated by still larger species of the same genus or those of other genera (Feinsinger, 1976). In Guatemala this may be the most common species in the Caribbean-slope foothills, where the birds are sometimes very abundant around various flowering trees, or at coffee plantations (Land, 1970).

Breeding Biology
Breeding of this species in Mexico reportedly occurs between February and May (Howell and Webb, 1995). Little information is available, but many other species of the widespread genus *Amazilia* have been studied, and major differences from these are unlikely. Nests of other better known species of *Amazilia* hummingbirds are cup-like structures placed over the top of a horizontal branch or supported by a fork (Sick, 1993). For example, in the congeneric but slightly larger glittering-throated emerald of eastern South America, nests are usually placed on horizontal branches no higher than 4 meters, and nest-building requires 6 to 12 days. There is the usual clutch of two eggs, laid on alternate days. Incubation lasts 15 to 17 days, and the nestling period lasts 19 to 22 days. Several broods may be produced in a single season, with new nests normally built for each nesting effort (Haverschmidt, 1968).

Evolutionary and Ecological Relationships
This is clearly a close relative of species belonging to the large genus *Amazilia*, such as the azure-crowned hummingbird and the green-fronted hummingbird, which are both very similar in most of their plumage traits.

AZURE-CROWNED HUMMINGBIRD

Amazilia cyanocephala (Lesson)

Other Names
Red-billed azure-crown. Colibri coronizul, Chupaflor cabeciazul (Spanish).

Range
Resident in Mexico from Tamaulipas south through the Isthmus to Chiapas, thence through Central America to Honduras and northern Nicaragua; also reported in northern Costa Rica as a very rare resident or stray.

North American Subspecies
A. c. cyanocephala. Resident from Tamalipas south to Nicaragua.

Measurements
Of *cyanocephala:* Wing, males 57.5–62 mm (ave. of 10, 59.7 mm), females 54–61 (ave. of 8, 57.9 mm). Culmen, males 19–22 mm (ave. of 10, 20.3 mm), females 19.5–23 mm (ave. of 8, 21.2 mm) (Ridgway, 1911). Eggs, 14.2 × 9.0–9.2 mm (Rowley, 1984).

Weights
An unsexed sample of 13 birds had a mean mass of 5.8 g (*SD* 0.63 g) (Brown and Bowers, 1985). Four males from Mexico in the Field Museum had a mean mass of 7.1 g (range 6.4–8.6 g).

Description (After Ridgway, 1911)
Adult male. Forehead and crown bright metallic blue, becoming metallic bronze-green on the rest of the nape, sides of the head, and other upperparts, with the bronzy tones becoming more olive on the tail-coverts and tail; metallic green feathers extend down to the sides of the neck and flanks, but the median portions of the chin, throat, breast and abdomen are all pure white, as are the femoral tufts; iris brown; feet dusky; bill dark red, with a dusky tip.

Adult female. Similar to the male, but the forehead and crown less brilliant blue and more greenish.

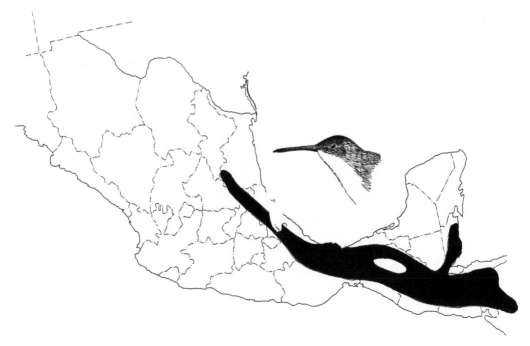

Residential range of the azure-crowned hummingbird. *(Adapted from Howell and Webb, 1995)*

Immature. The forehead and crown duller than adults, the throat and chest off-white, and the outer rectrices pale gray distally (Howell and Webb, 1995); in quite young birds, the bill is probably mostly blackish.

Identification

In the hand. This species closely resembles the somewhat smaller white-bellied emerald (culmen 19–22 mm vs. 16–19 mm respectively), but has a bluish rather than greenish crown. It is even closer in size to the violet-crowned and green-fronted hummingbirds, but these forms respectively have violet- and green-tinted foreheads and crowns.

In the field. This is a medium-sized hummingbird with immaculate white underparts, uniformly green to bronzy upperparts, and a bright, glittering blue forehead and crown. It is similar in size and general appearance to the violet-crowned hummingbird, but is apparently entirely allopatric with it. Except for the blue crown, it also closely resembles the somewhat smaller white-bellied emerald. Its vocalizations include hard, buzzy notes, repeated chipping calls, and prolonged trills or rattles.

Habitats

Associated with pine-oak woodlands and brushy habitats from near sea level to about 2500 meters' elevation. In Oaxaca the birds occupy humid, low-altitude oak and oak-pine forests, especially those adjacent to tropical evergreen and tropical semi-deciduous forests (Binford, 1989). In Honduras, breeding occurs in lowland pine savannas as well as montane pine forests (Monroe, 1968).

Movements

No specific movements have been documented, but birds have occasionally wandered as far as the base of the Yucatan Peninsula. In Honduras the birds apparently wander considerably during the non-breeding season, ranging in elevation from cloud forest to lowland rainforests (Monroe, 1968).

Foraging Behavior and Floral Ecology

There is little specific information on this species, which apparently forages on a variety of plants from low to high heights, but is especially attracted to flowering *Inga* species (Leguminaceae) trees, which also are highly attractive to insects.

Breeding Biology

Little specific information exists. In Belize, breeding occurs from January to July, and in Guatemala from July to September. In Mexico generally the breeding season extends from March to August. In Oaxaca, nests have been found in March and April (Rowley, 1984; Binford, 1989). Rowley described two nests, both found in Oaxaca during early April, at elevations of about 800 meters. One was in the vertical fork of a small shrub about 1.2 meters above ground, on a hillside covered with oaks and pines with a heavy understory. The other was about 3.6 meters above ground, in the fork of a small sapling oak. Both nests were of the open-cup type, but one had a long vegetational streamer and was about a third larger than the first. Two heavily incubated eggs were present in both nests.

Evolutionary and Ecological Relationships

This species appears to be fairly closely related to the slightly smaller white-bellied emerald, and it also is obviously close to the more comparably sized violet-crowned and green-fronted hummingbirds.

BERYLLINE HUMMINGBIRD

Amazilia beryllina (Lichtenstein)

Other Names
None in English use; Chupaflor de berilo, Chupaflor colicanelo (Spanish).

Range
Breeds in Mexico from southeastern Sonora and southern Chihuahua east to Veracruz and south through most of Mexico to western Honduras and El Salvador (Friedmann et al., 1950). Accidental in Arizona (Animas and Chiricahua Mountains).

North American Subspecies (Presumed)
A. b. viola (W. de W. Miller). Resident in the Sierra Madre Occidental from southeastern Sonora to Guerrero and east to eastern Michoacan.

Measurements
Of *A. b. viola*: Wing, males 52–57.5 mm (ave. of 14, 55.5 mm), females 50.5–55.5 mm (ave. of 10, 53.9 mm). Culmen, males 18–20.5 mm (ave. of 14, 19.1 mm), females 19–21 mm (ave. of 10, 20 mm) (Ridgway, 1911). Eggs, 13.5–13.7 × 8.5–8.8 mm (Rowley, 1966).

Weights
The average of 13 males was 4.87 g (range 4.4–5.7 g); that of 8 females was 4.37 g (range 4.0–4.8 g) (Delaware Museum of Natural History).

Description (After Ridgway, 1911)
Adult male. Above bright metallic green or bronze-green, passing into duller purplish bronzy on rump, the upper tail-coverts rather violet to violet-purple; middle rectrices metallic purplish, violet, to bronzy purple; the remaining rectrices chestnut, tipped, or broadly margined at tip, with purplish bronze (this sometimes wanting or obsolete on outermost rectrix); secondaries chestnut, or dull rufous-chestnut, broadly tipped with dusky, the innermost ones (tertials) mostly of the latter color;

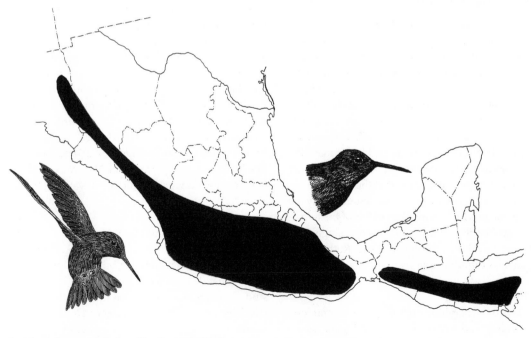

Residential range of the berylline hummingbird. (*Adapted from Howell and Webb, 1995*)

primaries chestnut or dull rufous-chestnut, with terminal portion extensively dusky, faintly glossed with purplish; malar region, chin, throat, sides of neck, chest, breast, sides, flanks, and upper abdomen bright metallic green (brighter and more yellowish than grass green), the feathers of chin and throat abruptly grayish white, those of under-parts of body dusky brownish gray, beneath surface; lower abdomen pale buffy gray, grayish cinnamon, or isabella color; femoral and lumbar tufts white; under tail-coverts pale chestnut broadly edged basally and (usually) narrowly margined terminally with white; maxilla dull black; mandible reddish basally, dusky at tip; iris dark brown; feet grayish brown or dusky.

Adult female. Similar to the adult male, but slightly duller in color, especially the underparts; nearly the whole abdomen being dull cinnamon-buffy; the feathers of chin and throat showing more or less the basal or subterminal white.

Identification

In the hand. Differs from the two similar *Amazilia* species in that the bill is not strongly broadened at the base. Further, the green flanks and underparts set it apart from the other species, although in females these areas are duller and more brownish.

In the field. Associated with forest edges, banana groves, and coffee plantations in Mexico; in drier areas, associated with wooded streams. Although similar to the rufous-tailed and buff-bellied hum-mingbirds, these two species never have bright green underparts. Males are also more chestnut-colored on the rump, and especially on the upper wing surface. Only the lower mandible is reddish, compared with reddish upper and lower mandible color basally in the other two species. Females are duller and more like these in that the lower abdomen is grayish or brownish, but generally are somewhat brighter below than either of the two other species. The male's call is a surprisingly loud *bob-o-leek!*, audible for some distance (Ruth Green, personal communication).

Habitats

In Mexico the berylline hummingbird is common and widespread among the wooded highlands between 900 and 3000 meters, especially in rather dense pine, oak, or pine-oak woodlands, fir forests, open areas having scattered trees or shrubs, and suburban gardens or vacant lots (Edwards, 1973). It is abundant in Colima, and its ecological range includes the thorn forest, parts of the tropical deciduous forest, oak woodlands, and both arid and humid pine-oak forests (Schaldach, 1963).

Movements

Except at the northern edge of its range, this species is probably essentially a permanent resident, subject only to seasonal altitudinal movements. However, in Sonora it is evidently only a summer visitor in the oak-pine zone of the southeastern mountains at about 1500 meters (van Rossem, 1945).

The U.S. records of this species are nearly all for Arizona from June 20 to the end of September. The first sighting was in Ramsey Canyon in the Huachuca Mountains, when an individual was seen from late June to early August 1967 (*Audubon Field Notes* 21:593). In 1971 a female was observed in Cave Creek Canyon of the Chiricahua Moun-tains from June 30 to August 1 (*American Birds* 25:890), and in 1975 a berylline was observed from June 27 onward through the summer in Ramsey Canyon. Finally, on July 13, 1976, a pair was discovered nesting there, and at least one young had hatched by July 22; however, the nest was found abandoned on August 16. The species was again found nesting in Ramsey Canyon the following year (Anderson and Monson, 1981). Finally, in 1979 the species was seen in Carr Canyon, adjacent to Ramsey Canyon, for the only Arizona record that year (*Continental Birds* 1:108). Since then, more than 30 records for Arizona have accrued (*American Birds* 48:972), and there also have been records for New Mexico's Guadeloupe Canyon and Big Bend National Park, Texas.

Foraging Behavior and Floral Ecology

According to Des Granges (1979), the berylline hummingbird is a wandering but not migratory species that defends feeding territories. In his study area, Des Granges noted that it was a relatively generalist type of forager. During the summer months the social dominance of the amethyst-throated hummingbird prevented it from foraging at the best tubular flower (*Malvaviscus arboreus*) in the arid pine-oak habitat, so the berylline fed at *Calliandra anomala* instead. During the winter months, however, the amethyst-throated hummingbird became less common, allowing the berylline access to the *Malvaviscus*. In the riparian gallery forest habitat the birds also fed on tubular flowers such as *Psittacanthus calyculatus*, whereas in the arid thorn forest they were attracted to *Ceiba aesculifolia* and *Lemairocereus*.

Breeding Biology

Very few observations on the breeding of this species have been published. Rowley (1966) located nine nests in Oaxaca between July and October, five of them during September. An apparently favorite nesting site there is the shrub *Wigandia caracasana*. Its outer dead seed stalks offer ideal nesting sites, especially where the plant's large leaves provide overhead protection from heavy rains. However, other nests were found in different flowering shrubs, one almost 5 meters up in an oak and another at least 15 meters above ground on the horizontal branch of a pine.

All the nests were abundantly covered by lichens, producing a nearly solid pattern. With a single exception they also always had a "streamer" of grass blades attached to the bottom with spider webbing, resulting in a very distinctive appearance. One of the nests measured nearly 50 mm in outside diameter, with an inside cup diameter of slightly more than 26 mm;

it was about 50 mm in depth, exclusive of the streamer.

The first U.S. nesting in Cave Creek Canyon was in a riparian habitat at 1634 meters elevation; the nest was on a slender branch of an Arizona sycamore (*Plantanus wrightii*), about 7.5 meters above the ground. The second nest was in a similar habitat, at 1722 meters' elevation in Ramsey Canyon, again located about 5.5 meters above ground in an Arizona sycamore. Both nests were covered with green leaf-like lichens and measured about 40 × 50 mm, with a cavity depth of 15 mm. The Ramsey Canyon nest is believed to have hatched between August 10 and 13, and the two young fledged on August 20 and September 1, suggesting a fledging period of about 20 days (Anderson and Monson, 1981).

Evolutionary and Ecological Relationships

This species is obviously a close relative of the buff-bellied and rufous-tailed hummingbird, and also of such forms as the blue-tailed hummingbird. Land (1970) noted that a Guatemalan population of the last-named species seems to represent an intermediate form, and thus perhaps the two should be considered conspecific. No definite hybrids are known, but the form described as *Amazilia ocai* may represent one involving the berylline hummingbird and the red-billed azure-crown (Berlioz, 1932).

Not enough is known of its foraging ecology to comment on interspecific relationships of this species.

CINNAMON HUMMINGBIRD

Amazilia rutila (DeLattre)

Other Names
Cinnamomeous hummingbird, Coral-billed hummingbird (*corallirostris*), Grayson's hummingbird (*graysoni*); Chupaflor rojizo, Colibri canelo (Spanish)

Range
Resident of Mexico from Sinaloa south along the Pacific slope (including the Tres Marias islands) to southern Chiapas, and again along the coast of Yucatan Peninsula to Belize, with additional Central American populations extending south to Costa Rica.

North American Subspecies
A. r. rutila. Resident from Sinaloa to Guatemala and Honduras.

A. r. graysoni (Lawrence). Resident on the Tres Marias islands.

A. r. corallirostris (Boucier and Mulsant). Resident in southern Chiapas.

Measurements
Wing, males 52.5–60 mm (ave. of 33, 56.5 mm), females 51–58 mm (ave. of 17, 55.3 mm). Culmen, males 19.5–23.5 mm (ave. of 33, 21.2 mm), females 20–23.5 mm (ave. of 17, 21.6 mm) (Ridgway, 1911). Eggs, 13.3–14.1 × 8.6 mm (Rowley, 1966).

Weights
Ten males averaged 5.17 g, and 11 females averaged 4.92 g (Stiles and Wolf, 1970). The mean of a mixed-sex sample of 14 birds was 5.0 g (range 3.3–5.7 g) (Dunning, 1993), and the mean of 5 birds of both sexes in the Field Museum was 5.1 g (range 4.6–6.0 g). Stiles and Skutch (1989) reported a mean mass of 4.8 g for the species; Arizmendi and Ornelas (1990) reported a mean weight of 4.3 g (*SD* 0.1).

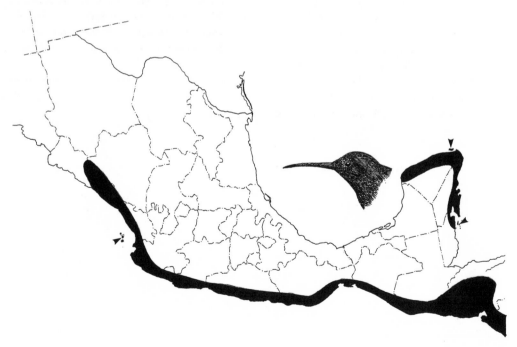

Residential range of the cinnamon hummingbird. The arrowheads represent insular breeding sites. (*Adapted from Howell and Webb, 1995*)

Description (After Ridgway, 1911)

Adults (sexes alike). Upperparts metallic bronze, feathers on the upper tail-coverts margined with rufous; rectrices mostly rufous, with broad tips and edgings of metallic bronze; underparts deep cinnamon to rufous-cinnamon, becoming paler on the chin, and with white femoral tufts; iris brown; feet dusky; bill carmine with a dusky tip.

Immatures. Apparently very similar to adults, but probably the bill is less reddish and more blackish, with parallel striations on the upper mandible (see species account for long-billed hermit); in especially young birds, the upperpart feathers and rectrices may be tipped with paler or buffy edges.

Identification

In the hand. This is the only medium-sized hummingbird of Mexico that has uniformly cinnamon underparts and a moderately long tail (31–37 mm) that is rich cinnamon, the rectrices narrowly edged and broadly tipped with metallic bronze.

In the field. This medium-sized hummingbird is rather easily identified by its cinnamon-colored underparts and the mostly cinnamon-colored tail (the latter trait also present in the sympatric berylline hummingbird, which, however, is glittering green on the throat and breast). The birds often forage around flowering shrubs, and their vocalizations include squeaking chipping notes, trills, and a territorial or courtship "song" by males consisting of thin and rather squeaky *chips* uttered in variable phrasing.

Habitats

Associated with pastures, arid scrub, brushy habitats, second-growth woodlands, and forest edges, from sea level to 1600 meters' elevation. In Oaxaca the species commonly occurs in arid tropical scrub, tropical deciduous forest, and the lower edges of tropical semideciduous forest (Binford, 1989). Hutto (1985) observed the species only in tropical deciduous forest in a distributional survey of western Mexican habitats. In a Yucatan habitat survey the species was encountered 25 times in coastal scrub, 11 times in old-field habitats, 8 times in dry forest, 8 times in field-and-pasture habitats, and twice in moist forest (Lynch, 1985). In Guanacaste

Province of Costa Rica, the species is most common along scrubby forest edges of tropical dry forest and does not enter primary forest (Stiles and Wolf, 1970).

Movements

No migrations or local movements have been documented. There is a recent sight record for Dona Ana County, New Mexico (*American Birds* 48:138).

Foraging Behavior and Floral Ecology

In a detailed study, Stiles and Wolf (1970) found that the favorite foraging resource for hummingbirds during most of their fieldwork was *Genipa americana* (Rubiaceae), a fairly small tropical tree that produces large numbers of white nectar-bearing flowers in a corolla tube measuring about 15 millimeters. The taller trees (6–15 m) seemed favored for foraging over lower ones. The birds were territorial at favored trees, but also tended to be observed around some trees supporting rather few flowers. As to daily use, the birds exhibited a rapid rise in activity after dawn, with a foraging peak occurring about a half hour after dawn. After that territorial activities increased rapidly and continued for much of the day. Sometimes they visited the orange-red flowers of the perennial scrophulariaceous herb *Lamorouxia viscosa* when it was in bloom.

Wagner (1946a) observed that individuals of this species and the green-breasted mango would divide the resources of a nectar-rich flowering tree, with the cinnamon hummingbird controlling the lower part of the tree, and the mango the top of the tree. In their Guanacaste study area, Stiles and Wolf (1970) found this species to occur in similar territorial competition for foraging sites with the steely-vented hummingbird, although none of the birds were then breeding. Each individual of these two species would occupy part of the tree, defending it from other hummingbirds (mainly) but sometimes also defending it against large black bees. Females evidently held territories on a fairly basis with males; this apparent selection for equal dominance might help account for the lack of sexual dichromatism in these species, in the view of these authors. Generally, the territories of cinnamon hummingbirds were located in the lower foliage areas and outside of the crown, whereas those of the steely-vented were smaller, more closely packed, and were usually situated in the interior of the crown. Territories were most strongly maintained during mid-day hours, and cinnamon hummingbirds were dominant to the steely-vented. However, by mid-afternoon the birds would begin leaving the tree, and all would be gone by 6:00 P.M.

Field studies in Jalisco indicate that this species is territorial and able to dominate all the other resident hummingbirds except for the larger plain-capped starthroat. There it is mostly a subcanopy forager, with progressively less use of the middle understory, the lower understory, and the canopy. Highly visited plant taxa included (in descending frequency) the herbaceous *Hamelia versicolor* (Rubiaceae), the vines *Combretum farinosum* (Combretaceae) and *Ipomoea bracteata* (Convolvulaceae), and the arboreal cactus *Opuntia excelsa* (Arizmendi and Ornelas, 1990).

Laboratory experiments suggest that this species can extract nectar from plants having corollas as deep as 57 to 57 millimeters, or about twice the bill length. Generally, however, the birds favor flowers with the shortest corollas that offer either high nectar volume or high sugar concentrations (Montgomerie, 1979, 1984).

Breeding Biology
Little specific information is available. In Mexico the records of breeding extend through much of the year; on the Yucatan peninsula breeding occurs from February to April, and again from August to November. Elsewhere in western Mexico there are breeding records for November to February, June to July, and September (Howell and Webb, 1995). Breeding in Jalisco occurs from January to April (Arizmendi and Ornelas, 1990).

Nests of this species are essentially like those of other *Amazilia* species, and presumably their breeding biology is very similar, but detailed studies have not been undertaken.

Evolutionary and Ecological Relationships
This species is clearly a close relative of the buff-bellied hummingbird, which is probably its nearest living relative. The buff-bellied hummingbird also occurs sympatrically in areas such as the Yucatan Peninsula, but there its habitat preferences are mostly moist forest and wet forest (Lynch, 1985), so the degree of interspecific ecological overlap must be quite limited.

BUFF-BELLIED HUMMINGBIRD

Amazilia yucatanensis (Cabot)

Other Names
Fawn-breasted hummingbird, Yucatan hummingbird; Chupamirto yucateco, Chupaflor vientre
castaño (Spanish).

Range
Breeds from the lower Rio Grande valley of Texas
south through eastern Mexico, Yucatan, Chiapas,
and Belize.

North American Subspecies (After Friedmann et al., 1950)
A. y. chalconota (Oberholser). Breeds from Cameron
and Hidalgo counties of Texas southward to
northeastern Mexico as far as San Luis Potosi
and northeastern Veracruz.

C. y. yucatanensis. Resident of the Yucatan
Peninsula.

C. y. cerniventris (Gould). Resident in Chiapas and
Veracruz.

Measurements
Wing, males 51–57.5 mm (ave. of 5, 53.8 mm),
females 52–52.5 mm (ave. of 2, 52.2 mm). Culmen,
males 19–21 mm (ave. of 5, 20.2 mm), females
21–21.5 mm (ave. of 2, 21.2 mm) (Ridgway, 1911).
Eggs, ave. 13.2 × 8.65 mm (extremes 11.8–15.3 ×
7.7–9.4 mm).

Weights
Seven males averaged 4.05 g (range 3.0–4.7),
7 females averaged 3.67 g (range 2.9–4.5) (Paynter,
1955, and U.S. National Museum specimens).

Description (After Ridgway, 1911)
Adults (sexes alike). Above metallic bronze-green
or greenish bronze, duller and darker on pileum;
the upper tail-coverts more or less tinged or
intermixed with cinnamon-rufous, sometimes
mostly of the latter color; middle pair of rectrices
mostly or with at least terminal fourth metallic
bronze, the basal portion chestnut; remaining

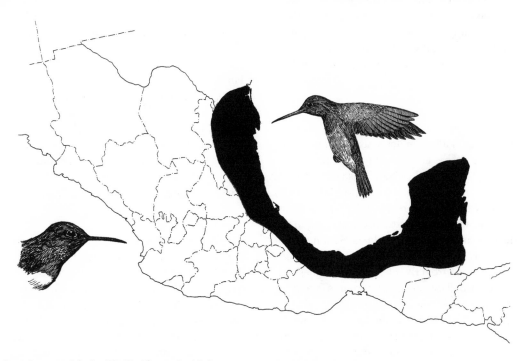

Residential range of the buff-bellied hummingbird. *(Adapted from Howell and Webb, 1995)*

rectrices chestnut, margined terminally with
metallic bronze; remiges dusky, faintly glossed
with violet; chin, throat, and chest bright metallic
yellowish emerald green, the feathers pale buff
basally or (on chest) subterminally; underparts of
body, including under tail- and wing-coverts and
axillars, pale cinnamon buff; femoral tufts white;
bill rosy reddish in life, dusky terminally; iris dark
brown; feet dull brown.

First-winter (sexes alike). Similar to first nuptial, but
pileum darker and duller; edgings of chin, throat,
and jugulum dark buff (Oberholser, 1974).

Identification
In the hand. Reddish bill, which is markedly
widened toward the base, and chestnut tail separate this from all other North American species
except the rufous-tailed hummingbird. In the buff-
bellied species the central tail feathers are somewhat shorter and are distinctly bronze-colored
toward the tip. Additionally, the flanks and underparts are more buffy or fawn-colored, as compared
with grayish in the rufous-tailed hummingbird.

In the field. Associated with semi-arid lowlands
dominated by woods or scrubby growth; sometimes occurs in citrus groves. Distinct from the
very similar rufous-tailed hummingbird by the
buff-bellied's slightly forked tail and more buffy
underparts. It has rather shrill calls that have not
been well described.

Habitats
In Texas, at the very northern limit of its range,
this species is found in dense thickets and among
flowering bushes and creeping vines along
streams, resacas, and gullies. It also occurs among
remnant patches of Texas palmetto (*Sabal texana*)
(Oberholser, 1974). More generally it is associated
with semiarid coastal scrub habitats along the
Mexican coastline. In Tamaulipas, Sutton and
Pettingill (1942) observed the race *A. y. chalconota*
to be common from river level to the highest
points reached on adjacent mountains (about
600 m), in brushy rather than deeply wooded
areas. *A. y. yucatanensis* of the Yucatan Peninsula
and Belize is most abundant in clearings within
high deciduous forest and moderately heavy rainforests (Paynter, 1955). *A. y. cerviventris* is associ-

ated with shaded woodlands (Sutton and
Burleigh, 1940).

Movements
Probably this species is relatively sedentary over
most of its range. In Texas it appears throughout
the year, but individuals apparently wander
north after the breeding season, rarely reaching
the central coast and the eastern edge of the
Edwards Plateau (Oberholser, 1974). There is also
a general movement out of the state during the
coldest months, probably involving a migration to
Tamaulipas and Veracruz, and the birds are less
frequent between October and April (Bent, 1940).
However, some birds do overwinter, and as
many as six have been counted on a single
Christmas count in Brownsville (Oberholser, 1974).
In addition, there are single extralimital sight
records for Midland (*Audubon Field Notes* 19:55),
Taylor (*Audubon Field Notes* 19:400), and the
vicinity of Beaumont (*Audubon Field Notes* 23:498).
Lowery (1974) summarized the species' status for
Louisiana, which at that time consisted of three
records, including one specimen record. Since then,
the species has been seen at least once more in
Louisiana (*American Birds* 32:364, 1020). There is
also one astonishing and highly questionable
record for Massachusetts (*Audubon Field Notes*
18:496).

Foraging Behavior and Floral Ecology
Buff-bellied hummingbirds use several native
flowers in Texas during spring and summer,
including Texas ebony (*Pithecellobium flexicaule*),
mesquite (*Prosopis glandulosa*), and anaqua (*Ehretia
anacua*). In fall and winter they share the flowers of
the giant Turk's cap (*Malvaviscus grandiflorus*) with
other hummingbirds (Oberholser, 1974).

Breeding Biology
Breeding in Texas was once much more common
than it is at present; 30 early egg records range
from March 24 to July 15, with 15 records between
May 9 and June 9, indicating the peak of the nest-
ing season (Bent, 1940). In Mexico, breeding has
been reported during April in San Luis Rotosi and
Tamaulipas and at the end of March in Nuevo
Leon (Friedmann et al., 1950). On the Yucatan
Peninsula the nesting season apparently begins in

late January and extends at least until mid-April (Paynter, 1955).

Nests are often built near woodland roads or paths and placed only a meter or two above the ground (Sutton and Burleigh, 1940; Sutton and Pettingill, 1942). According to Bendire (1895), they are usually saddled on a small, drooping limb or on the fork of a horizontal twig, between 1 and 3 meters from the ground. Small trees and bushes are the usual nest sites, including anachuita (*Cordia boissieri*), ebony, and hackberry, but sometimes nests are in willows. The nests are composed of shreds of vegetable fibers and thistle down, and are distinctively covered by dried flower blossoms, shreds of bark, and small pieces of light-colored lichens. Most are lined with thistle down, but some contain vegetable material resembling brown cattle hair. (They are about 40 mm wide and 32 mm deep; and the cup is usually about 22 mm wide and 16 mm in depth.) Bent (1940) described an exception to this general configuration, involving a nest that evidently had been used for three seasons, which was thus much higher than normal.

No information is available on incubation periods, incubation behavior, or brooding behavior and fledging periods. However, these details probably are similar to the information reported for the rufous-tailed hummingbirds.

Evolutionary and Ecological Relationships

Mayr and Short (1970) suggested that this species is part of a superspecies that includes *A. rutila*. I believe that *A. tzacatl* is at least as closely related to the former as is *A. rutila,* and might on morphological grounds be considered part of the same superspecies. The two forms are rather widely sympatric over the Yucatan Peninsula, however, and thus have clearly attained full species status. There are no known hybrids involving the buff-bellied hummingbird.

RUFOUS-TAILED HUMMINGBIRD

Amazilia tzacatl (De la Llave)

Other Names
Reiffer hummingbird; Chupamirto De la Llave, Chupaflor colirufo, Chupaflor pechigris (Spanish).

Range
Breeds from southern Tamaulipas in eastern Mexico southward through Middle America to Colombia, western Ecuador, and Gorgona Island.

North American Subspecies (After Friedmann et al., 1950)
A. t. tzacatl (De la Llave). Breeds from southern Tamaulipas south through eastern Middle America to Colombia (except southwestern) and east to the Andes of Merida, Venezuela. Accidental in southern Texas.

Measurements
Wing, males 46–61 mm (ave. of 60, 58.3 mm), females 52–58 mm (ave. of 36, 54.9 mm). Culmen, males 18–23 mm (ave. of 60, 20.8 mm), females 20–24 mm (ave. of 36, 21.3 mm) (Ridgway, 1911). Eggs, ave. 14 × 8.7 mm (extremes 13.5–14.2 × 8.6–9.1 mm).

Weights
The average of 12 males was 5.4 g; that of 10 females was 4.72 g (Hartman, 1954). An unsexed sample of 105 birds had a mean mass of 5.0 g (*SD* 0.28 g) (Brown and Bowers, 1995). Ten males had a mean mass of 5.22 g (*SD* 0.28 g); 9 females had a mean mass of 4.91 g (*SD* 0.19 g) (Stiles, 1995).

Description (After Ridgway, 1911)
Adult male. Above metallic bronze-green or greenish bronze, the pileum darker and duller; upper tail-coverts and tail chestnut, the rectrices margined terminally with dusky bronze, the coverts sometimes partly bronze or bronze-green; remiges dusky, faintly glossed with violet; lores

Residential range of the rufous-tailed hummingbird. *(Adapted from Howell and Webb, 1995)*

chestnut; malar region, chin, throat, chest, upper breast, and sides of lower breast bright metallic yellowish emerald-green, the feathers of chin and upper throat pale buff or buffy white basally, this much exposed on chin; abdomen and median portion of lower breast brownish gray; sides and flanks bronze-green; under tail-coverts cinnamon-rufous; femoral tufts white; bill reddish, dusky terminally, the maxilla sometimes mostly (rarely wholly?) blackish; iris dark brown; feet dusky (in dried skins).

Adult female. Similar to the adult male and perhaps not always distinguishable, but usually with the green of underparts more broken by whitish margins to the feathers; gray of abdomen paler; and cinnamon-rufous loral streak less distinct, sometimes obsolete.

Immature. Essentially like adults, but anterior underparts much duller metallic green; the chin and upper throat (at least) sometimes grayish brown or brownish gray, with little if any metallic gloss; and feathers of pileum, rump, etc., tipped (more or less distinctly) with rusty.

Identification

In the hand. Best distinguished from the quite similar *A. yucatanensis* by the more grayish underparts of *A. tzacatl* and its brownish rather than iridescent greenish central tail feathers. Additionally, the lores of *A. tzacatl* are chestnut, whereas those of *A. yucatanensis* are more definitely greenish, and the central rectrices are distinctly shorter than the outermost ones.

In the field. Found in thicket-like edge habitats, somewhat overgrown clearings, planted areas, and humid forested lowlands; likely to be confused only with the very similar buff-bellied hummingbird, which has a more buffy or tawny abdomen and a somewhat forked rather than nearly square-tipped tail. The rufous-tailed hummingbird utters a descending trill of rapid *ts* notes, and its song consists of a series of piercing *tss* or *tsip* notes that are opposingly accented and uttered at a rate of about 12 notes per 5 seconds (*Condor* 59:254). It also utters a *tchup*, a somewhat reedy and nasal *ca-ca-ca*, and burred or buzzy morning song that is sometimes uttered in flight (Slud, 1964).

Habitats

In Mexico, this species occurs in relatively open lowland habitats, especially along forest edges or clearings. In Belize, it resides in both semi-open and wooded areas, except for pine ridges. It occurs in heavy rainforest, but is less common there than at its edges (Russell, 1964). In Costa Rica it is the most widespread hummingbird species, and it is especially common in overgrown open and semi-open country and along thickety borders and woodland edges. Although uncommon in heavily forested areas, it is sometimes abundant at forest openings and in cultivated lands, often occurring around houses (Slud, 1964). In that country it is found from as low as sea level to at least 1800 meters' elevation, and it is often prevalent around citrus or other orchards (Carriker, 1910). In Honduras it ranges up to the lower edges of montane rainforest at about 1200 meters, but is most common at lower elevations in open forest, forest edge, and second-growth habitats (Monroe, 1968).

In Panama, the birds reside along streams, beaches, and more open border habitats in general. Less often they are found in forests, foraging from near the ground to canopy-level. They also are commonly observed along the borders of cultivated fields, in flower beds around houses, and in town parks planted to flowers (Wetmore, 1968). Skutch (1931) also noted that in Panama and other parts of its range, this species prefers open country to dark, humid forests and is the only species of hummingbird in the region that is truly characteristic of lawns, gardens, orchards, and plantings.

Movements

There is no evidence that this species undertakes any significant migrations or seasonal movements, which is not surprising considering the essentially tropical climate characteristics of its breeding range. Records of vagrants reaching the United States are very few. Two specimens, obtained (but apparently not preserved) at Fort Brown, Texas, in June and July 1876, constitute the only authenticated records for the species in the United States (AOU, 1957; Oberholser, 1974). There have been two more recent sight records for southern Texas: a sighting on November 11–12, 1969, of a bird at a hummingbird feeder at LaPorte, Texas (*American*

Birds 24:68) and a sighting on a golf course in Brownsville on August 20, 1975 (*American Birds* 29:84).

Foraging Behavior and Floral Ecology

Skutch (1931) has described the foraging of this species in some detail. He noted that the birds spend much time probing for insects or nectar in the large red blossoms of *Hibiscus simensis*, the blue trumpet-like flowers of *Thunbergia*, the blue-flowered *Clitoria*, and coral vines (*Antigonon*). They also hover beside the hanging flowers of banana (*Musa*) plants, where they probe among the white blossoms in company with stingless bees. In other areas they seem fond of citrus blossoms as well as those of guava (*Psidium*) and wild plantain (*Heliconia*) (Carriker, 1910). The species is evidently a generalist forager (Snow and Snow, 1980).

Stiles (1975) reported that in Costa Rica this species is one of nine that regularly visit the flowers of nine species of *Heliconia*. This species concentrates especially on *H. latispatha*, which is found in sunny habitats. Reproductive isolation among the species of *Heliconia* probably involves both spacial and temporal patterns of niche partitioning among the numerous available pollinators.

In Colombia, the rufous-tailed is common in the city of Medellin, where it commonly forages on *Thunbergia grandiflora*. The bananaquit (*Coereba flaveola*) also visits this plant, and the hummingbird typically obtains its nectar by using a hole in the corolla made previously by the bananaquit (Borerro, 1965).

Breeding Biology

The breeding season of this species seems quite extended and may include the entire year in some areas. In Mexico, active nests have been reported in April and July (Edwards and Tashian, 1959). In Belize occupied nests have been found in January, February, May, August, and September (Russell, 1964). Likewise in Panama the species probably nests from at least mid-December to mid-September, with an absence of records for June, October, and November (Borrero, 1975). In Colombia, nesting in the Cauca Valley and the vicinity of Medellin extends at least from April through December, and there is a February record for the Corillera Occidental. The birds probably breed twice per year at Medellin (Borrero, 1965, 1975).

In a study area at La Selva, Costa Rica, the breeding season was greatly prolonged, but peaked during the dry season when there was maximum flower availability. Fifteen nest records extend from February to December, with a peak in March and April and with no records for August, September, November, or January. Singing by males was noted for all months except August (Stiles, 1980). Likewise in Panama the majority of the nesting apparently occurs during the dry season, which there extends from January to May (Skutch, 1931).

The nest is an open cup, about 44 millimeters in diameter and about 32 millimeters in height. The exterior is composed of weathered grass strips, leaves, and bits of vegetation, decorated on the surface with lichens and mosses, which sometimes hang down festoon-like beneath it. The interior is lined with soft plant down, and the entire structure is bound together with cobwebbing (Skutch, 1931).

Nest locations are highly variable, ranging from 1 to 6 meters above ground, almost invariably on a rather slender support. Sometimes the nest lies between an upright branch and the base of a large thorn, in the angle between a horizontal and vertical stem, or in the axil of a slender leaf-stalk. Occasionally a lone leaf may be used as a foundation. The only apparent requirement is a horizontal support slender enough for the bird to grasp with its feet, from which building operations can begin on some nearby vertical or oblique surface. Frequently the nest is located close to or above a footpath (Skutch, 1931).

The nests are normally constructed so that they blend closely with the foliage around them. One nest, started on the 19th of December, was not completed until 12 days later, when the first egg was laid in it. In a second case, a week elapsed between the start of the nest and the first egg's appearance. On another occasion, a female spent 31 days attempting to produce a nest, only to have the attempts fail repeatedly. At least 12 fresh beginnings were made, involving 8 different locations. Most of these efforts were spoiled by the stealing of nesting materials by other rufous-tailed hummingbirds or tody flycatchers (*Todirostrum cincereum*). Generally when a nest is destroyed, a new one will be begun a short distance away, sometimes less than a week after the destruction of the first one (Skutch, 1931).

Eggs are usually laid on alternate days, but sometimes two days may pass between the laying of the two eggs. Incubation apparently begins shortly after the laying of the first egg and normally requires 16 days. The growth rate of the nestlings is fairly rapid. By the 6th or 7th day after hatching the eyes have opened and the pinfeathers are breaking open. By the time the young are 13 to 16 days old they are well feathered, and their combined weight causes a collapse of the nest cup. Fledging periods have ranged from 18 to 23 days, frequently to 19 days (Skutch, 1931). A. F. Skutch (1981) has provided additional observations on the reproductive biology of this species.

The mother continues to feed the young for a time after they leave the nest, but the exact period of such parental dependence is not known.

Evolutionary and Ecological Relationships

This species is a very close relative of *A. yucatanensis*, and the two forms may comprise a superspecies. Mayr and Short (1970) consider *A. yucatanensis* and *A. rutila* to comprise a superspecies, with *A. txacatl* a "closely related" species that replaces *A. yucatanensis* in wetter regions. The only reported hybrid combination seems to be a possible wild hybrid with *A. amabilis* (Butler, 1932).

Stiles (1985) has reported on the relationship of nine Costa Rican hummingbird species to the flowering phenology and pollination of eight *Heliconia* species. He found that this and other nonhermit hummingbirds are habitat-separated and pollinate *Heliconia* species that bloom fairly early in the rainy season. Besides such spacial and temporal isolating mechanisms, structural and ethological (flower-choice differences among pollinators) factors facilitate reproductive isolation. In the temperate zone of eastern Colombia the rufous-tail occurs with at least 13 other species of hummingbirds. There it is a generalist forager, defending favored flowers and also gleaning and hawking insects (Snow and Snow, 1980).

VIOLET-CROWNED HUMMINGBIRD

Amazilia violiceps (Gould)

Other names

Azure-crown, Salvin's hummingbird; Chupamirto corona azul (Spanish).

Range

Breeds in extreme southeastern Arizona and north-western and western Mexico from Sonora and Chihuahua south to Chiapas. Northern populations winter in Mexico.

North American Subspecies (After Friedmann et al., 1950)

A. v. ellioti (Berlepsch). Breeds locally in extreme southeastern Arizona (Guadalupe Mountains) and regularly (between 300 and 1800 meters' elevation), through northeastern and eastern Sonora, across Sinaloa and extreme southwestern Chihuahua (from sea level to 2250 meters), south on the Pacific slope of the Sierra Madre Occidental,

through western Durango to Colima, and thence east to Hidalgo.

Measurements

Wing, males 53–59.5 mm (ave. of 9, 57 mm), females 52–57 mm (ave. 6, 54.9 mm). Culmen, males 21–23.5 mm (ave. of 9, 22.6), females 21.5–24.5 mm (ave. of 6, 23.1 mm) (Ridgway, 1911). Eggs, ave. 13.7 × 8.8 mm (Rowley, 1966).

Weights

The average of 10 males was 5.78 g (range 5.5–6.2 g); that of 6 females was 5.19 g (range 5.0–5.5 g) (Delaware Museum of Natural History).

Description (After Ridgway, 1911)

Adult male. Pileum bright metallic blue or violet-blue; hindneck, upper back, scapulars, and wing-coverts dull bronze-green or olive glossed with

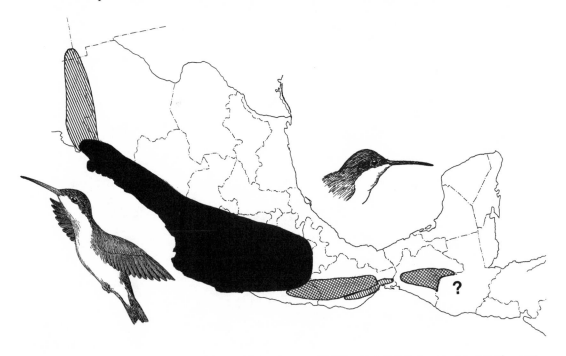

Breeding (hatched) and residential (inked) ranges of the violet-crowned hummingbird's nominate taxon. The residential ranges of two closely related forms, the Oaxacan endemic "cinnamon-sided" taxon *wagneri* (hatching), and the green-fronted hummingbird *viridifrons* (cross-hatching), are also illustrated. *(Adapted from Howell and Webb, 1995)*

bronze-green, the first usually more or less tinged or intermixed with blue on upper portion and along junction with white of foreneck; lower back, rump, and upper tail-coverts grayish brown or olive faintly glossed with bronze-greenish, the feathers sometimes narrowly and indistinctly paler on terminal margin, especially the upper tail-coverts, which are usually slightly more bronze-greenish; tail rather dull metallic greenish bronze; remiges dusky, faintly glossed with violet; rictal and malar regions and underparts (including under tail-coverts) immaculate white, the sides mostly light olive glossed with bronze-greenish; bill light rosy or carmine red, dusky at tip; iris dark brown; feet dusky (in dried skins).

Adult female. Similar to the adult male and not always distinguishable, but usually with slightly duller coloration.

Immature. Similar to adults, but feathers of upper-parts (including pileum) narrowly tipped or terminally margined with buffy, these markings broadest on rump and upper tail-coverts, some-times obsolete on back.

Identification
In the hand. The reddish bill (20–25 mm), distinctly widened near the base, and the white underparts provide for simple recognition of this species.

In the field. Favors riverside groves in mountain canyons, forest edges, and various plantations. The pure white underparts provide an excellent field-mark for both sexes; the bright red bill is also a helpful guide. The calls are similar to those of the broad-billed hummingbird, but are somewhat louder and less grating.

Habitats
In Mexico this species occupies a fairly broad vertical range, from 300 to 2250 meters. In Sonora it is most numerous in the foothills of the tropical zone above 300 meters, but it locally extends to about 1800 meters in the transition zone during summer months. Its favorite habitats in Mexico consist of scrubby riparian groves in deserts or foothills, forest edges, wooded parks or planta-tions, and scrub oak vegetation (from various sources).

In the United States, it has bred only in the Guadalupe Canyon area of extreme southeastern Arizona and adjacent New Mexico. Guadalupe Canyon drains southwestward from its source in New Mexico's Peloncillo Mountains, then cuts across the corner of Arizona before entering Mexico. It contains water in years of normal moisture and is lined by cottonwoods, sycamores, other trees, and various shrubs (Ligon, 1961). There the violet-crowned hummingbird has nested among fairly tall sycamores (*Platanus wrightii*) (Levy, 1958; Zimmerman and Levy, 1960).

Movements
Nothing is known of the seasonal movements of these birds, but certainly those in Arizona and New Mexico are migratory. Most U.S. records are for July and August. In adjacent Sonora, records are available from as early as March and April in the lowlands (about 90 meters' elevation) in the extreme southern part of the state (van Rossem, 1945). The latest record for the United States is probably the individual observed at a Tucson feeder from late November until late February, providing the first indication of overwintering in Arizona (*American Birds* 25:610). Also, an individ-ual visited a feeder at Santa Paula, Ventura County, California, from July 6 until September 19, 1976, representing the first California record of the species (*American Birds* 30:1004, 31:223). Two more California records have since been reported (*Western Birds* 9:91–92; *American Birds* 46:477).

Foraging Behavior and Floral Ecology
Scarcely anything has been written on this subject. Levy (1958) observed an adult violet-crown that perhaps was feeding young as it appeared every five to ten minutes from a sycamore grove to feed on the blossoms of an agave. The birds often feed at flowering trees 6 to 9 meters above the ground (Edwards, 1973).

Breeding Biology
The first account of possible nesting of this species in the United States was provided by Levy (1958), who observed as many as six birds in Guadalupe Canyon during July 1957. At that time he collected an adult female with a recent brood patch, but did not find a nest. However, during 1979, fieldwork in

the canyon by Zimmerman and Levy (1960) discovered nesting in both Arizona and New Mexico. The first nest was found under construction within Arizona on June 20, and on June 28 a completed nest was located nearby. On July 5, two more nests were discovered on the New Mexico side of the boundary. Another possible nesting pair was also seen repeatedly entering a large sycamore, but no nest was found. Of the four nests that were located, all were in sycamores at 7, 9, 10, and at least 12 meters above ground. In one case the nest was near the tip of a horizontal branch, and another was saddled on a horizontal branch about 1.5 meters from the tip and about 7.5 meters out from the trunk. A third was placed in the sharp angle of a "V" in a malformed, semipendant branch.

Few nests of this species have been described in detail, but Moore (in Bent, 1940) has provided one account. This nest was in the crotch of a dead twig 2 meters above the ground, at the end of the branch of a thorny bush overhanging a creek in Sinaloa. Most of the nest was composed of whitish cottony material from a paloblanco (*Celtis laevigata?*) tree. This was bound together by fine webbing resembling that of a spiderweb, but also strongly resembling the same cottony material from which the body of the nest was made. Three small twigs were attached to the external surface, as well as a number of pale greenish lichens, which provided the chief means of camouflage. At least two of the Arizona nests were also decorated with a few green lichens, and appeared quite white (Ligon, 1961).

Two nests of this species were found by Rowley (1966) among scrub oaks in Oaxaca. One was a little more than 1 meter above ground, and the other about 2 meters up; both were placed in small sapling oaks. Both were relatively crudely made

and, according to Rowley, might easily have passed for old nests of the previous year. One was about 38 millimeters in external diameter and about 25 millimeters in total depth, with a cup 12 millimeters in diameter. A photograph of one of the nests indicates that it was decorated externally with large lichens, and apparently held to the branch by webbing.

Nesting and fledging success rates for 16 nests, as estimated by Baltosser (1986), included an incubation success rate of 68 percent, a nestling success rate of 81.8 percent, and an overall nesting success (i.e., nests producing at least one fledged young) of 56.3 percent. The estimated seasonal productivity per female was 1.43 fledged young per season. Nearly all nest losses were attributed to predation.

Evolutionary and Ecological Relationships
Clearly the closest relative of *Amazilia violiceps* is *A. viridifrons*, which ranges from central Oaxaca and central Guerrero south to Chiapas. It is thus apparently partially sympatric with *A. violiceps* and cannot be considered conspecific on the basis of present knowledge.

The only possible hybrids so far reported involving this species are two birds, an adult male and a probable young female. The male, originally described as a new species (*Cyanomyia salvini*), was later believed to represent a possible hybrid involving *Cynanthus latirostris* (Bent, 1940). Later it was interpreted as simply an extreme example of *A. v. ellioti* (Friedmann et al., 1950), but Phillips et al. (1964) evidently regarded it as a hybrid with the broad-billed hummingbird. No identification could be made on the second specimen.

Ecological relationships between this species and other hummingbirds or plants are still unstudied.

GREEN-FRONTED HUMMINGBIRD

Amazilia (violiceps) viridifrons (Elliot)

Other Names

Cinnamon-sided hummingbird (*wagneri*); Colibri corona-verde, colibri flanquicanelo (Spanish).

Range

Consists of two similar taxa, distributed in three apparently parapatric populations. The first (*viridifrons*) extends from eastern Guerrero (where possibly locally sympatric with *violiceps*) to western Chiapas. A second population occurs in extreme eastern Oaxaca and interior Chiapas. These two disjunctive populations are geographically connected by a second taxon (*wagneri*) of uncertain taxonomic affiliation with the others (and possibly best associated with *violiceps* rather than *viridifrons*) that extends along the Pacific slope of the Oaxaca Mountains east approximately to the Chiapas border.

North American Subspecies

A. v. viridifrons. Resident from Oaxaca to eastern Chiapas; possibly also continuing south into Guatemala. Sometimes regarded as conspecific with *violiceps* (e.g., Phillips, 1964), but evidently widely sympatric with this taxon in Guerrero and considered as specifically distinct by most authors (Binford, 1989; Howell, 1993b; Howell and Webb, 1995).

A. v. wagneri. Resident on Pacific slope of Oaxaca. Sometimes associated with *violiceps*, but Binford (1989) relates it to *viridifrons*. Although it geographically separates two apparently identical populations of this taxon, this procedure seems safest given the present uncertain knowledge.

Breeding (hatched) and residential (inked) ranges of the violet-crowned hummingbird's nominate taxon. The residential ranges of two closely related forms, the Oaxacan endemic "cinnamon-sided" taxon *wagneri* (hatching), and the green-fronted hummingbird *viridifrons* (cross-hatching), are also illustrated. (*Adapted from Howell and Webb, 1995*)

Measurements

Wing, males 58–61.5 mm (ave. of 3, 60.1 mm), females 55–60.5 mm (ave. of 6, 58.5 mm). Culmen, males 22–24.5 mm (ave. of 3, 23.2 mm), females 20–24.5 mm (ave. of 6, 22.7 mm) (Ridgway, 1911). Eggs (of *wagneri*), 13.6–13.8 × 8.8 mm (Rowley, 1966).

Weights

A mixed-sex sample of 10 birds had a mean mass of 5.4 g (range 3.2–6.7 g) (Dunning, 1993).

Description (After Ridgway, 1911)

Adult male. Forehead and anterior crown indigo blue, becoming metallic green on the sides and rest of the head, and this color extending back over the upperparts to the rump and tail-coverts, which are more coppery-toned. Tail metallic bronze, the feathers narrowly tipped with whitish. The sides of the head below the eye, throat, and all underparts pure white, and the sides partly grayish brown glossed with bronze (in the nominate form; this area cinnamon-brown from the sides of the neck to the flanks in *wagneri*); iris brown; feet dusky; bill red, with a dusky tip.

Adult female. Like the male, but the forehead and crown dull dusky green, and the rectrices without buffy tips.

Immature. Closely resembles the adult female, but the bill blackish above, and the forehead and crown dark bluish green, the feathers tipped with cinnamon (Howell and Webb, 1995).

Identification

In the hand. This species has nearly identical measurements to the violet-crowned hummingbird; therefore, color differences in the forehead and crown feathers are what distinguish them (green in this species, violet in the violet-crowned). The distinctive taxon and so-called "cinnamon-sided" hummingbird (*wagneri*) are further separable from the violet-crowned by the narrow band of cinnamon separating the white underparts from the metallic-colored upperparts. Finally, the tail of the violet-crowned is slightly more grayish green (in males) than is the more bronzy-rufous tail of male green-fronted and cinnamon-sided hummingbirds, but these minor color differences

may be rather hard to detect without side-by-side comparison.

In the field. This medium-sized hummingbird is very similar to the locally presumably sympatric violet-crowned hummingbird, but instead has a green forehead and crown in both sexes. Otherwise the two taxa as nearly identical, and they occupy similar habitats. Vocalizations also are likely too similar (dry chattering notes have been reported by Howell and Webb, 1995) to be useful in distinguishing them.

Habitats

Associated with dry woodlands (especially thorn forests), brushy, or scrubby areas with scattered trees, from near sea level to about 1400 meters' elevation. In Oaxaca the birds are associated with tropical scrub, tropical deciduous forest, and tropical semideciduous forests, with breeding records occurring between 250 and 640 meters' elevation (Binford, 1989).

Movements

No movements of any kind have been documented.

Foraging Behavior and Floral Ecology

There are no available studies of the ecology of this species, which is seemingly extremely similar structurally to the violet-crowned hummingbird and is likely to be very similar ecologically.

Breeding Biology

There are no studies on this inadequately documented species, which is especially unfortunate inasmuch as local sympatric breeding without hybridization between it and the violet-crowned hummingbird remains to be established. Thus, the apparent sympatry in Guerrero that was documented by Phillips (1964) may be the result of seasonal movements into the area by the violet-crowned hummingbird (Binford, 1989). Breeding records for *viridifrons* in Mexico extend from December to June (except for March). Those of *wagneri* are scattered from January to October (Howell and Webb, 1995). Rowley (1966), who illustrated a nest of *wagneri*, stated that both of two nests were rather crudely made, and were placed on sapling oaks on steep hillsides. One was

about 1.2 meters above ground, the other about
1.8 meters. At least one was supported laterally
by a forking branch and was mostly covered by
lichens, judging from the published photograph.

Evolutionary and Ecological Relationships
Relationships among the so-called green-fronted,
cinnamon-sided and violet-crowned hummingbirds are still to be established; local studies probably are the only means by which the interactions
and interrelationships among these three taxa will
eventually be resolved.

STRIPE-TAILED HUMMINGBIRD

Eupherusa eximia (DeLattre)

Other Names
Blue-capped hummingbird (*cyanophrys*), Blue-fronted eupherusa (*cyanophrys*), Oaxaca humming-bird (*cyanophrys*), White-tailed hummingbird (*poliocerca*); Chupaflor colirraydo, Colibri colirrayado, Colibri Guerrerense, Colibri Oaxaqueño (Spanish).

Range
Resident in Mexico as three taxonomically recognizable but apparently allopatric populations. The first taxon (*eximia* group) extends from northern Oaxaca and southern Veracruz eastward to Chiapas, and southward throughout Central America. The second taxon (*poliocerca*) is endemic to Guerrero and Oaxaca. The third taxon (*cyanophrys*) is endemic to the Pacific slope of Oaxaca. Both of the latter taxa may be specifically distinct.

North American Subspecies (or Allospecies)
E. e. eximia. Resident from southern Veracruz south through Chiapas to Nicaragua.

E. (e.) poliocerca (Elliot). Resident in Guerrero and Oaxaca. Considered a distinct allospecies by Binford (1989), Howell and Webb (1995), and the American Ornithologists' Union (AOU, 1983).

E. (e.) cyanophrys (Rowley and Orr). Endemic to Pacific slope of Oaxaca (Rowley and Orr, 1964). Considered a probably distinct species by Howell and Webb (1995), Binford (1989), and the AOU (1983).

Measurements
Of *E. e. eximia:* Wing, males 56.5–61 mm (ave. of 11, 58.7 mm), females 50.5–54.5 mm (ave. of 8, 53.1 mm). Culmen, males 16–19 mm (ave. of 11, 17.6 mm), females 16–19.5 mm (ave. of 8, 17.6 mm)

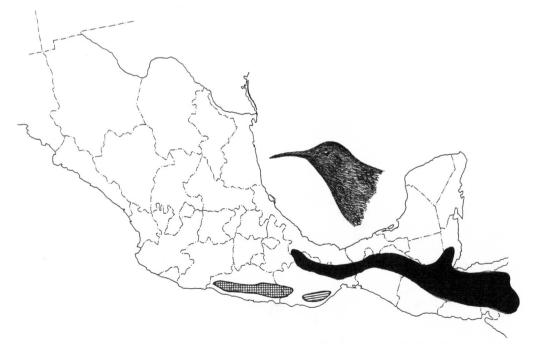

Residential ranges of the stripe-tailed hummingbird group, including the "white-tailed" *poliocera* (cross-hatching), the "blue-capped" *cyanophrys* (hatched), and the nominate ("stripe-tailed") *eximia* (inked). (*Adapted from Howell and Webb, 1995*)

(Ridgway, 1911). Eggs, 13–13.5 × 8.6–8.7 mm (Rowley, 1966).

Weights
An unsexed sample of 126 birds had a mean mass of 4.3 g (Brown and Bowers, 1985). The mean of 50 *eximia* males was 4.48 g; that of 31 females was 3.97 g (Feinsinger, 1976).

Description (After Ridgway, 1911)
Adult male. Upperparts and most underparts bright metallic green (but in *cyanophrys* the crown is bright metallic violet-blue), becoming more bronzy posteriorly, and the middle rectrices bronzy green to greenish bronze; lateral rectrices highly variable (in nominate *eximia* with black outer webs and the inner webs white tipped with black; in *poliocerca* the four outer pairs of rectrices white on the inner web, with the outer webs and tips of the inner webs pale grayish; in *cyanophrys* these rectrices entirely white); secondaries and inner primaries cinnamon-rufous with dusky tips; the under tail-coverts and femoral tufts white; iris brown; feet brown; bill dull black.

Adult female. Similar to the male, but the secondaries not tipped with dusky, the rectrices less strongly patterned with white, and the underparts light brownish gray, with some spotting of metallic green on the sides and flanks.

Immature male. Similar to the adult male, but the bright green underparts more restricted, and these areas mostly grayish brown; the outer rectrices may be tipped with white.

Identification
In the hand. This is one of the easiest of all Mexican hummingbirds to identify; the all-white or strongly white-striped tail distinguishes it from all others except for the much larger white-naped jacobin. The cinnamon-colored secondaries and inner primaries are also distinctive but not unique, as they also are present on the berylline hummingbird.

In the field. This small hummingbird is easily identified in the field by its flashing tail pattern; the mostly white to variously black-and-white striped rectrices are almost constantly flashed as the bird hovers before a food source or threatens another nearby individual. Otherwise males appear to be almost entirely glittering green (the endemic Oaxacan taxon *cyanophrys* has a distinctive blue crown), and females are green above and grayish white below. Both sexes also exhibit conspicuous cinnamon-colored patches on the secondaries and their major coverts. Calls include the usual assortment of buzzy and rattling notes, and the male's "song" is a series of high-pitched warbles made up of liquid or squeaky notes lasting 2 to 8 seconds (Howell and Webb, 1995). In Oaxaca individuals of all three known taxa may potentially occur, but breeding sympatry among the three is unproved and needs further field study.

Habitats
Associated with humid evergreen forests and forest edges, cloud forests, and montane pine-evergreen forests, from about 600 to 2300 meters' elevation. In Oaxaca, nominate *eximia* is a common permanent resident in cloud forest and upper tropical forests between 1300 and 1500 meters; *cyanophrys* is a common resident in cloud forest and the upper parts of tropical semideciduous forest between 1300 and 2600 meters; and *poliocerca* is very uncommon in tropical semideciduous forests and cloud forests between 900 and 1450 meters (Binford, 1989).

Movements
No migrations or other movements have been documented. Some non-breeding birds have been recorded at lower altitudes than those just mentioned for Oaxaca, suggesting that some altitudinal migration may occur.

Foraging Behavior and Floral Ecology
In Feinsinger's (1976) study at Monteverde, Costa Rica, nominate *eximia* was perhaps the most abundant hummingbird in that area, and at least males were territorial and highly aggressive in forested locations. The birds were especially attracted to flowering bushes (*Hamelia patens*, Rubiaceae) and *Inga brenesii* (Leguminaceae) trees, but were often displaced from these rich food resources by larger *Amazilia* hummingbirds. The birds usually visited the lower parts of flowering *Inga* trees and were generally forest-understory inhabitants. They also exploited the generally

scattered blossoms of various herbaceous or vine-like plants such as *Malvaviscus* (Malvaceae), *Psittacanthus* (Loranthaceae), *Lobelia* (Campanulaceae), *Kohleria* (Gesneriaceae), and *Manettia* (Rubiaceae). Using artificial feeders and sugar solutions, Carroll and Moore (1993) determined that the birds were able to distinguish among feeders and selectively choose the feeders that contained dissolved sugar alone, as well as those that included high and low concentrations of vitamins and minerals. They also chose feeders containing a high concentration of these supplements over feeders with a lower concentration.

Breeding Biology

Nesting of nominate *eximia* reportedly occurs from April to August, whereas *cyanophrys* nesting records are for May, plus September to November (Binford, 1989; Howell and Webb, 1995). Rowley (1966) described and illustrated the first four nests of *cyanophrys* to be discovered. Two were attached to bushes; one was on the exposed roots of a pine at a road cut; and one was attached to a branch of a small tree. The nests ranged in height from about 1.2 to 5.5 meters above ground, but the latter extreme represented the root-attached nest found at a steep road cut that was sheltered from above by an overhang; therefore, the maximum above-ground height (5.5 m) is rather misleading. The nests were cup-like structures made up of mosses, lined with plant down, and covered with varying amounts of lichens. Two of the nests had blossoms of *Salvia* incorporated into them.

Evolutionary and Ecological Relationships

These populations are part of a group of taxa that have ranges extending from Mexico to Panama and have received quite varied taxonomic treatment in the past. It is possible that as many as three species will be documented in the future, but lack of proof of sympatric breeding without hybridization makes the test of distinct biological species impossible to assess.

BLUE-THROATED HUMMINGBIRD

Lampornis clemenciae (Lesson)

Other Names
None in general English use; Chupamirto garganta azul (Spanish).

Range
Breeds from the mountains of southern Arizona, New Mexico, and Texas south through the mountains of Mexico to Oaxaca.

North American Subspecies (After Friedmann et al., 1950)
L. c. bessophilus (Oberholser). Casual to uncommon summer resident in the mountains of southern Arizona (Huachucas and Chiricahuas) and southern New Mexico (San Luis Mountains), becoming fairly common at elevations of 2400 to 3000 meters in the Sierra Madre Occidental of western Mexico to northwestern Durango and southeastern Sinaloa; winters at lower elevations.

L. c. clemenciae (Lesson). Fairly common summer resident in the Chisos Mountains of southern Texas and scarce in the Guadalupe Mountains of the same state, becoming common southward through the mountains of the Central Plateau and Sierra Madre Oriental to Oaxaca at elevations of 1800 to 3300 meters; winters at lower elevations.

Measurements
Wing, males 72–79 mm (ave. of 23, 76.7 mm), females 68.5–71 mm (ave. of 8, 69.7 mm). Culmen, males 21.5–25 mm (ave. of 23, 23.3 mm), females 24–27.5 mm (ave. of 8, 25.8 mm) (Ridgway, 1911). Eggs, ave. 16.3 × 9.9 mm (Bent, 1940, reports 16.23 × 12.45 mm for a single egg).

Weights
The average of 190 males in Oaxaca was 8.4 g (*SD* 0.4 g); that of 62 females was 6.8 g (*SD* 0.4 g)

Breeding (hatched) and residential (inked) ranges and migration zones or peripheral breeding areas (broken hatching) of the blue-throated hummingbird. (*Adapted from Howell and Webb, 1995*)

(Lyon, 1976). A sample of 195 males had a mean mass of 8.4 g (*SD* 0.24 g); 67 females had a mean mass of 6.8 g (*SD* 0.31 g) (Dunning, 1993).

Description (After Ridgway, 1911)
Adult male. Above rather dull metallic bronze-green, passing into olive-bronze or bronzy olive on rump, where the feathers have narrow terminal margins of pale brownish gray or buffy grayish; upper tail-coverts dusky (sometimes faintly glossed with greenish or bluish), narrowly and indistinctly margined with paler; tail black, faintly glossed with bluish, the outermost rectrix with terminal third (or less) abruptly white, the second less extensively tipped with white, the third usually with a small median white streak or mark (usually more or less fusiform or diamond-shaped) near tip; remiges dark brownish slate color or dusky, very faintly glossed with purplish; a conspicuous white postocular streak, extending obliquely backward and downward behind upper posterior margin of auricular region, the latter, together with the sub-orbital and loral regions, plain dusky; a more or less distinct malar streak of whitish (this sometimes obsolete); chin and throat metallic blue (varying from a greenish to a slightly violet hue), the feathers very narrowly and indistinctly margined with brownish gray and with concealed portion of the latter color; rest of underparts plain deep brownish gray or brownish slate color, the under tail-coverts broadly margined with white; femoral and anal tufts and tuft on each side of rump white; bill dull black; iris dark brown; feet dusky.

Adult female. Similar to the adult male, but blue of throat replaced by the general dull brownish gray of underparts.

Immature female. Similar to the adult female, but the pileum duller, the lower parts paler and more brownish, particularly on chin and upper throat, which are dull cinnamon (Oberholser, 1974).

Identification
In the hand. The large size (wing 68–80 mm) and the long, blackish, rounded bill (at least 21 mm long) separate the blue-throat from all other North American species except the magnificent hummingbird. Compared with that species, the blue-throated hummingbird has a more greenish tail,

which is rather square-tipped in males and somewhat rounded in females, and in both sexes has white markings at the tips.

In the field. Found in wooded streams of mountain canyons amid lush vegetation. The birds often utter loud, piercing seep notes while perched or in flight, and frequently fan the tail to reveal its green coloring and white tips. Although the female magnificent hummingbird also has whitish tips on the tail, the area in the blue-throat is larger and more pure white than in the magnificent.

Habitats
In Texas this species occurs in streamside vegetation of desert mountains, especially the Chisos Mountains, at elevations of 1500 to 2280 meters. There, it occurs among cypress (*Cupressus arizonica*), pines, oaks, and bigtooth maples in Boot Canyon, and moves out to drier canyon slopes to feed at agave plants (Oberholser, 1974). In Arizona it resides in moist canyons of the Huachuca and Chiricahua Mountains, and perhaps also the Santa Catalina and Santa Rita Mountains (Phillips et al., 1964). In New Mexico it is occasional in the southwest (San Luis Mountains) and the Guadalupe Mountains, and rarely reaches the northern highlands.

In Mexico the birds have a wide altitudinal range generally between 1800 and 3300 meters during the breeding season, locally descending to 300 to 900 meters during winter (Friedmann et al., 1950). Wagner (1952) noted that it sometimes reaches the limit of tree growth at about 3900 meters, and only in the coolest part of the year does it descend below 1800 meters at the southern end of its range.

Movements
The blue-throated hummingbird is a summer resident in the United States, with only rare winter occurrences north of Mexico. In Texas its spring migration is from March 18 to May 22, and its fall migration occurs between August 18 and October 24. There are a few December and early January records for the Rio Grande area (Oberholser, 1974). The earliest Arizona record is for early April, but nests with eggs have been found only a few weeks later.

Postbreeding vagrants probably account for most of the extralimital sightings of this species. These include several late summer or fall sightings along the coast of Texas, an early September sighting near Denver, Colorado (*American Birds* 25:85), and a possible fall sighting in Utah (*American Birds* 27:94).

In various parts of Mexico the species is either migratory or a resident form; Wagner (1952) summarized the details of their seasonal occurrence. In the Mexican highlands and adjoining valleys both sexes appear simultaneously between the last days of April and mid-May. Males immediately establish display territories; females fly about until the rainy season begins in late May or June and there is no longer a shortage of food in breeding areas. The sexes are somewhat separated during the breeding season: The males move into higher mountain levels above the firs and among the pines and bushgrasses, where flowering lupines are abundant; the females move to nesting territories with available overhanging nest sites, often in ravines or valleys. These tend to be fairly well-separated, but when conditions require, the birds will associate with one another and with other hummingbirds' groves that have an abundance of purple penstemons (*P. campanulatus*) and forest rims where mints (*Salvia cardinalis, S. elegans, S. genaereflora, Lamourouxia exerta*) occur.

Foraging Behavior and Floral Ecology

According to Wagner (1952), blue-throated hummingbirds live largely on insects, spiders, and plant lice. The particular color of the plant is not important to them, but the form of the flower determines to a great degree the kind of hummingbird that visits it. Thus, the small *S. mexicana*, although visited often by smaller hummingbird species, is utilized little by this one. On the other hand, *S. cardinalis* has a considerably larger blossom and is a highly favored food plant of the blue-throat in Mexico.

Marshall (1957) noted that in Arizona this species feeds at *Penstemon, Lobelia laxiflora*, and *Nicotiana*; in New Mexico it takes insects from the flowers of the shrubby honeysuckle, gilia, agave, and other plants (Bailey, 1928). On the basis of an analysis of three stomachs, Cottam and Knappen (1939) reported that the animal foods include

hemipterans, small beetles, flies, wasps, spiders, and daddy-longlegs. The stomachs also contained pollen grains and plant fibers.

In a study in the Chisos Mountains, Kuban and Neill (1980) found that this species was the most abundant hummingbird in the cypress-pine-oak habitat of their study area, and males defended elongated territories that paralleled Boot Creek. During the mornings they often foraged outside their territories in surrounding juniper woodlands, where nectar sources were more abundant prior to late June. When foraging on their territories in early June the birds ate only insects, which they captured by gleaning from vegetation as well as hawking them in the sage. When sage flowers became available in mid-July, the birds occasionally fed at them, but insects continued to dominate their diet throughout the summer.

In the Oaxaca area, Lyon (1976) observed that the blue-throat was the most successful of six major hummingbird species in terms of the number and sizes of territories maintained and the amount of preferred food plants (*Penstemon kunthii*) it dominated. However, in a year when magnificent hummingbirds were abundant, they effectively displaced the blue-throats in all open meadows, but were unable to do so in forest edge or open forest areas.

Breeding Biology

In Arizona, the breeding period of this species is relatively short; available egg records extend from April 22 to July 17 (Bent, 1940; Brandt, 1951). The timing of breeding in Mexico probably varies considerably by area. Eggs have been found in Veracruz in February; young have been found in Colima in March; and in the vicinity of Mexico City, the eggs are laid between late June and late September with a peak in July and August (Wagner, 1952).

To a greater degree than any other North American species of hummingbird, female blue-throats seek out nesting sites that are completely covered from above, such as in vertical-walled canyons, rock overhangs, or even under the roofs of structures such as bunkhouses or old barns. Brandt (1951) described a remarkable nest found in Ramsey Canyon, underneath a house built over a running stream. What was presumably the same female had occupied this nest for at least 10 years and produced up to three broods per season in it. In 1944 the female began renovating it on April 13. The first egg was laid on April 27, followed by a second egg two days later. When completed, the nest measured 127 millimeters high by 63 millimeters in diameter; when Brandt collected it on June 24 after the fledging of the young, he estimated that it consisted of some 24,000 kilometers of spider and insect thread! The female immediately began to construct a new nest, and by July 10 had deposited two eggs in it. Although this apparent 10-year lifespan is remarkable, even more impressive is a report of the apparent survival for 12 years of a single distinctive but unbanded male (Edgarton et al., 1951).

In Arizona, as apparently also in Mexico, males and females are not closely associated during the nesting period; the males occupy higher levels of the mountains. The nests also are fairly well spaced; Wagner (1952) diagramed the locations of five nests along a stream. The total distance between the two most widely separated nests was 250 meters, or an average inter-nest distance of about 60 meters. Foraging areas of the females did not apparently overlap.

When defending feeding areas, males often exhibit both intraspecific and interspecific aggressive behavior, and they nearly always dominate smaller species such as violet-crowned, black-chinned, and even magnificent hummingbirds. During such encounters the strongly patterned undertail surface is fanned and exposed, and this feature may be an important aspect of species recognition (Rising, 1965). The territorial call of the male is a loud *seep*, distinctly different from the *chip* call of the magnificent (Marshall, 1957). Evidently males are able to make clear distinctions between species during interspecific encounters (Lyon et al., 1977).

The shape of the nest varies considerably in this species, especially in its total depth. Wagner reported that the dimensions of the nest cup and outer nest diameter were fairly consistent, but that the total depth of 10 nests measured by him ranged from 4 to 11.5 centimeters. Depending on the size of the nest, it may take the female from 15 to 30 days to construct it, based on Wagner's observations.

The materials used in nest construction, other than spider webbing, also vary considerably. They may include oak blossom hulls, mosses, coarse straws, stems of weeds, and other materials. Sometimes the nests are decorated externally with lichens or other vegetation. They often are supported mostly from one side or the rim, and rarely are suspended from above, with a nearly lateral opening. The two eggs, laid approximately 48 hours apart, average 0.74 grams each (Wagner, 1952).

Incubation requires from 17 to 18 days, and an additional 24 to 29 days are needed to bring the nestlings to fledging. By the 10th day after hatching the eyes are opening, and by the 12th day the feathers of the back are starting to emerge. The young are alternately brooded and fed until about the 12th day, and after 16 days the weight of the young is typically greater than that of the adult. The young are fully feathered after about 24 days, but at that age the bill-length is appreciably shorter than that of the parent bird (Wagner, 1952).

Evolutionary and Ecological Relationships
Mayr and Short (1970) did not identify the blue-throated hummingbird's nearest relatives. Ridgway (1911) placed the species in a monotypic genus (*Cyanolaemus*), separating it from the more typical *Lampornis* by its bill shape, which is relatively less depressed and narrower basally. Certainly the species can be easily included within *Lampornis,* and probably it is a close relative of the amethyst-throated hummingbird, which ranges from Mexico to Honduras. Apparently the only known hybrid is a still unidentified specimen involving the Costa or possibly the Anna hummingbird (Mayr and Short, 1970).

The very large size of this species effectively allows it to dominate most other hummingbirds with which it comes into contact (Kuban and Neill, 1980; Lyon, 1976). On the other hand, it is unable to use some of the smaller flowers, such as those of *Salvia mexicana* (Wagner, 1952). Nor can it exploit those that are low in nectar production, such as *Penstemon perfoliatus* and *P. gentianoides* (Lyon, 1976), even though these may be used regularly by smaller species of hummingbirds.

AMETHYST-THROATED HUMMINGBIRD

Lampornis amethystinus (Swainson)

Other Names

Cazique hummingbird, Margaret's hummingbird, Violet-throated hummingbird (*margaritae*); Chupaflorvamatista, Colibri-serrano gorjiamatisto (Spanish).

Range

Resident in Mexico from Nayarit on the western slope and Tamaulipas on the eastern slope south to the Chiapas–Guatemala border, thence extending south to Honduras.

North American Subspecies

L. a. amethystinus. Resident from Nayarit and Tamaulipas to El Salvador and Honduras.

L. a. margaritae (Salvin and Goldman). Resident from Michoacan to Oaxaca. Sometimes considered to be specifically distinct.

L. a. circumventus (Phillips). Endemic to Oaxaca, but of doubtful validity (Binford, 1989).

L. a. brevirostris (Ridgway). Resident from Nayarit and Colima to State of Mexico. The validity of this race has also been questioned (Binford, 1989).

Measurements

Of *L. a. amethystinus:* Wing, males 65–71 mm (ave. of 9, 67.6 mm). Culmen, males 20–23 mm (ave. of 9, 21.5 mm) (Ridgway, 1911). Eggs, 14.5–15.4 × 9.2–9.6 mm (Rowley, 1966).

Weights

A sample of 21 males had a mean mass of 6.8 g (*SD* 0.4 g); 5 females a mean mass of 6.6 g (*SD* 0.13 g) (Lyon, 1976). Eight males in the Field Museum had a mean mass of 6.5 g (range 5.0–7.8 mm); 13 females had a mean mass of 5.3 g (range 5.0–5.8 mm).

Residential range of the amethyst-throated hummingbird. The arrowhead represents an extralimital breeding location. *(Adapted from Howell and Webb, 1995)*

Description (After Ridgway, 1911)

Adult male. Crown and upperparts metallic bronze-green to bronze, usually duller on the rump, and the upper tail-coverts dull black; tail black, with the lateral rectrices tipped brownish gray; the sides of the head from lores to auriculars mostly blackish, with a postocular streak of white above, and a very narrow cinnamon or cinnamon-buff streak below. This borders a gorget of metallic reddish purple (in *margaritae*) to pinkish amethyst (in *amethystinus*); these gorget feathers narrowly margined with grayish, producing a scale-like visual effect; rest of the underparts brownish gray, but with white lumbar and femoral tufts; iris brown; feet brownish; bill black.

Adult female. Similar to the adult male, but the crown darker and duller, the chin and throat dull cinnamon, the sides and flanks little if at all glossed with metallic feathers; the tips of the lateral rectrices more distinctly pale gray.

Immature male. Resembles the adult female, with a buff to gray (rather than cinnamon) throat; probably the upper mandible has the parallel striations typical of young hummingbirds as well.

Identification

In the hand. This species closely resembles the similar-sized garnet-throated hummingbird in several respects, but has a longer bill (culmen at least 20 mm) that is slightly decurved and lacks the cinnamon colors that are so evident on the wings of the garnet-throated hummingbird. Females closely resemble those of the magnificent hummingbird, but have gray rather than white tips on their outer rectrices.

In the field. This is a rather large hummingbird, and males have distinctive sooty-gray underparts, a dark face with a contrasting white eye-stripe, and a glittering purplish red gorget. Females have the grayish underparts and white eye-stripe, but instead of a glittering gorget have a pale-cinnamon-tinted throat. The species is most similar in size and color to the garnet-throated hummingbird, which is widely sympatric with it, but which has rich chestnut on the primaries and secondaries, a trait lacking in this species. It also resembles the blue-throated hummingbird, but lacks that species' conspicuous white tail markings (the tail instead

has grayish feather tips). Female magnificent hummingbirds are also similar, but likewise have white, not gray, feather tips on the outer rectrices. Vocalizations include rasping, buzzy notes, chattering calls, and the male's "song" is evidently a series of persistently repeated sharp and high-pitched chipping notes.

Habitats

Associated with cloud forests, high montane forests, pine-oak forests, and oak woodlands from 900 to 3000 meters' elevation. In Oaxaca it occurs commonly in cloud forests and humid pine-oak forests between about 1250 meters and 2900 meters (Binford, 1989).

Movements

No migrations or other movements have been documented.

Foraging Behavior and Floral Ecology

Lyon (1976) described this species as being non-territorial around its available food resources, but because of their larger size these birds were able to enter the territories of bumblebee and white-eared hummingbirds and chase them from favored foraging areas. However, when they entered the feeding territories of blue-throated or magnificent hummingbirds they were regularly chased away. The most favored plants in this meadow-like area within a pine forest were the scrophulariaeceous *Penstemon kunthii* and the iridaceous species *Rigidella orthantha,* the former blooming later in the summer as the latter species declined in importance as a foraging resource.

Wagner (1946a) reported that in the montane rainforests of Chiapas, the species hunts for insect prey by hovering near mossy tree trunks, slowly moving up and down nearby, and occasionally stopping long enough to capture a small insect. In the moss-covered cloud forests of Guatemala, the birds were observed visiting various red-flowered plants, including various mints (*Salvia nervata, S. cinnaberina,* and *Satureja* species), the herbaceous *Centropogon affinis* (Campanulaceae), and the shrub-like *Fuchsia minutiflora* (Onagraceae). At times they also utilized the blue flowers of a mint (*S. cacaliaefolia*) that bloomed during the rainy season (Skutch, 1967).

Breeding Biology

Skutch (1967) observed that in the Guatemala highlands territorial singing by males of this species occurs during the latter months of the year. At that time territorial males perch about 3 meters above ground and utter a single, mournful note at a rate of about 75 to 85 times per minute for hours on end, without apparent variation. The singing season there corresponded closely to the blossoming period of the species' favorite food plant, the mint *S. nervata.*

Nesting of this species in western Mexico and Guatemala occurs from October to December, and from May to July (Howell and Webb, 1995). In Guatemala, Skutch (1967) found four nests: two in November and two in December. One of these was located in low, gnarled trees on a mountain summit; two were in cypress forest at somewhat lower elevation; and one was in the zone of mixed hardwoods and pines at a still lower elevation. The nests ranged from about 1.2 to 2.4 meters above ground. All were open-cup nests, three of which were made mainly of mosses, decorated with lichens around the exterior. The fourth had a mossy exterior surface, without lichens or other decorations. At least two of the nests were attached to pendant or dangling vegetation. Two of the nests contained eggs when they were found, and one contained two newly hatched nestlings.

Evolutionary and Ecological Relationships

This species is a close relative of the blue-throated hummingbird, with which it is distinctly sympatric, both species occurring in humid pine-oak forests of Oaxaca (Binford, 1989) and doubtless also elsewhere in their extensively overlapping ranges. The marked interspecific differences in male gorget color and in the extent of white in the tails of females suggest these might be important isolating mechanisms in such areas.

GARNET-THROATED HUMMINGBIRD

Lamprolaima rhami (Lesson)

Other Names

None in general English use; Chupaflor alicastaño, Colibri alicastaño (Spanish).

Range

Resident in Mexico from Guerrero in the west and Puebla in the east, southeast to the Chiapas–Guatemala border, thence extending south in Central America to El Salvador and Honduras.

North American Subspecies

None recognized here. The validity of a race (*occidentalis*) described by Phillips (1966) is questionable (Binford, 1989).

Measurements

Wing, males 73.5–81 mm (ave. of 23, 77.2 mm), females 64–67.5 mm (ave. of 9, 66.2 mm). Culmen, males 15.5–18.5 mm (ave. of 23, 17.9 mm), females

15.5–18.5 mm (ave. of 9, 16.7 mm) (Ridgway, 1911). Eggs apparently undescribed.

Weights

Three birds of both sexes had a mean mass of 6.1 g (Dunning, 1993).

Description (After Ridgway, 1911)

Adult male. Upperparts bright metallic green to bronze-green becoming more bronzy posteriorly, and the tail entirely purplish black to violet-purple; the remiges and their greater wing-coverts rufous to chestnut, with dusky purplish feather edgings; sides of the face from the lores to the auriculars and the sides of lower throat velvety black to purplish bronze; the chin and throat bright metallic reddish purple to purplish red; a small white post-ocular spot, and below the blackish throat the breast rich metallic violet to metallic green, depending on angle of view; breast, sides, and

Residential range of the garnet-throated hummingbird. *(Adapted from Howell and Webb, 1995)*

flanks also glossed with bronze-green; the lower underparts more slate, with grayish white femoral and rump tufts; Iris brown; feet dusky; bill black.

Adult female. Similar to the adult male, but the metallic green more bronzy, the sides of the head less glossy black and more grayish; only a few scattered metallic feathers on the mostly buffy brown throat; the other underparts mostly grayish dusky to deep sooty brown; the lateral rectrices all tipped with brownish gray, and the postocular spot of the male poorly developed and sometimes extended into an eye-stripe.

Immature male. Resembles the adult female, but with varying amounts of metallic blue feathering present on the throat.

Identification

In the hand. This hummingbird has an unusually short and straight bill (culmen maximum 18.5 mm) relative to its body size, which together with its chestnut-colored wing feathers fully separates it from the otherwise rather similar-appearing and comparably sized blue-throated, amethyst-throated, and magnificent hummingbirds.

In the field. This large hummingbird has distinctive chestnut-colored patches on the flight feathers and their greater coverts—the only comparatively large Mexican hummingbird having this trait. The male's magenta-colored gorget is fully outlined posteriorly in black, unlike the similar amethyst-throated hummingbird, which has dusky coloration confined to the sides of the head. The white postocular spot of the male shows up well against the surrounding black area of the auricular region, and the female has a somewhat less conspicuous whitish spot, or more a linear streak behind the eye. She may also exhibit a few iridescent violet feathers on the throat where the male's gorget is located. The black bill is also relatively short and straight in this species.

Habitats

The garnet-throated hummingbird is associated with cloud forests and humid montane pine-oak forests, as well as forest edges and scrubby second-growth thickets, from about 900 to 3000 meters' elevation. Nesting records range from about 1500 to 2300 meters. In Honduras the birds are limited to cloud forests above 1600 meters' elevation (Monroe, 1968).

Movements

No migrations or other movements have been documented.

Foraging Behavior and Floral Ecology

No specific information is available. The birds are said to favor various flowering trees as a primary food resource, and to feed at locations ranging from low to high above ground level. Possible foraging-site territoriality is still unreported. The short, straight bill of this species suggests it must be something of a generalist, with a very limited ability to extract nectar from flowers with deep-tubed corollas.

Breeding Biology

Breeding records in Mexico extend from December to March on the Pacific slope, and from April to May on the Atlantic slope and Oaxaca (Howell and Webb, 1995). Rowley (1966) reported on two nests that he found in Oaxaca. One was located along a washed-out creek bed where the roots of trees and bushes had been exposed below an overhanging bank. The nest was attached to the exposed roots, and was composed of green moss, with a plant fiber inner lining; and the entire structure was attached to the root by spider webbing. A large leaf and some pine needles also had been worked into the nest, which was of a large size (about 80 mm in external diameter at the base) and shape as to resemble the nest of a wood-pewee (*Contopus*) rather than a hummingbird. It was situated on a nearly level exposed section of root, with rootlets dangling below the nest. A second nest was later found in a similar situation, and it too was quite large, perhaps because it had been built on the remains of a previously used nest. A long "streamer" of pine needles hung below the nest.

Evolutionary and Ecological Relationships

This is the only member of its genus, but there seems little reason for maintaining this genus rather than merging it with *Lampornis*. The somewhat shorter and straighter bill of this species is practically the only feature that can be identified as a presumed generic trait separating it from typical *Lampornis* species.

MAGNIFICENT HUMMINGBIRD

Heliodoxa fulgens (Swainson)
(*Eugenes fulgens* of AOU, 1983)

Other Names
Rivoli hummingbird; Chupamirto verde montero, Chupaflor magnifico (Spanish).

Range
Breeds from the mountains of southern Arizona and southwestern New Mexico south through Middle America to western Panama. Northern populations winter in Mexico.

North American Subspecies (After Friedmann et al., 1950)
H. f. fulgens (Boucard). Breeds in the mountains of southern Arizona (north to the Grahams and Santa Catalinas), southwestern New Mexico (San Luis Mountains), and southwestern Texas (Chisos Mountains), south through the mountains of Mexico (usually 1500–3300 meters) to the Isthmus of Panama. Has bred rarely in Colorado.

Measurements
Wing, males 69.5–76 mm (ave. of 31, 73 mm), females 66.5–70.5 mm (ave. of 14, 68.7 mm). Culmen, males 25.5–31 mm (ave. of 31, 27.4 mm), females 27–30.5 mm (ave. of 14, 29.1 mm) (Ridgway, 1911). Eggs, ave. 15.4 × 10.2 mm (extremes 14–16.5 × 9.4–11.4 mm).

Weights
The average of 119 males was 7.7 g (*SD* 0.4 g); that of 24 females was 6.4 g (*SD* 0.5 g), in Oaxaca (Lyon, 1976).

Description (After Ridgway, 1911)
Adult male. Pileum rich metallic violet or royal purple, the forehead (at least anteriorly) blackish, usually glossed with green or bluish green; hindneck sides of occiput, and auricular region velvety black in position *a* (see note at beginning of Part

Breeding areas (hatched), residential areas (inked), and migration zones or peripheral breeding areas (broken hatching) of the magnificent hummingbird. *(Adapted from Howell and Webb, 1995)*

Two for position descriptions), metallic bronze, bronze-green, or golden green in position *c;* rest of upperparts metallic bronze, bronze-green, or golden green, including tail, the latter sometimes with rectrices passing into pale grayish at tip; remiges dark brownish slate or dusky, faintly glossed with purple or purplish bronze; chin and throat brilliant metallic emerald green (more yellowish in position *a,* more bluish in position *b*), this brilliant green area extending much farther backward laterally than medially; chest and upper breast velvety black in position *a* (bronze or bronze-green in position *c*), passing into dusky bronze or bronze-green on lower breast, this into grayish brown or sooty grayish on abdomen and flanks; femoral and anal tufts white; a small white postocular spot or streak (sometimes a whitish malar streak also); under tail-coverts light brownish gray (sometimes glossed with bronze or bronze-green), margined (more or less distinctly) with whitish; bill dull black; iris dark brown; feet dusky.

Adult female. Above, including four middle rectrices, metallic bronze, bronze-green, or golden green, the pileum duller (sometimes dull grayish brown anteriorly); three outer rectrices (on each side) with basal half more or less bronze-green, then black, the tip brownish gray or grayish brown, this broadest on outermost rectrix, much smaller on third; remiges as in adult male; post-ocular streak of white and below this a dusky auricular area; underparts brownish gray or buffy grayish, glossed laterally with metallic bronze or bronze-green, the feathers of chin and throat margined with paler or with dull grayish white, producing a squamate appearance; femoral and anal tufts white; under tail-coverts brownish gray (sometimes glossed with bronze-green), margined with pale brownish gray or dull whitish; bill, etc., as in adult male.

Immature male. Intermediate in coloration between the adult male and the adult female, the crown partly violet, the throat only partly green, and the chest slightly intermixed with black; the tail exactly intermediate, both in form and color.

Immature female. Similar to the adult female, but feathers of upperparts narrowly margined termi-

nally with pale grayish buffy, and underparts slightly darker and suffused, more or less, with pale brownish buffy.

Identification
In the hand. A large hummingbird (wing at least 65 mm) notable for its long, blackish bill (25–30 mm), which is rounded rather than widened at the base, and its fairly long, greenish-black tail, which is slightly forked in males and rather square-tipped in females. The similar-sized blue-throated hummingbird has a more bluish tail, which is squared-off in males and slightly rounded in females.

In the field. Associated with oak or pine forests and (in Middle America) cloud forests, especially near forest edges or thinned woods, where they perch low in shrubby or thickety growth mixed with trees. Male's small white postocular spot resembles a staring eye (Slud, 1964). It may be distinguished from the blue-throated hummingbird by its tail, which is blackish and slightly forked in males, and greenish with white corners in females. Its call is a thin, sharp *chip.*

Habitats
In Arizona, magnificent hummingbirds occur commonly during summer among the maples of the lower mountain streams, and extend upward from 1500 to 2250 meters on mountain slopes to the ponderosa pine zone, with males occasionally reaching the edge of the fir zone (Bent, 1940; Phillips et al., 1964). In the Huachuca Mountains the birds are probably most common at the lower edge of the pine belt, where flowering agaves are relatively abundant. Likewise in the Chiricahua Mountains they are common in more open parts of the ponderosa pine forest, such as where fire has killed some of the trees and where flowering plants such as penstemons and honeysuckles (*Lonicera involucrata*) are in blossom (Bent, 1940).

In New Mexico, the species has been found during summer at about 2000 meters in the San Luis Mountains (Ligon, 1961). It also locally summers in mountainous areas of the northern and western portions of the state, from the fir forest zone down into evergreen and adjacent riparian wooded habitats. During migration it also occurs in intervening lowland and residential areas (Hubbard, 1978). Very rarely the species also summers in the mountains of Colorado, and there is a single nesting record for Boulder County (Bailey and Niedrach, 1965).

In Texas it occurs uncommonly between 1800 and 2300 meters' elevation in the Chisos Mountains during summer and probably breeds there, although nests have not been found. In these mountains the species favors the pinyon-juniper-oak zone and occurs in somewhat higher and drier habitats than those typical of blue-throated hummingbirds (Oberholser, 1974; Wauer, 1973).

In Mexico, magnificent hummingbirds generally range from 1500 to 3300 meters' elevation, occasionally descending as low as 900 meters (Friedmann et al., 1950). In Guatemala they occur from 900 to 2600 meters in cloud forest and woodlands of oaks or pines (Land, 1970). In Honduras, they inhabit cloud forests above 1000 meters, and occasionally pine-oak vegetation or scrubby growth (Monroe, 1968). In El Salvador they occur between 2100 and 2400 meters among oaks, pines, and rocky areas containing agaves (Dickey and van Rossem, 1938).

In Costa Rica the species extends from timberline down to 2100 meters, and occasionally as low as 1500 meters. There it is mainly associated with forest edges, breaks in woodlands, parklike pastures, and thinned woods, and is scarce in dense forests (Slud, 1964). At the southern end of its range in western Panama, it is found on volcanic slopes from 1550 to 2300 meters' elevation (Wetmore, 1968).

Movements

In Arizona, magnificent hummingbirds have been recorded as early as April 5 and as late as November 11 (Phillips et al., 1964), but they probably normally arrive in the state in the latter part of April. They usually occur in Texas from early April until late September, but have been reported from March 30 to October 12 (Oberholser, 1974). They have been observed in New Mexico from May through October, but have also rarely been seen in the Sandia Mountains during winter months (Hubbard, 1978; Ligon, 1961).

Post-breeding movements of adults or immatures are probably responsible for many of the extralimital records of this species in the United States. Thus it has been reported at least once during late summer from Nevada (*American Birds* 30:103) and Utah (*American Birds* 25:885) and several times from Colorado throughout the summer months from late June to September 6 (Bailey and Niedrach, 1965).

Farther south in Mexico and Central America the species probably is not markedly migratory, but at least in some areas the birds evidently move to lower altitudes during the non-breeding season (Land, 1970).

Foraging Behavior and Floral Ecology

According to Des Granges (1979), this species is "trapliner" rather than territorial. In Colima it was found to forage largely on thistles (*Cirsium species*), but visited isolated clumps of flowers along its trapline as well, especially the tubular flowers of *Penstemon roseus*. It also is apparently associated with *Lobelia laxiflora* in more arid environments.

In a study area in Oaxaca, Lyon (1976) found that males of this species occupied feeding territories that averaged about 720 square meters in area, compared with slightly smaller territories for females. These territories were second in size only to those held by male blue-throated hummingbirds, and collectively the two species dominated the areas containing the richest nectar sources. They partitioned such areas on the basis of habitat. In one year a large population of magnificent hummingbirds managed to displace the slightly larger blue-throats from all their territories in open meadows, but were unable to do so in meadow–forest edge habitats or in open forest areas. Both species foraged primarily on *P. kunthii*, which blooms from May to October and provides the richest single source of nectar available in the area.

In a Costa Rica tropical highland habitat, Wolf et al. (1976) found that individuals of this species spent 84 percent of their foraging effort at the blossoms of only two species (*Centropogon talamanensis* and *Cirsium subcoriaceum*); these plants accounted for 99 percent of the territoriality records. A small amount of foraging was done at the flowers of *Bomarea costaricensis* and *Fushsia splendens* as well. In this area, as in Oaxaca, the species was territorial rather than trapliner.

Marshall (1957) reported that this species often captures insects near tree foliage, especially sycamores and Apache pines. Although riparian woodland is its preferred habitat, Marshall observed that it is sometimes found far from water, and when necessary can be independent of nectar sources. It is especially attracted to agaves (*A. americana* and *A. parryi*), the flowers of which are rich in insect life. Moreover, it forages on penstemons, honeysuckles, scarlet salvias, irises, and planted scarlet geraniums, as well as blossoms of *Erythrina corallodendrum* (Toledo, 1974). A sample of three stomachs included a large variety of insect life, including leaf bugs, plant lice, leafhoppers, parasitic wasps, beetles, flies, moth fragments, and a considerable number of spiders (Cottam and Knappen, 1939).

When not actually feeding, the birds seem to spend an unusual amount of time perched relatively high in tall or dead-topped trees, where they are prominently outlined against the sky (Dickey and van Rossem, 1938). However, Slud (1964) stated that the birds often perch and forage fairly low in shrubby or thickety growth, and Edwards (1973) stated that they often perch on inner tree branches within 4 to 6 meters of the ground.

Breeding Biology

In Arizona the nesting season is fairly short. Bent (1940) listed 24 egg records extending from May 6 to July 28, with half of these between June 14 and July 14. Birds either breeding or in breeding condition have been found in Mexico from March to November (Friedmann et al., 1950). Similarly, in Guatemala the nonbreeding season extends from December to February (Land, 1970). In El Salvador, however, breeding probably occurs during the winter months, although year-round breeding is also a possibility (Dickey and van Rossem, 1938).

Magnificent hummingbirds place their nests fairly high in trees. Of six nest sites mentioned by Bent (1940), the total range was from 3 to 16 meters above ground, and the average was 9 meters. A variety of trees are used as substrates, including cottonwood, mountain maple, sycamore, alder, walnut, pine, and Douglas fir, with none seemingly preferred over the others.

The nests are usually saddled on horizontal branches that may be smaller in diameter than the nest itself. They are usually slightly more than 50 millimeters in outside diameter and 50 millimeters in total depth, with a cup slightly more than 25 millimeters wide. They resemble the nests of the ruby-throated hummingbird, but are generally larger, proportionately broader, and not so high. The nests are composed of mosses and other soft vegetation, lined internally with plant down or soft feathers, and externally decorated with lichens that sometimes nearly "shingle" the entire outer surface. In one observed case, the birds required a week to build a nest and lay two eggs (Bent, 1940).

There is no information available on the incubation period, the development of the young, or the fledging period.

Evolutionary and Ecological Relationships

Although the magnificent hummingbird is usually considered a monotypic genus (or sometimes of two species, with the Middle American form *H. f. spectabilis* separated specifically), Zimmer (1950–53) has urged the merger of *Eugenes* into the fairly large genus *Heliodoxa*. I have followed this procedure here, and it seems entirely consistent with a broad generic concept. However, Wetmore (1968) has suggested that such a merger is premature on the basis of available information.

In any case, it seems likely that the nearest relatives of *H. f. fulgens* are among the species of the genus *Heliodoxa*. Further, the genus *Lampornis* is probably not far removed from this cluster of species. Short and Phillips (1966) have also suggested that *Amazilia* and *"Eugenes"* are probably not as distantly related to each other as most current classifications suggest. The only known hybrid involving the magnificent hummingbird is with the broad-billed hummingbird (Short and Phillips, 1966).

Nothing specific can be said about the ecological relationships of the magnificent hummingbird. Certainly it locally overlaps with the blue-throated hummingbird and presumably competes with it. But the blue-throated is a slightly larger species than the magnificent which would probably place the latter at a competitive disadvantage.

PLAIN-CAPPED STARTHROAT

Heliomaster constantii (DeLattre)

Other Names
Constant star-throat; Chupamirto ocotero, Chupaflor pochotero (Spanish).

Range
Breeds in the arid tropical zone of the Pacific Coast from southern Sonora, Mexico, south to Costa Rica (Friedmann et al., 1950). Accidental in Arizona.

North American Subspecies (Presumed)
H. c. pinicola (Gould). Confined to the western slope of the Sierra Madre Occidental, from sea level to about 1200 meters, from extreme southern Sonora to western Jalisco (Friedmann et al., 1950).

H. c. leocadiae (Bourcier and Mulsant). Resident of central and southwestern Mexico, from Michoacan and Guerrero south to Guatemala.

Measurements
Wing, males 63–70.5 mm (ave. of 15, 66.9 mm), females 64–68.5 mm (ave. of 10, 65.7 mm). Culmen, males 33.5–36.5 mm (ave. of 15, 34.5 mm), females 34–37.5 mm (ave. of 10, 35.5 mm) (Ridgway, 1911). Eggs, apparently undescribed. Those of the slightly smaller *H. longirostris* average 13.2 × 8.65 mm (Rowley, 1966).

Weights
The average weight of 6 males was 7.43 g (range 6.5–9.0); that of 3 females was 7.2 g (range 6.8–7.8 g) (J. L. Des Granges, personal communication). A sample of live-captured and museum specimens (sex and number unstated) averaged 7.5 g (*SD* 0.9 g) (Arizmendi and Ornelas, 1990).

Description (After Ridgway, 1911)
Adult male. Above metallic bronze or bronze-green, somewhat duller on pileum, especially on forehead; rump with longitudinal median patch or broad streak of white; middle rectrices usually dusky terminally, the other retrices extensively blackish terminally, the inner web tipped with a

Residential range of the plain-capped starthroat. *(Adapted from Howell and Webb, 1995)*

spot of white; remiges brownish slate or dusky, faintly glossed with purplish; a postocular spot and a conspicuous malar stripe dull white; the auricular, suborbital, and loral regions dusky; chin sooty or blackish; throat bright metallic red or purplish red (varying from orange-red or scarlet to rose-red), the feathers narrowly margined terminally with pale grayish or dull whitish (invisible except when viewed from behind); underparts of body brownish gray (deep smoke gray or nearly mouse gray), fading to white on abdomen and anal and femoral regions; under tail-coverts pale gray basally, dusky subterminally (in form of a V- or U-shaped bar), broadly white terminally; a large and conspicuous tuft of silky white feathers on sides, between flanks and back; bill dull black; iris dark brown; feet dusky.

Adult female. Very similar to the adult male and not always distinguishable, but usually (?) with the blackish of chin slightly more extended.

Immature. Similar to adults, but greater part (sometimes whole) of throat dark sooty brown or dusky, the feathers margined terminally with grayish white.

Identification
In the hand. Very large size (wing more than 60 mm) and long bill (culmen 33–38 mm) eliminate all other North American species; moreover, the tail feathers are broadly tipped with white.

In the field. Associated with arid scrubby lowlands, woodland borders, dry woods, plantations, and the like. Its very large size, long bill, and broadly white-tipped tail are distinctive. There is also a streak of white on the rump, a white malar stripe, and a throat that is sometimes glittering red (males) or spotted with dusky (females). Both sexes have dark grayish underparts. These birds often hawk for insects. While they are in flight, and sometimes while perched, a tuft of white flank feathers may be visible near the posterior edge of the wing.

Habitats
In Mexico this species occurs in scrubby, rather arid woodlands, woodland edges, partially open country with scattered trees or thickets, and scrubby riparian woodland, from sea level to 1500 meters (Edwards, 1973). In Colima it is found from the thorn forest zone to the upper edges of the tropical deciduous forest, but is most common in the denser thorn forest zone (Schaldach, 1963).

Movements
Since the plain-capped starthroat is essentially tropical in distribution, it probably undergoes relatively few seasonal movements. In Sonora it is essentially a resident species, but van Rossem (1945) suggested that there may be a partial exodus from there during the winter months.

Post-breeding wandering has perhaps accounted for the few records of this species in the United States, most of which have occurred in Arizona. The first reported individual visited a feeder in Nogales during late September 1969. In 1978 an individual was seen at a feeder in Patagonia, Arizona, between July 15 and 20. Earlier that year a possible sighting of the species was made on June 17 in Sycamore Canyon, west of Nogales, and on June 24 a similar sighting was made southwest of Patagonia. These sightings all occurred close to the Mexican border, about 500 kilometers north of the species' known range limits in Sonora. Finally, in Phoenix, well to the north of these observations, an individual frequented a yard feeder from October 17 to November 28, 1978 (Witzeman, 1979), and on June 28, 1980, an individual was observed near Sierra Vista (*American Birds* 34:919). There are also two recent New Mexico sight records (*American Birds* 48:138).

Foraging Behavior and Floral Ecology
Almost nothing is known of the foraging ecology of the plain-capped starthroat. According to Des Granges (1979), it resides in the Volcan de Colima area and has exhibited feeding territoriality there. It showed a moderate degree of social dominance with other species, but almost throughout the entire year it fed nearly exclusively on aerial arthropods; during the short blooming period of *Ceiba aesculifolia*, however, nectar-feeding occupied nearly all of its time. This species frequently catches insects while flying close to the ground above roads or wide trails, and otherwise tends to perch quietly 4 to 6 meters above ground (Edwards, 1973).

Breeding Biology

There is apparently no description available of the nest or breeding activities of this species.

Evolutionary and Ecological Relationships

The two Central American species of starthroats are so similar to each other that they may constitute a superspecies, with the plain-capped starthroat occupying more arid and westerly portions of Mexico and Central America and the long-billed starthroat adapted to more humid lowlands, extending southward to Brazil.

The unusually long and decurved bills of the starthroats approach those of the hermit hummingbirds, and perhaps these species compete locally to some extent. However, at least the plain-capped starthroat is strongly insectivorous and does not normally defend feeding territories (Des Granges, 1979).

LONG-BILLED STARTHROAT

Heliomaster longirostris (Audebart and Vieillot)

Other Names
Pale-crowned starthroat (*pallidiceps*); Chupamirto picudo, Picolargo coroniazul (Spanish).

Range
Resident in Mexico from Guerrero on the western slope and Veracruz on the eastern slope to the Chiapas–Guatemala border, continuing on through Central America and in South America south to Peru, Bolivia, and Brazil, and east to Venezuela, Trinidad, and the Gulanas.

North American Subspecies
H. l. pallidiceps (Gould). Resident from Guerrero and Veracruz south to Honduras.

H. l. masculinus (Phillips). Resident of Pacific slope of mountains in Guerrero and Oaxaca.

Measurements
Of *H. l. pallidiceps*: Wing, males 57–61.5 mm (ave. of 11, 58.9 mm), female 57 mm. Culmen, males 30–33.5 mm (ave. of 11, 31.8 mm), female 32.5 mm (Ridgway, 1911). Eggs, 12.8–13.6 × 8.5–8.8 mm (Rowley, 1966).

Weights
Stiles and Skutch (1989) reported 7 g as a mean species mass. Two Trinidad males weighed 6.5 g each; a female weighed 7.5 g; and 2 unsexed birds weighed 6 and 6.5 g (ffrench, 1991). The collective mean of these 5 birds is 6.6 g.

Description (After Ridgway, 1911)
Adult male. The forehead and crown bright metallic blue, the occiput dark bronze to coppery bronze; bronze to bronze-green upperparts to the rump and central tail feathers, except where there is an elongated white patch in the middle of the rump; the lateral rectrices mostly metallic bronze, but the outer pairs of rectrices with white terminal spots; a conspicuous white malar stripe extends back from the lores, above which is a dark brown auricular

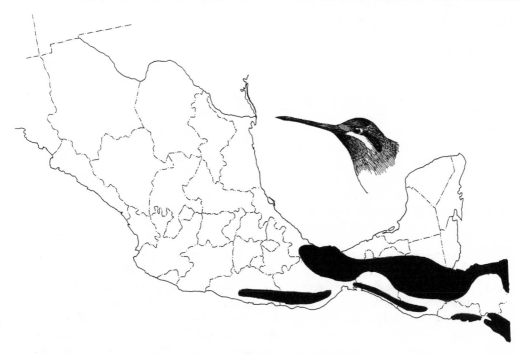

Residential range of the long-billed starthroat. (*Adapted from Howell and Webb, 1995*)

patch and a small white postocular spot; below the white malar stripe a metallic reddish purple gorget, bounded anteriorly by a blackish chin and below by a greenish breast, which gradually fades to a white central abdomen, bounded above by metallic bronze flanks and sides; a large tuft of silky white feathers between the flanks and rump; iris brown; feet dusky; bill black.

Adult female. Similar to the male, but the forehead and crown dull metallic green, the chin mostly blackish, with scattered reddish purple feathers, and the underparts generally paler.

Immature male. Similar to the adult female, but without any purple feathers on the dull white throat. Very young birds of both sexes may have the upperpart feathers margined with pale rusty brown or buffy.

Identification
In the hand. This is the only species of Mexican hummingbird with a straight bill that exceeds half the length of the wing and is as long or longer than the tail.

In the field. The extremely long bill of this species and strongly patterned face (a conspicuous white malar stripe) identifies it as a starthroat. Adults are easily identified by the glittering blue forehead and white postocular spot, and readily distinguished from the plain-capped starthroat, which has a green forehead and a postocular streak. Immatures of the two species are more similar, but the long-billed has a rather sooty gray upper breast rather than being greenish in this area, and the differences in the postocular markings should also help to separate them. Vocalizations include sharp *pik* or *peek* notes (Howell and Webb, 1995) and various "dry syllables" (Skutch, 1972), but no definite male song seems to have been discerned.

Habitats
Associated with humid evergreen forests, forest edges, and coffee plantations, from sea level to about 1200 meters' elevation in Mexico, 1500 meters in Guatemala, 750 meters in Honduras, and 1400 meters in Costa Rica. It is common in areas where bananas and coral bean (or "poro"; *Erythrina*) trees have been planted and is generally attracted to gardens and flowering shrubbery.

Movements
No migrations have been documented.

Foraging Behavior and Floral Ecology
The long bill of this highly insectivorous species allows it to forage at long-tubed blossoms such as coral bean trees, where it also can capture insects similarly attracted to these nectar-rich flowers. Individuals often try to take territorial possession of such trees, expelling others of their own species as well as many other hummingbirds, for as long as the tree remains in bloom (Skutch, 1972). It similarly likes to forage in the canopy of other profusely flowering tropical trees such as *Tabebuia* species (Bignoniaceae), the blossoms of woody vines such as *Mandevilla hirsuta* (Apocynaceae), and in gardens planted to *Ixora* species (Rubiaceae) (ffrench, 1991). It is also attracted to the white flowers of banana plants, around which insects are usually abundant.

Wagner (1946a) described these birds hawking for small flying insects that fly in swarms. The birds would fly out to a swarm and remain below it for a few seconds, then quickly fly up through the swarm, catching insects mid flight. After passing through the swarm three to five times, it would return to its lookout post.

Breeding Biology
This species' display behavior is still undescribed, but in the blue-tufted starthroat of South America, the male flies before the female as if ascending an invisible stairway, holding its body in an erect position and erecting its lateral blue necktufts (Sick, 1993).

Rowley (1966) found two nests of the long-billed starthroat in Oaxaca during November; other Pacific-slope breeding records extend from November to March. These two nests were quite different in appearance and location: One was attached to a *Cecropia* tree and was situated on the elbow of a small, sharply bending branch elevated about 15 meters above ground. The other was only 2.4 meters above ground, placed on a slanting branch of a shrub among understory vegetation. Both were open-cup nests, but one had an abundance of bark strips attached to the sides of the nest, and the other had only a few lichens attached.

Skutch (1972) also described two Costa Rican

nests, found during November and February. One was placed about 6 meters above ground on a small, dead tree. It was attached to the upper side of a bend in a small, horizontal stub of a branch and was composed mainly of mosses and liverworts, decorated externally with a few lichens and lined with vegetable down. This nest was in initial construction when discovered, and Skutch determined that about six more days were spent in its completion. Eight days intervened between the completion of the nest and the laying of the first egg; a second egg was laid two days later. The second nest was found about 10 meters up in a dead tree in a pasture. This one was attached to the upper side of a branch that was also sharply bent in the form of an inverted *L*, with the nest attached just above the angle in the branch. Although fully exposed to view, its brown color closely matched that of the dead wood around it. However, this nest disappeared about a week after its discovery.

Skutch (1972) closely followed the incubation and brood-rearing activities in the single successful nest. Incubation during 15 sessions ranged from 3 to 137 minutes, averaging 34.3 minutes, and recess intervals averaged 10 minutes. All told the female was on the nest about 77 percent of the total observation periods. The two eggs hatched on successive days, with the incubation period estimated at 18.5 to 19 days. When the chicks were 5 to 10 days old they were fed an average of 1.2 times per hour, as compared with about twice per hour when they were 16 to 23 days old. The birds were well feathered by two weeks of age, and at 19 days of age their remiges were quite long. They fledged at 25 to 26 days, but their mother continued to feed them to some extent for at least another 23 days.

Evolutionary and Ecological Relationships

This species and the plain-capped starthroat are very close relatives and quite distinct from the two South American species of this genus.

MEXICAN SHEARTAIL

Tilmatura (Calothorax?) eliza (Lesson and DeLattre)

Other Names
None in general English use; Tijereta yucateca (Spanish).

Range
Endemic resident of Mexico along the northern coast of the Yucatan Peninsula (including Holbox Island). Local breeding has also been reported from central Veracruz.

North American Subspecies
None recognized.

Measurements
Wing, males 34.5–36 mm. (ave. of 7, 37.1 mm), females 37.5–40 mm (ave. of 4, 38.7 mm). Exposed culmen, males 20.5–22.5 mm (ave. of 7, 21.3 mm), females 22–23 mm (ave. of 4, 22.4 mm) (Ridgway, 1911). Eggs, no information.

Weights
One male weighed 2.3 g, and 3 females had a mean mass of 2.6 g (range 2.5–2.7 g) (Paynter, 1955).

Description (After Ridgway, 1911)
Adult male. Upperparts metallic bronze to bronze-green, this color extending to the deeply forked tail; the three longest and most lateral pairs of rectrices are purplish bronzy black, with the inner webs of the progressively shorter second and third pairs strongly edged with pale cinnamon; chin and throat bright metallic purple, this gorget bounded behind by white on the lower throat and sides of the head, and the white extending down the middle of the breast and underparts to the under tail-coverts; sides and flanks metallic bronze-green; iris brown; feet dusky; bill black.

Adult female. Similar to the adult male, but the tail much shorter (double-rounded rather than forked

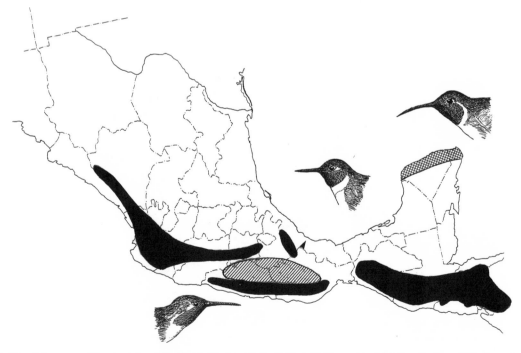

Residential ranges of the sparkling-tailed hummingbird (inked), beautiful hummingbird (hatched), and Mexican shear-tail (cross-hatching, plus an arrowhead showing an extralimital breeding location). *(Adapted from Howell and Webb, 1995)*

in profile); the three outer rectrices light cinnamon rufous, with a black subterminal band, and the two outermost rectrices broadly tipped with white; the third may also be white-tipped; the throat and underparts grayish white, with no metallic gorget feathers.

Immature male. Similar to the adult female, but the throat flecked with glittering pink feathers, and the tail with a broader black subterminal band (Howell and Webb, 1995).

Identification

In the hand. The long, decurved bill of this species is more than half the length of the tail in males, and as long or slightly longer than the tail in females. Males are further easily recognized by their forked tail (the central rectrices are about 30 percent as long as the outermost pair, which are 35–38 mm). Females have a uniformly short tail that is double-rounded (its central rectrices average only about 4 mm shorter than the outermost pair).

In the field. Within its limited range, this small species may be easily distinguished by the long, decurved bill of both sexes (longer than the head) and, in males, the deeply forked tail and brilliant metallic red gorget. Females and young males have rather short tails, but no other hummingbird in the area has such a long, curved bill. Vocalizations include the usual chipping notes of hummingbirds, and buzzy sounds are produced during male display flights and actually may be mechanical wing noises rather than vocalizations.

Habitats

Associated with arid scrub habitats, especially those with scattered trees or high perches, from sea level to about 300 meters' elevation. The birds also occur in coastal mangrove woods, forest edges, and city gardens, but generally seem to be fairly closely confined to coastal environments.

Movements

No definite migrations have been documented, although vagrants have been reported south to northeastern Quintana Roo. Reported disjunctive breeding in central Veracruz (600 km away from its nearest known breeding locations on the Yucatan Peninsula) is also notable.

Foraging Behavior and Floral Ecology

No detailed information is available. The species' long bill should allow it to forage on flowers having relatively deep corollas.

Breeding Biology

During aerial display, the male makes buzzing shuttle flights back and forth in front of the perched female, with his tail spread and twisted almost at right angles to the plane of his body, then steeply ascends to about 30 meters or more before suddenly diving and perching beside the female. Nesting in the Yucatan Peninsula reportedly occurs from August to April; breeding-condition birds have been obtained in Veracruz during April (Howell and Webb, 1995). However, the nest and eggs of this species are apparently still undescribed.

Evolutionary and Ecological Relationships

The two species of sheartails are the only members of the traditional genus *Doricha* (Ridgway, 1911; Sibley and Monroe, 1990), but this poorly defined genus (based largely on male tail configuration) might easily be included with the general woodstar assemblage in an expanded genus *Calothorax*, as was done in the earlier edition of this book, and as treated by Howell and Webb (1995). The similarities in bill length and bill shape of the two species of sheartails are substantial, suggesting marked similarities in foraging strategies between these two. There are also similarities in bill shape and length with the lucifer-beautiful hummingbird group, which are also seemingly fairly closely related taxa.

SPARKLING-TAILED HUMMINGBIRD

Tilmatura dupontii (Lesson)

Other Names
Dupont's hummingbird, Sparkling-tailed wood-star; Chupaflor colipinto, Colibri colipinto, Chupamirto de Dupont (Spanish).

Range
Resident in Mexico from Sinaloa and central Veracruz south and east to the Chiapas–Guatemala border, thence southward in Central America to northern Nicaragua.

North American Subspecies
None recognized.

Measurements
Wing, males 33–35 mm (ave. of 10, 34 mm), females 33.5–36 mm (ave. of 8, 34.7 mm).

Exposed culmen, males 12.5–13.5 (ave. of 10, 12.9 mm), females 12–14.5 mm (ave. of 8, 13.6 mm) (Ridgway, 1911).

Weights
Twelve birds of unspecified sex had a mean mass of 2.2 g (Brown and Bowers, 1985). Three males in the Field Museum had a mean mass of 2.7 g (range 2.4–3.3 g), and 2 females averaged 3.4 g (2.8 and 4.1 g). One male weighed 2.2 g, and 2 females weighed 2.45 and 2.6 g (Binford, 1989). Collectively, these 4 males averaged 2.6 g; the 4 females averaged 3.0 g.

Description (After Ridgway, 1911)
Adult male. The upperparts metallic bronze to bronze-green, the central pair of the highly

Residential ranges of the sparkling-tailed hummingbird (inked), beautiful hummingbird (hatched), and Mexican sheartail (cross-hatching, plus an arrowhead showing an extralimital breeding location). *(Adapted from Howell and Webb, 1995)*

forked rectrices also metallic bronze-green, but the next pair with purplish black on the inner web, and the remaining pairs mostly purplish black, tipped with white; the two outermost pairs of rectrices crossed by white and rufous bands, and the third with a white spot on the middle of its inner web; a conspicuous white patch on either side of the rump; chin and throat metallic violet-blue, this gorget bounded posteriorly by a grayish white breast-band; other underparts, including the flanks and sides, dark metallic bronze-green; iris brown; feet dusky; bill black.

Adult female. Similar to the male, but the forehead duller, and the chin, sides of head, throat, and most underparts cinnamon rufous; a pale cinnamon buff or white patch of feathers on either side of the rump, and the tail much shorter and slightly double-rounded rather than deeply forked; the central pair of rectrices metallic bronze-green, with blackish tips, and the lateral rectrices with blackish subterminal bands and tipped with pale cinnamon-rufous.

Immature male. Similar to the adult female, but with the outer rectrices tipped with white, and with variable green mottling on the underparts; some show a white median band when the tail is closed (Howell and Webb, 1995).

Identification

In the hand. This hummingbird has a uniquely short, straight, and black bill (culmen 12–15 mm), and (in adult males) a tail that is deeply forked (outer rectrices 38–47 mm) and spotted with white. This trait is lacking in the short-tailed females, but both sexes have conspicuous white patches on either side of the rump.

In the field. Both sexes of this tiny hummingbird have very short, straight blackish bills and white lateral rump patches that are interrupted in the middle of the rump by green. Adult males have white-spotted, deeply forked tails (not usually evident, since the rectrices are held tightly together in flight), and females are a rich pale cinnamon on the underparts. As with other woodstars, the tail is slightly cocked and is often pumped vertically while the hovering bird forages, the wings then making a loud buzzing sound. Vocalizations include high-pitched chips, and the male's "song" is a series of thin squeaking notes uttered from an exposed perch.

Habitats

Associated with open woodlands, especially oak woodlands, brushy areas, thickets, and second-growth habitats, generally ranging in Mexico from a minimum of 500 to about 2500 meters' elevation. In Oaxaca the birds range from about 900 to 2400 meters' elevation and occupy humid pine-oak and some adjacent arid pine-oak forests (Binford, 1989). In Guatemala the species ranges from 550 to about 1950 meters' elevation and is fairly common in open woodlands and second growth (Land, 1970). In Honduras the species is common in cloud forest and pine-oak habitats above 1000 meters, but descends during the non-breeding season into tropical evergreen forest just above sea level (Monroe, 1968).

Movements

No movements have been specifically documented, but there is apparently some altitudinal migration downward during the non-breeding season.

Foraging Behavior and Floral Ecology

Surprisingly little information exists on the species, which is evidently a generalist, feeding on a variety of plants in and away from the forest and forest edges, sometimes hawking insects during extended flights from exposed perches.

Breeding Biology

No information on the breeding biology of this species is available, but breeding-condition birds have been collected during August (Howell and Webb, 1995). The nest and eggs are apparently still undescribed. In the woodstar genus *Calliphlox*, the nest is of the usual hummingbird type, consisting of an open cup-like structure, saddled over a more or less horizontal branch or twig.

Evolutionary and Ecological Relationships
The questionably valid genus *"Philodice"* has been
erected to include this species and the magenta-
throated woodstar of Central America, its obvious
nearest relative. Like the other woodstar species,
they seem to play the role of small, insect-like
traplining generalist in the ecology of tropical
hummingbirds.

LUCIFER HUMMINGBIRD

Calothorax lucifer (Swainson)

Other Names
None in general English use; Chupamirto morada grande (Spanish).

Range
Breeds in southwestern Texas (Chisos Mountains), southward through central and southern Mexico to Guerrero, mostly at elevations of 1200 to 2250 meters. Migratory at northern end of the range. Has bred in southern Arizona; probably a regular breeder in adjacent Sonora.

North American Subspecies
None recognized.

Measurements
Wing, males 36–39 mm (ave. of 10, 37.6 mm), females 39–44 mm (ave. of 8, 41.2 mm). Exposed culmen, males 19.5–22 mm (ave. of 10, 21.1 mm), females 20–22.5 mm (ave. of 8, 21.2 mm) (Ridgway, 1911). Eggs, ave. 12.7 × 9.7 mm (extremes 12.0–13.8 × 9.2–10.1 mm).

Weights
Thirteen adult males from Arizona averaged 3.2 g (range 2.9–3.7 g); 5 females averaged 3.3 g (range 3.2–3.5 g). Nine females from Texas averaged 3.5 g (range 3.1–3.9 g); 3 males averaged 3.4 g (range 3.1–3.6 g) (Scott, 1994).

Description (After Ridgway, 1911)
Adult male. Above metallic bronze, bronze-green, or golden green, usually duller on pileum, especially on forehead; remiges dull brownish slate or dusky, faintly glossed with purplish; four middle rectrices metallic green or bronze-green, the rest of tail purplish or bronzy dusky or blackish; a small postocular spot (sometimes also a rictal

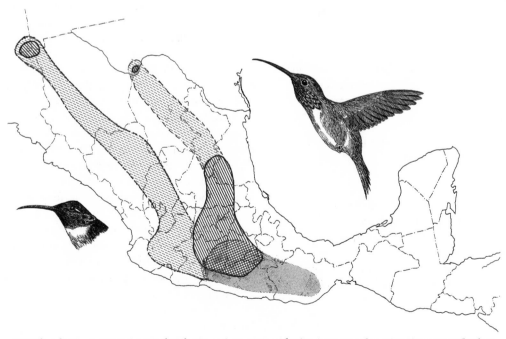

Breeding (hatched), residential (cross-hatched), and wintering (shading) ranges, plus migration routes (broken hatching) of the lucifer hummingbird.

spot) of dull whitish; chin and throat brilliant metallic solferino or magenta-purple, changing to violet, the posterior feathers of sides of throat much elongated; chest dull white; sides and flanks mixed light cinnamon and metallic bronze or bronze-green, the median portion of breast and abdomen pale grayish or dull grayish white; under tail-coverts dull white with central area of pale brownish gray; femoral tufts white; bill dull black; iris dark brown; feet dusky.

Adult female. Above as in adult male but lateral rectrices much broader, the three outermost (on each side) with basal half (approximately) light cinnamon-rufous, then (distally) purplish black, the two outermost broadly tipped with white, the black terminal or subterminal area on second and third separated from the cinnamon-rufous of basal portion by a narrow space of metallic bronze-green; fourth rectrix (from outside) mostly metallic bronze-green but terminal or subterminal portion blackish and outer web edged basally with light cinnamon-rufous; a postocular spot or streak of cinnamon-buff, and beneath this a narrow auricular area of grayish brown; malar region and underparts dull light vinaceous-cinnamon or cinnamon-buff, passing into dull whitish on abdomen; the under tail-coverts mostly (sometimes almost wholly) whitish; femoral tufts white; bill, etc., as in adult male.

Immature male (first winter). Similar to adult female, including broader tail feathers, but darker above and below, size smaller, throat with some metallic purple feathers (Oberholser, 1974).

Immature female (first winter). Similar to adult female (Oberholser, 1974).

Identification

In the hand. Unique among North American hummingbirds in having a blackish (culmen 19–22 mm) bill that is more than half as long as the wing and distinctly decurved. The underparts are buffy to pale buff. The tail of the male is deeply forked; that of the female is rounded, with a tawny base and white tips.

In the field. Inhabits open desert-like country, often where agaves are abundant, and the long and decurved bill is the best fieldmark for both sexes.

The male is about the same size as the Costa hummingbird, but has a deeply forked tail and lacks red color on the forehead. The female resembles several other hummingbirds but, apart from the longer bill, the pale cinnamon-buff underparts and the tawny color at the base of the tail are fairly distinctive. The birds utter shrill, piercing shrieks when defending their nest; other calls have not been described.

Habitats

In the Chisos Mountains of the Big Bend area of Texas, lucifer hummingbirds occur from May through early July on the open desert or on the slopes of the mountains up to about 1500 meters. During late July and early August, after breeding, the birds begin to move upward into various canyons such as Boot Canyon at 1900 meters (Wauer, 1973). They sometimes occur at higher altitudes as well; as many as 10 individuals have been seen near the South Rim of these mountains at about 2200 meters in mid-July (Fox, 1954). Actual nests in the area have been found from 1100 to 1500 meters. The first Texas nests were found in a desert-like habitat dominated by agave (*A. lechuguilla*), sotal (*Dasylirion leophyllum*), and ocotillo (*Fouquieria splendens*) (Pulich and Pulich, 1963). Associated plants included candelilla (*Euphorbia antisyphilitica*), catclaw (*Acacia greggii*), mormon tea (*Ephedra*), yuccas (*Yucca* species), and cacti. Blooming cacti provided the principal source of food during the unusually late nesting observed by the Pulichs.

In northern Sonora, 30 females were found along a 3-kilometer stretch of Arroyo Cajon Bonito at about 1200-meter elevations. There they nested among sycamores and hackberries and foraged from streamside upward along the drier upper slopes, where agaves were in flower (Russell and Lamm, 1978). More generally in Mexico the species is found in open country having scattered trees and shrubs and among brushy vegetation in arid areas. There is no special marked preference for any particular species of plant nor for any specific altitudinal zone (Wagner, 1946b).

Movements

In Texas, this species arrives as early as March 7 and has been reported as late as November 12 (one winter record exists for January 4). Typically it arrives during March, with males preceding females by a few days. By late August the birds are moving toward lower canyons, where they remain until about the second week of September. Between then and November the birds move back to their winter territories in the Mexican desert (Oberholser, 1974).

Vagrant birds in Texas sometimes reach the Edwards Plateau, more rarely the Gulf Coast (Rockport, Aransas County). Other than Texas, the species is essentially confined to Arizona, where it is very rare but where nesting has been reported (*American Birds* 27:804). There is also a single state record for Williamsburg, New Mexico (*American Birds* 33:796).

In the Valley of Mexico, seasonal movements of these birds are associated with variable availability of foods; there are few plants during the winter months (Wagner, 1946b).

Foraging Behavior and Floral Ecology
The relatively few observations of lucifer hummingbirds suggest that it consumes a fairly high proportion of insects in its diet, and its attachment to flowering agaves is probably in large part associated with the abundance of small insects usually found around these plants. In Texas, the overwhelming favorite plants from May through September are the yellow-green blossoms of several agaves (*A. americana, A. chisoensis, A. harvariana,* and *A. lechuguilla*). In early spring the birds also visit Chiso bluebonnet (*Lupinus havardii*) and ocotillo (*Fouquieria splendens*), and late fall migrants often concentrate on tree-tobacco (*Nicotiana glauca*) that has been planted along the Rio Grande (Oberholser, 1974). The species also has been observed foraging at the blossoms of *Erythrina coralloides* and E. *"corallodendrum"* (Toledo, 1974).

Wagner (1946b) reported that 11 stomach samples that he examined invariably contained insects, spiders, and other small animals, particularly dipterans. These insects are extracted from flowers of *Erythrina americana, Salvia mexicana, Loesalia mexicana, Lupinus elegans,* and other plants, including eucalyptus and *Opuntia* cactus. Sometimes the birds also resort to removing entangled flies from spiderwebs (Bent, 1940), but they do not hawk them in flight.

Breeding Biology
Although few actual nests have been found in Texas, the breeding season there probably extends from May to July. Nests have been found from May 8 to August 2, and very recently fledged young have been observed between June 8 and the first week of August. There are now at least six nesting records for the Big Bend area (*American Birds* 32:1026). The single-reported Arizona nesting of this species was in Guadalupe Canyon, Cochise County, in May 1973 (*American Birds* 28:920).

In Mexico, breeding likewise seems to occur in the summer months. Bent (1940) reported six Mexican egg records from June 15 to July 4; most were from the state of Tamaulipas. In the Valley of Mexico, the principal breeding period is between May and September, with extremes of April and October (Wagner, 1946b).

Feeding territories in the Chisos Mountains are sometimes associated with the distribution of flowering agaves. The territory of one male contained two agaves along a cliff edge, and from them extended 6 meters on each side to some low trees, and 4.5 meters backward from the edge to another small tree (Fox, 1954). All but two of six observed males defended feeding territories, which usually consisted of circular areas having 12-meter radii. In one of the two exceptional cases, one male shared some agaves with black-chinned hummingbirds, and another male partially shared its territory with the first. Of two females, one showed partial possession of her area, while the other apparently visited undefended agaves. Two immature birds were evidently non-territorial. Wagner reported that in Mexico the calculated size of a male breeding territory was 30 to 50 meters in diameter—considerably larger than the feeding territories reported by Fox.

The first two nests discovered in Texas were both placed in agave (*A. lechuguilla*) stalks, about 2 to 4 meters above the ground (Pulich and Pulich, 1963). Four nests found in Tamaulipas were all placed in shrubs, again only a few meters above ground (Bent, 1940). Four nests observed by Wagner were in low shrubs (*Senecio salignus*) about 1 to 2 meters above ground. The two nests found in the Chisos Mountains were similarly constructed. Both were placed on two or three dried pods that were attached to the stalk. On these pods the birds had made a foundation and constructed the nests of plant fibers, grass seeds, and pieces of leaves. Both were lined with plant down and feathers, and one was decorated externally with small leaves. The other nest, still unfinished, was undecorated. Neither contained lichen decorations, although this seems to be typical of nests found in Mexico (Bent, 1940; Wagner, 1946b). A third nest found in the Big Bend area was located almost 2 meters up in a dead shrub, but with flowering individuals of *Agave lechuguilla* in the vicinity (Nelson, 1970).

Nests in Texas are usually placed on the live or dead stalks of cane cholla, on dead lechuguilla stalks, or on leafy branches of ocotillo. Sometimes nests are placed on the old nests of the previous year, but females that begin a second clutch while still feeding nestlings build their second nest at a new site. Male lucifer hummingbirds are the only North American hummingbirds known to display to females that are sitting on their nests, including not only incubating females but also those tending nestlings. Such courted females may remain on the nest or may take off and ascend with the male, either chasing him or facing and hovering upward with him. The most typical displays performed by males in Texas are shuttle flights lasting 30 to 45 seconds, and having associated wing or tail sounds that sound like card-shuffling noises and that can be heard at least 100 meters away (Scott, 1994). Presumably the courting of nesting females is an adaptive behavior for any regularly double-brooding species.

According to Wagner (1946b), courtship has two phases. The first is to attract the female and induce her into copulation, and the second is simultaneously performed by both partners. The first of these is a display flight by the male, performed daily in the same place, usually in the first 5 hours after sunrise. Its form is variable, but in part consists of repeated lateral flights between two perches on a tree or bush. At greater levels of excitement the male may ascend vertically upward in a spiraling flight and then pitch downward again to a perch. The more rarely seen second phase was observed once by Wagner a short time after dawn. In this case a male performed before a perched female, beginning his display with a sharply ascending and somewhat

spiraling flight upwards for about 20 meters. At the peak of the ascent the bird hovered in place, and then began a rapidly descending swoop toward the female, pulling out of the dive while still some distance above her, and terminating the descent with a series of pendulum-like swings of decreasing amplitude as he slowly descended toward her.

The flight terminated with the male again gaining altitude. The dive was accompanied by a sound resembling that of an electric fan, but apparently lacked vocalizations (Wagner, 1946b). This display has some components, especially the pendulum-like swinging, very much like that of the Bahama woodstar; it also resembles the display dives of *Archilochus* and *Selasphorus* to some degree.

Details of incubation and brooding behavior are not well known, but Wagner (1946b) estimated that the incubation period is 15 days from the laying of the second egg to the hatching of the first egg, or a total incubation period of 15 to 16 days. Incomplete observations indicated a fledging period of 22 to 24 days. Similarly, Pulich and Pulich (1963) estimated the fledging period at 21 to 23 days, also on the basis of incomplete information.

Recent observations by Scott (1994) support the 15-day estimate for incubation, and a nestling period of 19 to 23 days. The fledglings remain near the nest and are fed for 13 to 19 days after leaving the nest. The female's care of the nestlings may be combined with nest construction and incubation of a second clutch.

Evolutionary and Ecological Relationships

The unusually long and decurved bill of this species is not associated with obtaining nectar from unusually long tubular-blossomed plants, but instead may principally serve for extracting insects from plant blossoms and perhaps similar recesses. It thus seems somewhat comparable to the starthroats, which also have unusually long bills and apparently feed largely on insects rather than nectar. The starthroats often hawk insects during flight, however, which is rare for lucifer hummingbirds.

Taxonomists have generally maintained the genus *Calothorax* exclusively for the lucifer hummingbird and a very closely related species—the slightly shorter billed beautiful hummingbird of southern Mexico. These two forms clearly constitute a superspecies. They are also obviously very close relatives of the sheartails (*"Doricha"*) of the same general area, differing from them only in the longer and more deeply forked tails of the males. I can see no reason for maintaining these two genera as distinct, nor for excluding some of the "woodstars" from the same genus. Other Central and South American genera such as *Acestrura, Chaetocercus, Philodice,* and *Calliphlox* may be part of the same assemblage, but no detailed comparisons of these types have been made in the present study.

The very small size of this species probably places it at a competitive disadvantage with many other sympatric species, and Fox (1954) noted that black-chinned and broad-tailed hummingbirds sometimes trespass on the territory of a male lucifer. However, a female drove away an immature magnificent hummingbird and some black-chinned hummingbirds from one agave, but another was driven away from an agave by a male black-chinned. There are apparently no plants that this species has specifically become dependent upon, although several observers have commented on the attraction of lucifer hummingbirds to flowering agaves.

BEAUTIFUL HUMMINGBIRD

Calothorax pulcher (Gould)

Other Names

None in general English use; Chupamirto morada chico, Tijereta Oaxaqueña (Spanish).

Range

Endemic resident of Mexico from central Guerrero and southern Puebla east to eastern Oaxaca.

North American Subspecies

None recognized.

Measurements

Wing, males 35–39 mm (ave. of 8, 36.7 mm), females 36.5–40.5 mm (ave. of 3, 38.5 mm). Exposed culmen, males 17–18.5 mm (ave. of 8, 17.9 mm), females 18.5–19 mm (ave. of 3, 18.7 mm) (Ridgway, 1911). Eggs, 12.4 × 8.2–8.3 mm (Rowley, 1984).

Weights

Three birds of unspecified sex had a mean mass of 2.9 g (Brown and Bowers, 1985).

Description (After Ridgway, 1911)

Adult male. Upperparts metallic bronze to bronze-green, this color extending to the four middle rectrices, which are much shorter than the outer three pairs; the rest of the rectrices plain purplish bronzy black, the second from the outermost the longest, and the outermost with rounded rather than pointed tips (compare lucifer hummingbird); a small white postocular stripe, above a gorget of metallic magenta-purple feathers; Gorget bounded below by a white throat, the white extending down the median portion of the breast to the under tail-coverts; Sides and flanks metallic bronze or bronze-green, with some cinnamon-rufous on the flanks; iris brown; feet and bill black.

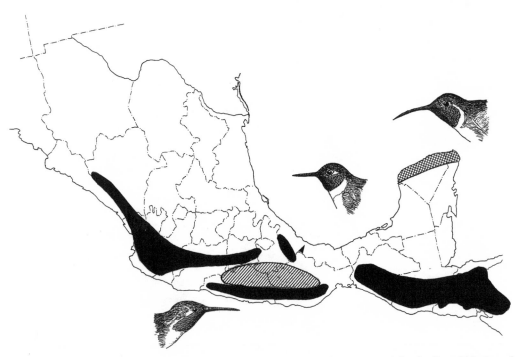

Residential ranges of the sparkling-tailed hummingbird (inked), beautiful hummingbird (hatched), and Mexican shear-tail (cross-hatching, plus an arrowhead showing an extralimital breeding location). *(Adapted from Howell and Webb, 1995)*

Adult female. Upperparts similar to the male, but somewhat duller throughout, especially on the forehead, and without a metallic gorget; tail shorter and double-rounded rather than forked in profile, with the middle pair of rectrices bronze-green, the second pair blackish near the tips, and the three lateral pairs cinnamon-rufous basally and with subterminal black bands and white spots or tips; underparts including the throat dull grayish buffy, which becomes more cinnamon-toned on the flanks.

Immature male. Initially similar to the female, but gradually developing a forked tail and a metallic gorget. Young birds of both sexes probably have parallel striations on the upper mandible.

Indentification
In the hand. Adults of both sexes differ mainly from the lucifer in their slightly shorter bills (17–19 mm, vs. 19–22 mm), which are less decurved. Males have outermost rectrices that are round-tipped and nearly as wide as the adjacent pair, rather than pointed and much narrower. Females have shorter wings (35–39 mm, vs. 39–44 mm), and shorter middle rectrices (12–16.5 mm vs. 16–21 mm) than is the case with the lucifer.

In the field. This tiny hummingbird closely resembles the lucifer hummingbird and is only incompletely geographically separated from it. These two species can be safely distinguished only in the hand: Both species are unique in their long, slightly decurved bills and (in adult males) their magenta-red iridescent gorgets and somewhat forked tails. Females of the two species are even more similar than males, but as with the male the female's bill of this species is somewhat shorter and is only very slightly decurved. Vocalizations

include the usual high-pitched chirps of hummingbirds, and thin and high-pitched squeaky notes that are produced during aerial courtship display.

Habitats
This species is associated with arid brush, thorn forest, and scrubby open habitats with scattered trees or elevated perches, from about 1000 to 2000 meters' elevation (Howell and Webb, 1995). In Oaxaca the species is an uncommon permanent resident in arid subtropical scrub, at elevations of about 1000 to 2200 meters (Binford, 1989).

Movements
No migrations or local movements have been documented.

Foraging Behavior and Floral Ecology
No specific information is available, but the ecology of this species is unlikely to differ substantially from that of the lucifer hummingbird.

Breeding Biology
Rowley (1984) found a nest with two eggs during early May, in a garden in the city of Oaxaca. At the time of the nest's discovery, the female was sitting on two fresh eggs. The nest was situated approximately 2.5 meters above ground, and it was attached to the forking branch of a *Bougainvillea* vine. The lining was composed of very small plant down, perhaps from a milkweed, and the outside was covered by brown composite pappi. Incubation intervals by the female lasted 2 to 15 minutes.

Evolutionary and Ecological Relationships
This species is a very close allopatric relative of the lucifer hummingbird, scarcely separable from it at the species level.

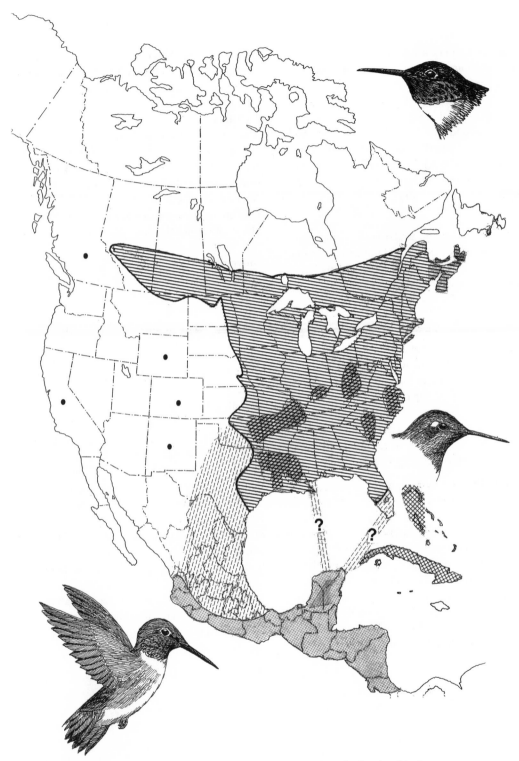

Breeding (hatched) and wintering (stippled) ranges and migration routes (broken hatching) of the ruby-throated hummingbird. Areas of greater breeding densities are shown by cross-hatching. Dots represent states or provinces with extralimital sightings. The Florida–Yucatan migration is speculative; occasional overwintering occurs on Cuba. The residential range of the Cuban emerald is also shown (cross-hatching).

RUBY-THROATED HUMMINGBIRD

Archilochus colubris (Linnaeus)

Other Names
None in general English use; Mansoncito garganta de fuego, Chupaflor rubi (Spanish).

Range
Breeds throughout the eastern half of North America west to the middle of the Great Plains, from Southern Saskatchewan east to Nova Scotia, and south to southern Texas, the Gulf of Mexico, and Florida. Winters from middle Florida through southern Mexico and the rest of Middle America to Panama; casual in Cuba, Hispaniola, the Bahama Islands, and Bermuda (Friedmann et al., 1950).

North American Subspecies
None recognized.

Measurements
Wing, males 37–40 mm (ave. of 10, 38.5 mm), females 43.5–45.5 mm (ave. of 10, 44.5 mm). Exposed culmen, males 15–17 mm (ave. of 10, 15.9 mm), females 17–19.5 mm (ave. of 10, 18.2 mm) (Ridgway, 1911). Eggs, ave. 13.0 × 8.4 mm (extremes 11.5–14.5 × 7.8–9.1 mm).

Weights
The average of 419 adult females (May–September) was 3.34 g (range 2.7–4.8 g); that of 202 adult males (April–September) was 3.03 g (range 2.5–4.1 g) (Clench and Leberman, 1978).

Description (After Ridgway, 1911, with Modifications)
Adult male. Above metallic bronze-green, including middle pair of rectrices; remiges dark brownish slate or dusky, faintly glossed with purplish; tail (except middle pair of rectrices) dark bronzy purplish or purplish bronzy black; chin, malar region, suborbital region, and auricular region velvety black; a small postocular spot of white; whole throat brilliant metallic red (nearest geranium red in frontal light, changing to golden or even greenish in side light); chest dull brownish white or very pale buffy brownish gray, passing gradually into deeper brownish gray on breast and abdomen, the sides and flanks darker and overlaid by metallic bronze-green; femoral tufts and tuft on each side of rump white; under tail-coverts brownish gray (sometimes glossed with greenish bronze) centrally, broadly margined with dull white; bill dull black; iris dark brown; feet dusky.

Adult female. Above metallic bronze-green, golden green or greenish bronze, including middle pair of rectrices; three outer rectrices, on each side, broadly tipped with white (the white tip on third rectrix smaller and mostly confined to inner web), metallic bronze-green for basal half (more or less), the intervening portion black; remiges dark brownish slate or dusky, faintly glossed with purplish; a small postocular spot of dull white; auricular region deep dull grayish; lores dusky; malar region and underparts dull grayish white or very pale brownish gray (usually more decidedly whitish on chin, throat, and malar region); the flanks and shorter under tail-coverts usually more or less tinged with pale buffy brownish; femoral tufts and tuft on each side of rump white; bill, etc., as in adult male.

Immature male. Similar to the adult female, but feathers of upperparts very narrowly and indistinctly margined terminally with pale grayish buffy, throat with small mesial streaks of dusky, and underparts usually more strongly tinged with buffy brownish, especially on sides and flanks. Also, the sixth primary is narrower (outer web 0.3 mm or fewer subterminally) and more abruptly angulated (Norris et al., 1957).

Immature female. Similar to the young male, but throat without dusky streaks. Furthermore, the outer web of the sixth primary (counting from the inside) is at least 1 mm wide throughout most of its 20-mm length, rather than being very narrow toward the tip. Separation from adult females is difficult, but young females tend to have a brighter yellow to orange gape, more brown on the flanks, a more streaked throat, and unworn edges on the back and crown feathers (Leberman, 1972).

Identification

In the hand. Adult males are best identified and
separated from black-chinned hummingbirds by
the combination of having inner primaries with
small notches near the tips of the inner webs, the
tail slightly forked, and a purplish red gorget.
Females of both species have notched primaries,
but the ruby-throated has a double-rounded
tail, with the middle rectrices slightly shorter
(22–24.5 mm) than the more lateral ones. The
exposed culmen (17–19.5 mm) is also very slightly
shorter than that of the black-chinned hummingbird, and the wing is also slightly shorter (43.5–
45.5 mm vs. 46–48.5 mm).

In the field. A widely distributed species that occurs
in deciduous woodlands during the breeding
season, but also is found in suburban parks and
gardens. Males produce humming sounds with
their wings in flight, and utter a mouse-like
squeaking call. The adult male's brilliant red
gorget serves for identification in most areas, and
except where the black-chinned hummingbird
might be present the female's dull-white throat
and otherwise greenish upperparts provide
sufficient basis for recognition. Her tail is rounded
and white-tipped, and lacks any rufous color at
the base. The typical display flight of the male
consists of a series of long, arching flights in a
pendulum-like manner, apparently without
associated vocalizations.

Habitats

On its extensive breeding grounds of the eastern
United States, this species is found in mixed woodlands and eastern deciduous forests rich in flowering plants; it sometimes also breeds in city parks or
other areas planted with flowers. In Canada, its
northern breeding limits approximately coincide
with the southern limits of boreal forest; there it is
generally associated with woodland clearings and
edges, gardens, and orchards (Godfrey, 1986). Near
the western edge of its breeding range in North
Dakota it is associated with brushy margins or
openings of tracts of deciduous forest, including
river floodplain and upland forests (Stewart, 1975).
At its southwestern limits in Texas it occurs in
open coniferous and mixed woodlands, meadows
with scattered groves and flowering vegetation,

and urban areas (Oberholser, 1974). On their
wintering areas of Costa Rica, the birds have foraged where thickets and woodlands alternate with
pastures (Wetmore, 1968). In El Salvador, the
species winters in a wide variety of habitats, but
appears to prefer open, sunny areas such as thin
second-growth gallery forest and the edges of
clearings (Dickey and van Rossem, 1938).

Movements

Ruby-throated hummingbirds are present in their
wintering areas of Central America from late
October until late March or early April (Dickey
and van Rossem, 1938; Wetmore, 1968). In Texas
spring migration occurs from middle or late March
to mid-May, and fall migration lasts from late July
or early August to late October. A few birds winter
in coastal Texas, especially in and near the Rio
Grande delta, and they occur casually to the eastern edge of the Edwards Plateau (Oberholser,
1974). The species sometimes winters as far north
as southern Alabama, and at least twice has been
reported as far north as North Carolina in December (Hauser and Currie, 1966).

Although early estimates of the energy costs
of migration made ornithologists question the
nonstop flight range of ruby-throats, newer data
indicate that individuals carrying at least 2.1 grams
of fat should be able to fly more than 950 kilometers, or easily far enough for a nonstop flight
across the Gulf of Mexico (Norris et al., 1957).
Such flights are evidently regular in spring
(Sargent and Sargent, 1996).

During the spring migration, birds usually reach
the southern parts of the Gulf Coast states in late
February or early March. They generally move
northward at the same approximate rate of movement as the 1.7° C isotherm, which closely correlates with the flowering times of several important
spring foods such as *Aesculus pavia, Ribes odoratum,
Aquilegia canadensis,* and similar species (Austin,
1975). As the birds reach their summer ranges,
there is a gradual increase in the numbers of
important flower species. Thus, in southern
Florida, six important species are in bloom in
January (and five continue through December), in
the Carolinas five species bloom through April,
and in the northeastern states nine species are
flowering by May (Austin, 1975). Blooming in the

northeastern states terminates in October, by which time the southern migration is well underway.

Foraging Behavior and Floral Ecology

The most complete survey of ornithophilous flowers in the breeding range of the ruby-throated hummingbird is that of Austin (1975), who reported that this species forages on and presumably helps pollinate at least 31 plant species in 21 genera, including 18 plant families. In Austin's view, at least 19 species of eastern North American plants have been influenced by selection for hummingbird pollination, and the birds also assist in the reproduction of at least 11 more species, many of which are of southwestern or West Indian affinities. R. I. Bertin (*Can. J. Zool.* 60:210–219, 1982) has concluded that the ruby-throated hummingbird and its North American food plants are facultative mutualists, with the plants relatively more labile in that they are also adapted for pollination by various insects.

Ruby-throated hummingbirds are probably adapted to several primary foraging species in various parts of their extensive breeding range. Thus, in New York their nesting distribution has been reported to be governed locally by the occurrence of *Monarda didyma*, whereas in North Dakota *Impatiens* has been assigned equal importance. Red is the predominant color of most hummingbird-adapted flowers in most areas and particularly in western states. A study on the ruby-throated hummingbird in Saskatchewan, near the northern limit of the species' breeding range, provided some information on color and feeding preferences in this species (Miller and Miller, 1971). This study found that foraging on artificial feeders occurred more or less equally throughout the day, beginning as early as half an hour after sunrise and terminating as late as an hour or more after sunset. Usually the birds concentrated on only one or two feeders, but at times they also explored other known or potential food sites as well as other colorful objects not associated with feeders. Apparently they quickly learned to recognize food sources by association with color and location, and they often investigated potential food plants before they were in full flower. Experimental coloring of the feeding solutions indicated that the birds made fewer visits

to blue fluids than expected by chance, but significantly more visits to clear fluids. No definite preference was shown for red fluids initially, but training the birds to associate red fluids with sugar and the others with unsweetened water quickly taught them to make the appropriate association; they soon were visiting the "correct" feeder with about 90 percent accuracy. Evidently a distinctive color is a definite aid in remembering the location of a food source.

In observations extending over two summers in a South Carolina garden, Pickens and Garrison (1931) noted that only 4 percent of the common wildflowers were red, but hummingbirds predominantly visited plants of reddish colors. The authors also noted that, of 300 types of garden plants, red and orange colors occur among species of Western Hemisphere origin at a rate three times greater than among species of the Eastern Hemisphere, suggesting the importance of hummingbirds in the evolution of the Western Hemisphere flora. All eight species listed by James (1948) as the most important flowers for ruby-throated hummingbirds are red or partially red.

Breeding Biology

In general, male ruby-throated hummingbirds precede the females during spring migration, and usually they either migrate singly or occur locally in groups of as many as several individuals gathered temporarily around favored food sources. However, the birds established territories as soon as they arrived on their breeding grounds. Territory sizes have not been accurately defined, but Pickens (1944) noted that in some cases nests may be located as close as about 63 meters apart. However, since males are promiscuous, this does not provide any indication of actual territorial limits of a single male. Pitelka (1942) reported that the feeding territory of a male in early summer consisted of about 970 square meters and was centered on a food supply, with mating a possible secondary function.

The usual aerial display has been well described by a variety of early writers, all summarized by Bent (1940). It generally consists of flying back and forth along the arc of a wide circle, as if the bird were supported by a swaying wire, and with the swings "so accurate and precise that they suggest a geometric figure drawn in the air rather than the flight of a bird." Frequently the male passes very close to the head of the female at the bottom of the arc, with the wings and tail producing the loudest buzzing at that point. In another variation both the male and the female hover in the air, a short distance apart while facing each other, ascending and descending vertically over distances of 1.5 to 3 meters. In one case these vertical oscillations were done out of synchrony, so that when the male was at the top of his flight the female was at the bottom of hers. Evidently the interest of the female in any male is limited to the few days immediately prior to egg-laying; likewise the male loses interest in a female after egg-laying is completed (Bent, 1940). Copulation has not been well described, but on one observed instance it occurred on the ground after a display flight by the two birds similar to that just described (Whittle, 1937).

The construction of the nest is usually completed by the female alone and prior to egg-laying, with the exception of bits of lichens and additional lining that may be added later. The nest is typically constructed mostly of bud scales, about 25 millimeters deep and 25 millimeters in diameter, lined with plant down, and covered on the outside with lichens. It usually resembles a knot saddled on a fairly small limb that slants downward from the tree. It is always sheltered from above by other limbs, and often is located directly over a brook or other open area. The height of the nest typically ranges from about 1.5 to 6 meters above the substrate, but rarely may be as high as 15 meters above ground.

Nest-building begins with leafy materials or bud scales, which are fastened to the limb by spider silk. Lichens are added around the outside before the lining of plant down is put in; in some cases the lining may not be added until the eggs have been laid and incubation has begun. Lining is added during the incubation period, and may even be added as late as two weeks after hatching. With the exception of a single observation by Welter (1935), there is no evidence that the male participates at all in nest construction.

A considerable variety of trees are used as nest sites. Pickens (1944) noted that the most commonly used sites in the upper piedmont and lower montane zones of South Carolina are lichen-covered post oaks (*Quercus minor*). However, in the lower piedmont where the oaks are more scrubby, pines and other trees species are more commonly used. In the Allegheny Park of New York, hornbeams commonly serve as nest sites, and in various other areas hickories, gums, tulip-poplars, junipers, and other species have been used. Probably species having a fairly rough and lichen-covered bark are favored over relatively smooth-barked species.

The length of time required for construction of the nest seems highly variable. Hinman (1928) noted that a nest begun on May 29 was completed by June 5, and by June 8 incubation of two eggs was underway. On the other hand, Welter (1935) observed that in one nest 10 days elapsed from its initiation until the first egg was laid, while a later nest was essentially completed in a single day, with an egg being deposited four days later. In this unusual case the male reportedly assisted in nest-building.

Almost invariably two eggs are laid; Bent (1940) reported no exceptions to this rule, but Welter (1935) reported that only one egg was laid in one of the nests he observed. There is a one-day interval between the laying of the two eggs, and incubation begins with the laying of the second egg.

Although some shorter estimates have been reported by Bent (1940), the normal incubation period is probably 16 days. Egg dates extend from late March to June 15 in Florida, from late May to early July in New York, and from June 1 to July 17 in Michigan. It may be double-brooded in some areas (Nickell, 1948) and certainly is known to renest following an initial nesting failure. Old nests are sometimes reoccupied for subsequent use, even in subsequent years, but frequently a new nest is built, often in the same tree or a nearby one (Bent, 1940).

Fledging periods vary considerably in different areas; records from New England range from 14 to 28 days, and in a few carefully observed cases have been 20 to 21 days (Bent, 1940; Hinman, 1928). During the first few days after hatching the female inserts her tongue into the throat of the nestlings and squirts in nectar and tiny insects. During feeding, the female typically stands on the edge of the nest, braces her tail against its side, and sometimes thrusts her entire bill down into the nestlings' throats. As the young birds become larger the bill may be inserted at right angles, and even later the food may be passed directly from beak to beak (Bent, 1940).

Little can be said of post-fledging survival rates, but among 2290 ruby-throated hummingbirds banded over an 11-year period by Baumgartner and Baumgartner (1992), 53 percent were immatures among this initial banding sample, and 46.7 percent of 334 total later returns were of birds that had been banded as immatures, suggesting a fairly high annual recruitment rate. Among 1268 banded males, two individuals subsequently survived at least five more years; and, among 1022 females, three survived at least six more years, two at least seven more years, and one at least nine more years.

Evolutionary and Ecological Relationships
Mayr and Short (1970) concluded that the ruby-throated and the black-chinned hummingbirds are very closely related geographic representatives, comprising a superspecies. They also included the Costa hummingbird in this species group, as it is especially similar to the black-chinned hummingbirds.

As summarized by Austin (1975), the plant relationships of the ruby-throated hummingbird are complex; it or other hummingbirds have been effective in the selection and evolution of at least 19 species of eastern North American plants. Pickens (1927) noted that *Macranthera "LeContei"* (= *flammea*) is the "most delicately adjusted" species of ornithophilous plant that he was aware of in its structural adaptations for pollination by the ruby-throat; it is endemic in the southeastern states from Louisiana to Georgia. Bené (1947) listed 28 species or genera of plants (including various garden or horticultural forms) reportedly visited by the species, several of which (e.g., cardinal flower, jewelweed, black locust, horse chestnut, Oswego tea) were uniquely listed for the ruby-throat.

In addition to their relationships to these plants, ruby-throated hummingbirds exhibit a close association with sapsuckers and their associated drilling behavior at various sap-producing trees. Foster and Tate (1966) observed hummingbirds feeding more frequently at sapsucker-drilled trees than the sapsuckers themselves. Although they came to these trees primarily to feed on sap, they also fed on insects similarly attracted to the tree. Other hummingbird species (Anna, broad-tailed, and rufous) also have been observed feeding at sapsucker drillings, but apparently mainly on the sap itself.

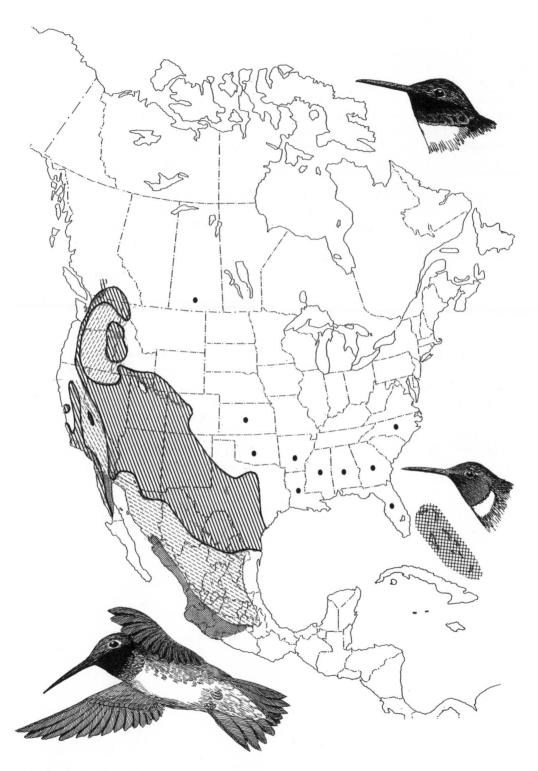

Breeding (hatched) and wintering (stippled) ranges, plus migration routes (broken hatching), of the black-chinned hummingbird. Dots represent states or provinces with extralimital sightings. The residential range of the Bahama woodstar is also shown (cross-hatching).

BLACK-CHINNED HUMMINGBIRD

Archilochus alexandri (Bourcier and Mulsant)

Other Names
Purple-throated hummingbird; Terciopelo
barbanegro (Spanish).

Range
Breeds from southwestern British Columbia and
northwestern Montana south through western
Montana, central Idaho, western Colorado, New
Mexico, and south-central and southwestern Texas
to northern Baja California, Sonora, and extreme
northwestern Chihuahua (AOU, 1957). Winters
from northern Baja California south to Guerrero,
Distrito Federal, and Michoacan.

North American Subspecies
None recognized.

Measurements
Wing, males 41.5–44 mm (ave. of 10, 42.7 mm),
females 46–48.5 mm (ave. of 10, 47 mm). Exposed
culmen, males 18–20.5 mm (ave. of 10, 19.2 mm),
females 19.5–22 mm (ave. of 10, 20.6 mm)
(Ridgway, 1911). Eggs, ave. 12.5 × 8.3 mm
(extremes 11.68–13.72 × 7.87–8.89 mm).

Weights
The average of 34 males was 3.09 g (range
2.7–4.1 g); that of 24 females was 3.42 g (range
3.3–3.7 g) (from various sources, including data
from Museum of Vertebrate Zoology).

Description (After Ridgway, 1911)
Adult male. Above rather dull metallic bronze-
green, darker and duller on pileum, the forehead
sometimes dull dusky; remiges dark brownish
slate or dusky, faintly glossed with purplish; tail
(except middle pair of rectrices) bronzy purplish
black; loral, suborbital, auricular, and malar
regions, chin, upper throat, and sides of throat uniformly opaque or velvety black; lower throat
metallic violet or violet-purple, changing to black
in position *b* (see note at beginning of Part Two for
position descriptions); chest dull grayish white or
very pale brownish gray; the underparts of body
similar, but usually more decidedly grayish medially; the sides and flanks darker and glossed or
overlaid with metallic bronze or bronze-green;
under tail-coverts brownish gray (sometimes
glossed with bronzy) centrally or medially, broadly
margined with white; femoral tufts and tuft on
each side of rump white; bill dull black; iris dark
brown; feet dusky.

Adult female. Above rather dull metallic bronze-
green, the pileum much duller, usually dull
grayish brown or brownish gray, at least on forehead and crown; remiges dark brownish slate or
dusky, faintly glossed with purplish; three outer
rectrices (on each side) broadly tipped with white,
the subterminal portion (extensively) black, the
basal half (more or less) metallic bronze-green
(sometimes grayish basally); underparts dull white
or grayish white (more purely white on abdomen
and under tail-coverts); the throat sometimes
streaked or guttately spotted with dusky; femoral
tufts and tuft on each side of rump white; bill, etc.,
as in adult male.

Immature male. Similar to the adult female, but
feathers of upperparts margined terminally with
pale grayish buffy; underparts more or less
strongly tinged or suffused with pale buffy brownish; throat streaked or spotted with dusky. Juvenile
males have a white tip to the third rectrix that is no
more than 12 millimeters long; in juvenile females,
the white area is at least 12.7 millimeters. In
subadult males, the white tip is no more than
9.45 millimeters; in subadult females, it is at least
10 millimeters. Subadults of both sexes differ
from those of ruby-throated hummingbirds in
that the maximum width of the tenth primary is
usually greater than 2.5 millimeters within 5 millimeters of the tip, rather than less than this width
(Baltosser, 1987).

Immature female. Similar to the young male, but
throat usually immaculate or with the dusky spots
or streaks smaller and less distinct.

Identification

In the hand. Males are best identified by the combination of the inner primaries having notches near the tips of the inner webs, a slightly forked tail, and a throat that is blackish, becoming violet posteriorly. Females also have notched primaries, but the tail is rounded, with the middle rectrices the longest (24–26 mm), and a dull-white throat. The exposed culmen (19.5–22 mm) of females is also very slightly longer than that of the otherwise similar female ruby-throated hummingbird, as is the wing (46–48.5 mm vs. 43.5–45.5 mm).

In the field. Occurs in mountain meadows and woodlands, canyons, and orchards. Males produce a dry buzzing sound with their wings when they fly, and utter a low *tup.* The blackish throat of the male is a distinctive fieldmark, but the female cannot be safely told from the female ruby-throat in a few areas (such as eastern Texas) where both might occur. Males do not sing while perched or in the air, but resemble the ruby-throat in that their courtship consists of long, swinging, pendulum-like swoops above the perched female, producing a long and drawn-out plaintive note at the bottom of the dive. They also perform a droning flight, resembling a horizontal figure eight (Bent, 1940).

Females closely resemble those of the Costa hummingbird, but have longer bills, are generally grayer below, and have throat markings ranging from nearly immaculate to extensively marked with dusky. Young males are also rather heavily marked with dusky on the throat and are usually grayer below than Costa males. Perhaps the best distinction of these two species is their voice: The chip call of the black-chinned is low pitched, softer, and a slurred *tew* or *tchew*, rarely run together in a series, and the birds are generally much less vocal. While hovering, both species often nervously flick the tail open and pump it up and down (Stiles, 1971).

Habitats

In breeding areas of California, breeding female black-chinned hummingbirds are characteristic of deciduous trees along stream bottoms, particularly in canyons, but they also occur in irrigated orchards. Evidently proximity to water is important in that area, and the associated trees include willows, cottonwoods, alders, sycamores, and valley oaks. Males frequent similar habitats, but often are found on drier canyon sides such as among live oaks and chaparral, or in desert washes where mesquite and catclaw thickets occur (Grinnell and Miller, 1944). In Texas the species is the most common nesting hummingbird and is associated with agave-cactus desert and subhumid juniper-oak woods (Oberholser, 1974). In Arizona black-chins nest primarily among willows, cottonwoods, and sycamores along streams and in olive trees in towns (Phillips et al., 1964). Their favorite habitats there seem to be the mouths of canyons where sycamores occur in association with water, but they also extend to the dry washes where small patches of willows are present (Bent, 1940).

Movements

Black-chinned hummingbirds winter almost entirely within Mexico, but occasionally have been reported during winter in southern California (once at Pacific Grove) and south-coastal Texas. Fall departure dates from Washington, Oregon, and California range from mid-August to late September. In Texas the majority of the birds are gone by mid-October, with a few persisting into December (Oberholser, 1974). Spring migration into Texas occurs from middle and late March to early May. The species usually arrives in Arizona in mid-March, and is common in some parts of southern California by late March. By mid-May it has reached northern Oregon, eastern Washington, and southern British Columbia (Bent, 1940). In Arizona the species is common through September, but apparently by midsummer most of the adult males have left the state, leaving only females and juveniles (Phillips et al., 1964).

There are a considerable number of extra-limital records, including locations as far east as Nova Scotia and Massachusetts. Among more than 500 wintering hummingbirds banded in Louisiana by Newfield (1966), 23 percent were black-chinned! Stray birds also have been reported from Georgia, Alabama, Mississippi, Arkansas, Tennessee, and Ontario.

Foraging Behavior and Floral Ecology

Much of the available information on the foraging behavior of black-chins comes from the work of

F. Bené (1947). He found that the timing of the bird's arrival in its southwestern breeding areas coincides with the flowering of important food plants. In the vicinity of Phoenix, the advent of irrigation and introduction of exotic flora evidently were responsible for the development of black-chin breeding in that area. When they arrive in late February or early March, the plants on which they feed are already in bloom, including agave, tree-tobacco (*Nicotiana glauca*), lantana, citrus trees, shrimp plant (*Beloperone guttata*), nasturtium, *Buddleia,* and others. The birds depart the area in June or July, presumably to move to higher elevations in the mountains where midsummer food is more abundant, and perhaps to undertake a second breeding cycle there.

Bené (1947) noted that black-chinned hummingbirds have been reported visiting at least 37 species or genera of flowers. He determined that they have no innate preference for red flowers, but that color preference could be conditioned by training. He also established that the form of a flower could serve as a conditioning stimulus, and that the memory of receptacle features or flower shape was apparently held for periods lasting several months. He also established that the birds could detect fairly small differences in concentrations of honey solutions, but judged that flower scent did not play any evident role in locating hidden nectar receptacles.

Bent (1940) summarized the many species of plants mentioned as foraging sites for black-chins, and noted that the exotic tree-tobacco (*N. glauca*) is very popular in California. Bené (1947) noted that the species is especially attracted to Hall's honey-suckle (*Lonicera japonica*); other species that have been specifically mentioned include *Delphinium cardinalis, Anisocanthus thurberi,* ocotillo, *Lycium andersoni,* palo verde, ironwood, Texas buckeye, Texas redbud, and Texas mountain laurel (Bent, 1940).

Breeding Biology
Although the data are not very clear, the sexes migrate separately in the spring, and there is an interval between their arrivals in the spring breeding areas, with males often arriving in advance. According to Bené (1947), black-chins establish three kinds of territories. One is a female nesting preserve in a breeding locality that includes a nest site, one or more perches, a roost site, and a feeding site. The second is a mating area, visited by both sexes. Last is a male feeding preserve that includes guarding perches, a feeding site, and perhaps a roost. Males sometimes visit the female's nesting preserve but do not feed on it. The male roosts near the female at the time of copulation and for a few days thereafter, but arrives only at nightfall, after his last meal of the day. On the preserve the nest and feeding sites are specifically defended, with the strength of defense growing as the breeding cycle progresses. Within the male preserve, only a relatively small area (3–6 m in diameter) may be defended. After the breeding season the sphere of active defense may be an area as small as 7 to 15 meters in diameter, containing a couple of feeders.

Although birds probably initially establish territories on the basis of local food supply, a later change in the territory may bring about a change in the nest site, perch, or roost, and thus a shift in the area guarded for food. For a male the roost is the central point in the territory, and he returns to it invariably except for the few days at the time of courtship, when he roosts beside the courted female. As the young birds become independent, they settle on the territories left by their parents at the end of the breeding cycle, and remain there until they depart for wintering areas (Bené, 1947).

With the appearance of females on the male's display areas, aerial display begins. It consists of a series of long, swooping, pendulum-like maneuvers about 30 meters in length, with the male passing very close to the female at the bottom of the arc, and ending about 5 meters higher (Bent, 1940). It sometimes takes the form of a narrow horizontal figure eight, and is accompanied by a loud whistling sound, presumably made by the wings or tail. According to Woods (1927), the dive may be accompanied by about four thin, vibrant notes, and the males utter a distinct courtship note as well, consisting of a long, drawn, pulsating, plaintive, liquid note. At the apex of each flight the bird may hover and call, or produce an apparently mechanical sound evidently caused by patting his wings together underneath him, sounding like the noises made by a bathing bird. Bené (1947) reported seeing as many as three swooping courtship sequences per day. These continued at least through the egg-laying period, but ceased shortly after egg-laying was finished, apparently through repeated rejections by the female. In the case of two females, the interval from the start of nest construction to the laying of the first egg was five days, whereas a third female probably laid her first egg six or seven days after nesting was started. The eggs are usually laid one or two days apart, but Demaree (1970) observed a case in which two eggs were laid in a single day, only three days after nest-building had been initiated.

The nest is very similar in size to that of the ruby-throated hummingbird, but typically is not covered by lichens on the outside. It often is composed almost entirely of the yellowish downy material from the underside of sycamore leaves, resulting in a distinctive yellowish cast. However, the outside is sometimes covered by various small materials attached to spider webbing, including lichens. These often include bud scales, stamens, flower, bark, or leaf fragments, and the like. Nests are usually 1 to 3 meters above ground and often overhang small or dry creek beds. They are located either in a fork or on a small, drooping branch of one of many trees, including alders, cottonwoods, oaks, sycamores, laurels, willows, apples, and oranges (Bent, 1940). All but 5 of the 26 nests tabulated by Pitelka (1951a) were in live oaks. Occasionally the nests are situated among woody vines or on taller herbaceous weedy plants. Rarely have they been found as high as 9 meters above ground, but the average height is less than 3 meters (Pitelka, 1951a).

Incubation is by the female alone, and has been estimated to last from 13 to 16 days. In a nest closely observed by Demaree (1970), incubation began the day after egg-laying was completed. One of the eggs hatched 16 days later; the other did not hatch. The young bird left its nest 21 days later, and remained in the ivy vines around the nest for 2 more days.

There have been several reported instances of females incubating eggs in one nest while feeding young in another (Cogswell, 1949), suggesting that double-brooding may be fairly frequent in black-chins. In one instance, a female was observed building a nest during the final week of brooding her earlier brood; thus, laying of the

eggs in the new nest apparently occurred two to seven days after the fledging of the first one. These observations were made in California, but double-brooding may also occur in Arizona (Phillips et al., 1964) and New Mexico (Bailey, 1928). In the Santa Barbara area of California the species is single-brooded, breeding from late April through June, at the same time that nesting occurs in the Costa hummingbird (Pitelka, 1951a).

The female continues to feed her young for several days after fledging, but the young birds gradually begin to learn the identity of suitable food plants. This occurs only gradually, after repeated visits to obnoxious species, probing fertilized flowers with dried nectaries, overlooking familiar fruitful species, and generally obtaining their food in an awkward fashion (Bené, 1947).

Nesting and fledging success rates for 157 nests of this species have been estimated by Baltosser (1986). He reported an overall rate of nesting success (percentage of nests producing at least one fledged young) as 34 percent among three study areas in Arizona and New Mexico. Later, apparently second, nestings averaged slightly more successful than early-season nestings. Average total seasonal productivity per female ranged from 1.89 young in Guadelupe Canyon to 0.53 in Rucker Canyon, Arizona. Predation was the primary cause of losses in this and other nesting hummingbirds, the major culprits apparently being avian predators such as jays.

Evolutionary and Ecological Relationships
Hybridization in the wild has occurred with at least the Allen, Costa, Anna, and broad-tailed hummingbirds (Lynch and Ames, 1970; Mayr and Short, 1970); a possible hybrid with the ruby-throated hummingbird has also been suggested (*Bulletin of Oklahoma Ornithological Society* 2:14–15). As Mayr and Short have noted, this species is closely related to the ruby-throated hummingbird—the two having essentially allopatric breeding ranges—and they comprise a well-defined superspecies. They suggested that the Costa hummingbird is also a close relative of this superspecies and also mentioned that the broad-tailed hummingbird may be more closely related to this superspecies than to other species now included in *Selasphorus*.

Although it has a relatively wide total breeding distribution, the black-chinned hummingbird is common only in the southwestern United States, where a maximum abundance of bird-adapted flowers occurs. Bené (1947) listed nearly 40 species of flowers frequented by the species, many of which are southwestern in their distribution patterns. Those listed exclusively for the black-chinned hummingbird include chuparosa (*Beloperone californica*), shrimp plant (*B. guttata*), catmint (*Nepeta species*), myrtle (*Vinca major*), palo verde (*Parkinsonia microphylla*), *Poinciana,* rose of Sharon (*Althea* species), garden balsam (*Impatiens balsamina*), and iris (*Iris* species). Most of these plants are, of course, garden or horticultural forms rather than native North American flowers, but at least chuparosa is a typical hummingbird-adapted species with a number of specific adaptations that facilitate probing without damaging the ovaries (Grant and Grant, 1968).

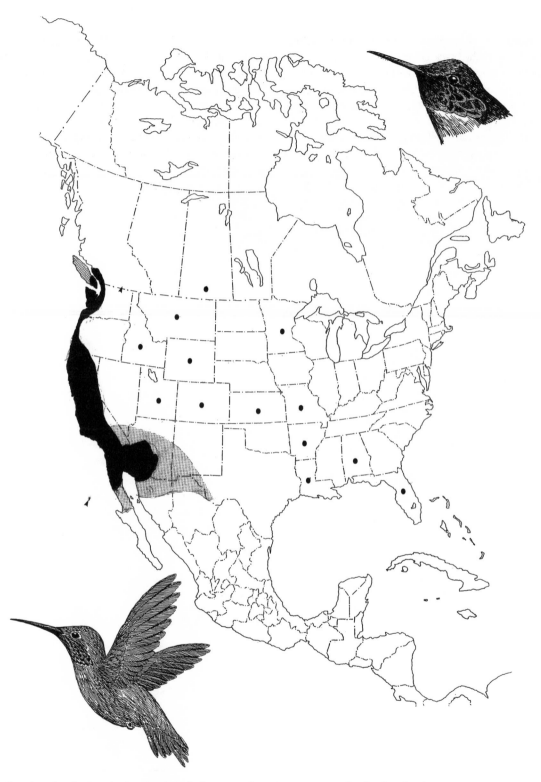

Breeding (hatched) and wintering (stippled) ranges, plus migration routes (broken hatching), of the Anna hummingbird. Dots represent states or provinces with extralimital sightings.

ANNA HUMMINGBIRD

Archilochus anna (Lesson)
(*Calypte anna* of AOU, 1983)

Other Names
None in general English use; Chupamirto cuello escarlata (Spanish).

Range
Breeds in California west of the Sierra Nevada and southern coastal mountains from Humboldt, Shasta, and Tehema counties south to the Sierra San Predo Martir and San Quintin in northwestern Baja California. Also reaches Santa Cruz Island, California, and Guadalupe Island, Baja California. Winters over the breeding range, northward to Humboldt Bay and to the islands of the coast of California and Baja California, and southward on the mainland of Baja California. Also winters eastward from southern California across southern Arizona to northern Sonora, and northward to southern Oregon. In recent years breeds increasingly in Arizona (Zimmerman, 1973).

North American Subspecies
None recognized.

Measurements
Wing, males 48.5–51 mm (ave. of 10, 49–7 mm), females 48–51 mm (ave. of 10, 49.6 mm). Exposed culmen, males 17.5–20 mm (ave. of 10, 18.2 mm), females 17–22 mm (ave. of 10, 18.8 mm) (Ridgway, 1911). Eggs, ave. 13.31 × 8.65 mm (extremes 11.3–14.3 × 7.7–9.4 mm).

Weights
The average of 81 males was 4.31 g (range 3.3–5.8 g); that of 40 females was 4.07 g (range 3.3–4.7 g) (from various sources, including specimens in Museum of Vertebrate Zoology).

Description (After Ridgway, 1911)
Adult male. Whole head except occiput and auricular region brilliant metallic rose red, changing to solferino and violet in certain lights—more golden or even greenish in position *b* (see note at beginning of Part Two for position descriptions)—the latero-posterior feathers of throat elongated; a small postocular spot or streak of white; occiput (except laterally, where at least partly metallic purplish red), hindneck, back, scapulars, wing-coverts, and rump metallic bronze-green, the upper tail-coverts and middle pair of rectrices similar but usually less bronzy, or more bluish, green; tail (except middle pair of rectrices) dark grayish, faintly glossed with greenish bronze, the rectrices blackish terminally and medially; remiges brownish slate or dusky, faintly glossed with purplish; chest pale brownish gray or dull grayish white, the feathers darker brownish gray beneath surface; rest of underparts deeper grayish, strongly glossed with metallic bronze-green laterally, the feathers more or less distinctly margined with paler grayish; femoral tufts and conspicuous tuft on each side of rump white; under tail-coverts brownish gray glossed with bronze-green or greenish bronze (especially on shorter coverts) and broadly margined with pale gray or grayish white; bill dull black; iris dark brown; feet dusky.

Adult female. Above metallic bronze or bronze-green, duller on pileum, the forehead sometimes dull grayish brown; middle pair of rectrices metallic green or bronze-green, sometimes dusky terminally; the next pair similar, but with terminal portion (broadly) blackish; third pair similar to second, but more extensively blackish terminally and narrowly tipped with white; two outer pairs with basal half (more or less) brownish gray, the tip pale brownish gray or dull grayish white (broader on outermost rectrix), the intermediate (subterminal) portion black; remiges brownish slate or dusky, faintly glossed with purplish; chin and throat pale brownish gray or dull grayish white, the center of throat usually with an admixture of metallic red or purplish red feathers, sometimes with a considerable patch of metallic reddish; the lower throat, at least, with mesial guttate spots or broad streaks of dusky grayish brown or dull bronzy; rest of underparts as in the adult male, but slightly paler and less extensively glossed with metallic greenish; bill, etc., as in adult male.

Immature male. Similar to the adult female, but tail less rounded; lateral rectrices with dark subterminal portion duller blackish and less sharply contracted with dull grayish of basal portion; feathers of upperparts very narrowly and indistinctly margined with pale buffy grayish; and (at least older individuals) with metallic purplish red feathers on crown as well as on throat. Subadult males have a white tip to the third rectrix that is no more than 0.6 millimeters long; in subadult females it is at least 1.5 millimeters. Juvenile females have a white tip of at least 1.4 millimeters; that of males is no more than 0.1 millimeters (Baltosser, 1987). See account of Costa hummingbird for distinguishing the young of these species.

Immature female. Similar to the adult female, but feathers of upperparts narrowly margined with pale brownish or dull buffy and throat without metallic red feathers.

Identification
In the hand. Like the Costa hummingbird, differs from typical *Archilochus* forms because the males lack notches near the tips of the inner webs of the primaries, and the lateral rectrices are distinctly narrowed, although not so much as in the Costa (males 2.5–3 mm, females 5 mm). Females also have wider central rectrices (at least 8 mm in width) than do Costa females and the wing length is slightly greater (48–51 mm vs. 43.5–46 mm).

In the field. Adult males are best separated from male Costa hummingbirds by their red rather than violet crown and gorget. Females of the two species are very similar, but females of the Anna hummingbird have a reddish patch on the throat and darker underparts than Costa females. When feeding the birds utter a heavy *chick,* which is repeated rapidly when alarmed; they produce a series of *ztikl* notes when chasing intruders. The song is a generally high-pitched assortment of squeaks, gurgles, and hissing noises, uttered while perched. The male also sings in flight: Typically he flies upward 22 to 44 meters, sometimes pausing to sing, and at the top of his climb he pauses a second time to sing before diving sharply downward over the female, making a sharp *peek* at the bottom of the dive. On Guadalupe Island the songs are distinctly different from those on the mainland

(Mirsky, 1976). The *peek* dive-note may be vocal (Baptista and Matsui, 1979). Female Annas are distinctly larger than other western hummingbirds except for the broad-tailed, and the individual chip notes are very hard, sharp, and explosive *tzip* or *kip* sounds. These are often run together into an excited chittering, which like the chatter calls are fairly loud and harsh. Besides being generally grayer below, females also have less conspicuous black on the tail than do most other species, but juvenile males are sometimes fairly pale below and have little or no red on the throat. These birds usually have buffy-gray feather edgings on the upper surface, and when hovering birds of all ages tend to hold the tail down, nearly in line with the body, with little pumping, spreading, or flicking (Stiles, 1971).

Habitats
Essentially confined to California during the breeding season, this species then typically occupies broken chaparral or woodland, or mixed woodland and chaparral in open stands. During the breeding period the sexes show some habitat segregation, with males in more open habitats—such as on canyon sides, hill slopes, or level washes—and females in tracts of evergreen trees, especially live oaks. Outside the breeding period and when foraging, the birds also occur in other habitats, often to 1800 meters' elevation in the mountains during late summer, especially those rich in preferred flower species. Limited wintering also occurs in southern Arizona, especially in the areas where nighttime temperatures do not fall below freezing. Late summer habitats are quite diverse, with the birds tending to move to higher elevations, occasionally into boreal forest communities. Likewise, in August and September they use lower altitude areas, sometimes extending out into stubble-fields in search of plants such as blue-curls (*Trichostema*) (Grinnell and Miller, 1944).

Movements
Unlike most hummingbirds, this species is relatively sedentary, although the birds withdraw from the northern parts of the California breeding range in the colder months and migrate in small numbers into southern Arizona in September or early October, remaining there until December or

(rarely) early March (Phillips, et al., 1964). They also do a fair amount of post-breeding wandering in California, with vagrant birds reaching the Farallon Islands, San Clemente, Santa Catalina, and most of the other islands of the Santa Barbara group during winter or spring months (Grinnell and Miller, 1944). In recent years the species has expanded its winter range considerably (Zimmerman, 1973).

There are a surprising number of extralimital records for Anna hummingbirds. For example, it has been reported several times in Alaska (*American Birds* 29:104; 31:211; 33:205; 43:1356), British Columbia (*American Birds* 25:618; 29:718), and Washington (including one nesting record) (*American Birds* 27:89; 29:718; 30:881). Since the late 1980s the species also has turned up in unlikely locations such as Florida, Alabama, Missouri, Minnesota, Kansas, and Arkansas. It is also becoming increasingly frequent in Alaska and has been seen as far west as Anchorage. It is rare but regular in western Oregon, mostly as a winter visitor, and nested there in 1981 (*American Birds* 35:855). There are two nesting records for British Columbia (*American Birds* 42:312).

Foraging Behavior and Floral Ecology
The chaparral habitats of California include a number of hummingbird-adapted flowers and, although there is some ecological overlap in the foraging habitats of Anna and Allen hummingbirds, the Anna commonly exploits flower species such as bush monkey-flower (*Diplacus* species), Indian pink (*Silena laciniata*), and Indian paintbrushes (*Castilleja martinii* and *C. foliolosa*) especially in spring (Grant and Grant, 1968).

Bent (1940) described several observations of Anna hummingbirds feeding on the wing for midges and gleaning the trunks and branches of trees for invertebrates. They regularly visit sapsucker holes (Bent, 1940), as does the ruby-throated hummingbird; the birds evidently consume both the exuding sap and the insects attracted to the sap. According to Bent, the very favorite plant species of Anna hummingbirds is the red gooseberry (*Ribes speciosum*), which is abundantly in flower when the Annas begin nesting activities. Also of importance in southern California is the exotic tree-tobacco (*Nicotiana*

glauca), and the fall blossoms of century plants (*Agave americana*) are widely available in the state as a result of planting. Likewise, the planted red-hot-poker (*Tritoma*) is common in gardens. Grinnell and Miller (1944) stated that perhaps the most important of the introduced plants are the many varieties of *Eucalyptus*, both for foraging and for nesting.

Stiles (1972b) stated that the tree-tobacco and *Eucalyptus globulus* are the most important introduced plants for hummingbirds in chaparral areas of California, and that citrus orchards also provide important spring nectar sources for breeding and migrating birds. However, the two most important native species in his study area (Santa Monica Mountains) were two species of *Ribes* (*malvaceus* and *speciosum*). In higher and somewhat cooler areas a species of manzanita (*Arctostaphylos glauca*) was frequently visited in early months. Stiles suggested that the reproductive strategies of the Anna hummingbird and *Ribes speciosum* have evolved together, resulting in an unusually early breeding season for the hummingbird.

Breeding Biology
In the Santa Monica Mountains, the breeding season of the Anna hummingbird begins in November or December, as males move onto breeding territories. The first heavy winter rains usually fall between November and January. The season ends in April and May when the males abandon their territories, and females cease their nesting activities between May and July. Thus, the female breeding cycle lags somewhat behind that of the male, sometimes by as much as a month (Stiles, 1972b).

Prior to breeding, adult and first-year males both hold feeding territories at localized rich sources of food, such as at eucalyptus or tree-tobacco plants. Females do not hold feeding territories during the prebreeding season, but either visit undefended flowers or poach from defended areas (Stiles, 1972b).

Breeding territories of males are larger and more energetically defended than are non-breeding ones, and are typically established around rich, dependable, and easily exploited food resources. Males usually establish territories where *Ribes malvaceus* is in flower, and may later shift slightly as *Ribes speciosum* begins to replace the other

species. Occasionally birds will also commute to
local food sources well outside their territories,
sometimes as far as 700 to 800 meters away and
rarely up to a kilometer away. Territories usually
have from five to ten regular perches within the
core area, which is 900 to 1000 square meters in
area and frequently about 50 to 100 meters in
diameter. Although Stiles was unable to mark
his birds, he had four cases of probable use of
the same territory by the same individual in
two successive years. Outside the core area is
an ill-defined buffer zone, which may range
from 2 to 6 hectares in size (Pitelka, 1951b; Stiles,
1972b).

Females begin nest-building a few weeks after
the males have established their territories. They
probably spend much more time than males in
search for insect food, but also need a reliable
source of nectar. Thus, the location of suitable
flowers is an important part of nest-site selection.
Prior to nesting, females often restrict their
activities to a single food source and will defend it,
although less vigorously than would males. The
nest site is likewise probably chosen with reference
to the food source, and that site is then also
defended, becoming a nesting territory. Several
perches are also chosen within the territory, for
resting, preening, or using to sally forth to chase
insects (Stiles, 1972b).

Nesting locations are extremely variable for
Annas, ranging in height from less than 1 meter to
9 meters from the ground and on nearly every
possible substrate, including electric wires, climb-
ing vines, and many kinds of trees, such as citrus,
eucalyptus, and many chaparral species, especially
oaks. The nests are similar to those of the Costa,
but average somewhat larger (nearly 50 mm in
diameter), often have fibers or stemmy materials in
the walls, and often are extensively ornamented
with lichens. The lining frequently includes small
feathers (Bent, 1940).

Nest-building usually takes seven days, but
the range may be from three days to at least two
weeks (Legg and Pitelka, 1956). Probably most
copulations occur during this time. Trousdale
(1954) observed copulation during a period when
a female was gathering tent caterpillar webs. The
gathering was interrupted when the female flew
to a clothesline and perched. A male that had

been perching and singing on the clothesline then
immediately left his perch to hover over and then
mount the female. The male's wings fluttered during copulation, and as soon as it was completed
the female returned to the oak tree and the male to
his previous perch. According to Schuchmann
(1979), copulation is preceded by the male flying
back and forth before the perched female four to
six times, producing a whistling sound with the
bill slightly open. The female follows the male's
movements with her bill, and when she opens
her bill slightly the male alights beside her and
copulation follows. Stiles (1982) also has described
courtship and aggressive displays of this species
in detail.

At the time of mating the nest is already at least
partly built, but materials may be added to it well
into incubation, which requires 14 to 19 days
(Stiles, 1972b). Eggs are apparently laid a day
apart, and sometimes the young hatch a day apart,
suggesting that incubation may begin with the first
egg (Kelly, 1955). Feeding during incubation occasionally occurs well away from the nest site, with
weather regulating the periods that the adult bird
is away from the nest. On cold or wet days, the
bird remains on the nest longer and returns early.
During incubation, the female does not go into
torpor at night, but rather maintains normal body
temperature (Howell and Dawson, 1954). Probably
unusually long foraging periods in the evening
hours are needed to compensate for this energy
drain, which is also affected by the relative thickness and thermal conductivity of the nest itself
(Smith et al., 1974).

By the time the birds are 6 days old they are
well-covered with down, and their eyes typically
open on the 5th day after hatching. Little or no
brooding occurs beyond the 12th day. The nestling
period ranges from 18 to 23 days, and the young
remain dependent on their mother for a few days
thereafter, but become independent within a week
or two (Stiles, 1972b). If a nesting effort fails, the
female usually begins to renest within about
10 days, and in many areas two broods are normally raised during a breeding season. In Stiles'
(1972b) study area, the first eggs were found in
December and the last eggs were laid in May, but
Bent (1940) indicated a total span of 86 California
nesting records that ranged from December 21 to

August 17, with a peak between February 22 and May 18.

Juveniles begin to show territorial behavior at an extremely early age, sometimes only shortly after leaving the nest. The young birds often move about in pairs, presumably siblings, which may often chase one another in a manner resembling play. The young birds gradually become less "playful," and by fall young males frequently are holding feeding territories alongside adult males. In Stiles' (1972b) study area Anna hummingbirds as well as rufous and Allen hummingbirds all occupied stands of tree-tobacco in large numbers, with Anna territories usually larger and more dispersed; the *Selasphorus* species territories tended to be smaller and more tightly packed. In individual encounters between the species, the Annas tended to be victorious, but they were unable to penetrate the areas occupied by *Selasphorus* because of the collective superiority of the latter.

Evolutionary and Ecological Relationships

According to Mayr and Short (1970), this species may be more closely related to the rufous superspecies (rufous and Allen hummingbirds) than to other species of *Archilochus*, including the Costa. Natural hybridization has occurred with black-chinned, calliope, and Allen hummingbirds (Banks and Johnson, 1961), and more recently it has been reported for the Costa (Wells et al., 1978).

Regardless of the closeness of their phyletic relationships, there is considerable ecological overlap and interaction between breeding Anna and Allen hummingbirds (Pitelka, 1951b). Their nesting cycles overlap considerably as well, although the Anna begins to breed earlier and often continues slightly longer in southern and central California (Pitelka, 1951a, 1951b). In the Berkeley area, the resident Anna hummingbird is territorial and breeding before the Allen males arrive; moreover, the Anna is larger, giving it a competitive advantage. Thus territories of Allen males tend to be peripheral to those of Anna, although (rarely) a male Allen is able to displace a territorial male Anna (Pitelka, 1951b).

The chaparral flora of California has several species adapted to winter growth and flowering (Grant and Grant, 1968), some of which have probably evolved in conjunction with the Anna hummingbird. As Stiles (1972b) noted, the *Ribes* species in particular is closely associated with the Anna, and the reproductive strategy of *R. speciosum* has probably evolved in association with it.

Breeding (hatched), residential (inked), and wintering (stippled) ranges of the Costa
hummingbird. Areas of denser breeding populations are shown by cross-hatching.
Dots represent states or provinces with extralimital sightings.

COSTA HUMMINGBIRD

Archilochus costae (Bourcier)
(*Calypte costae* of AOU, 1983)

Other Names
Coast hummingbird, Ruffed hummingbird;
Chupamirto garganta violeta (Spanish).

Range
Breeds in western North America from central
California, southern Nevada, and southwestern
Utah; south to the Santa Barbara Islands, southern
Baja California including all near-shore islands,
southern Arizona, and southwestern New Mexico;
south to Sonora and Sinaloa, including Tiburon
and San Esteban islands. Winters over most of the
breeding range, from southern California and
southwestern Arizona southward (AOU, 1983).

North American Subspecies
None recognized.

Measurements
Wing, males 43–45.5 mm (ave. of 13, 44.4 mm),
females 43.5–46 mm (ave. of 12, 44.7 mm). Exposed
culmen, males 16–19 mm (ave. of 13, 17.2 mm),
females 17–20 mm (ave. 12, 18.2 mm) (Ridgway,
1911). Eggs, ave. 12.4 × 8.2 mm (extremes 11.4–14 ×
7.6–9.4 mm).

Weights
The average of 33 males was 3.05 g (range
2.5–5.2 g); that of 27 females was 3.22 g (range
2.5–3.4 g) (various sources, including specimens in
Museum of Vertebrate Zoology). Twenty-five adult
males averaged 2.98 g (range 2.88–3.08, *SD* 0.25 g),
and 19 females averaged 3.25 g (range 3.12–3.38,
SD 0.26 g) (Baltosser and Scott, 1996).

Description (After Ridgway, 1911)
Adult male. Head, except postocular region, very
brilliant metallic violet or amethyst purple, changing to violet-blue or even greenish and more
reddish purple (magenta) in certain lights, the
latero-posterior feathers of throat much elongated;
rest of upperparts, including four middle rectrices,
rather dull metallic bronze-green or greenish
bronze; tail (except four middle rectrices) grayish

brown or brownish gray, faintly glossed with
bronze-greenish, the rectrices darker on shafts and
toward tip; remiges brownish slate or dusky,
faintly glossed with purplish; foreneck very pale
brownish gray or grayish white, passing into more
decidedly grayish on chest and median line of
breast and abdomen; rest of underparts metallic
bronze-green or greenish bronze, the feathers more
or less distinctly margined with dull grayish;
femoral tufts and conspicuous tuft on each side of
rump white; under tail-coverts light brownish gray
or bronzy centrally, margined with whitish; bill
dull black; iris dark brown; feet dusky.

Adult female. Above rather dull metallic bronze-
green or greenish bronze, much duller on pileum,
where (at least on forehead) sometimes dull grayish brown; middle pair of rectrices bronze-green;
the next pair similar but with terminal portion
black; third pair tipped with dull white or pale
brownish gray, extensively black subterminally
and dull brownish gray basally, the gray and black
separated (at least on outer web) by more or less
metallic bronze-green; fourth and outermost pairs
with whitish tip broader, basal grayish more
extended, and with little if any metallic greenish
between the gray and black; remiges brownish
slate or dusky, faintly glossed with purplish;
underparts pale brownish gray, paler (dull
whitish) on chin, upper throat, and under tail-
coverts; femoral tufts and tuft on each side of
rump white; bill, etc., as in adult male.

Immature male. Similar to the adult female but
feathers of upperparts more or less distinctly
margined with pale grayish buffy; tail double-
rounded instead of rounded; and throat with a
central patch of metallic purple or violet feathers
(in older individuals, similar feathers also on
crown). Juvenile males have a white tip to the third
rectrix that is no more than 5.6 millimeters long; in
juvenile females, the white tip is at least 6.3 millimeters. In subadult males, the white tip is no more
than 9.25 millimeters; in females, it is at least
9.8 millimeters. Juveniles of both sexes can be

distinguished from juvenile Anna by the maximum width of the outermost rectrix, which is not more than 4.45 millimeters in the Costa and at least 5.4 millimeters in the Anna (Baltosser, 1987).

Immature female. Similar to the adult female, but feathers of upperparts margined with pale grayish buffy.

Identification
In the hand. Differs from typical *Archilochus* forms in that the males lack notches near the tips of the inner webs of the primaries, and the lateral rectrices are unusually narrowed (outermost rectrix only 1.5 mm in males, 3.5 mm in females).

In the field. Found in hot desert areas where water is often lacking, but sometimes also occurs in chaparral and woodland areas. The male is the only North American hummingbird with a purple gorget and crown that has a lateral extension down the side of the head. Females and immature males lack this, and at most have a few reddish flecks on an otherwise grayish throat. They cannot be separated with certainty from female black-chinned hummingbirds, but are slightly smaller and more grayish on the crown. The absence of cinnamon on the tail helps to distinguish them from other western species, and the female Anna hummingbird often has a patch of red on the throat. The call-notes include soft *chip* or *chik* notes, and males utter two types of static (non-aerial) song: a single whistled note and a much shorter two- or three-noted whistle. The prolonged whistle is made during the diving display as the bird reaches the bottom of a deep U-shaped arc that may begin from a height of 30 to 60 meters. As the bird reaches the bottom of the dive, it begins to produce a shrill whistle that lasts about two seconds, after which it begins immediately to ascend and finally flies in a broad horizontal circle until it reaches its starting point. Beside these differences, the call is much longer than is the case with the Anna hummingbird. A similar but shorter call is produced while the male perches during static display (Wells et al., 1978).

Female Costas resemble black-chinned hummingbirds, but their *chip* call is very different—a very high, light and sharp *tik* or *tip,* which is often repeated to produce a rapid twitter. They are the palest in general coloration of the western hummingbirds, and the purest white below, usually having an immaculate white throat. Juvenile males likewise are very light below, with brownish feather-edgings above; often young males have some purple on the throat near the corners of the future gorget (Stiles, 1971).

Habitats
Throughout California, the breeding habitat consists of deserts or desert-like washes, mesas, or side-hills, particularly where sages, ocotillo, yuccas, and cholla cacti are abundant (Grinnell and Miller, 1944). In Joshua Tree National Monument, California, Costas are most commonly found during the nesting season along canyons or washes, where ocotillo, mesquites, or other flowering plants may be found (Miller and Stebbins, 1964). In Arizona the species favors habitats rich in ocotillos, chuparosa, and cacti (Phillips et al., 1964).

There apparently is no habitat separation of the sexes during the breeding season in this species (Pitelka, 1951a). Most observers stress that the Costa is relatively independent of water at this time, and thus extends into more xeric habitats than do any other North American hummingbirds. Probably the same is true of the wintering period, although nothing specific has been written of this aspect of Costa biology.

Movements
Like the Anna hummingbird, Costas undergo only limited seasonal movements. In California, wintering occurs on the Colorado Desert northwest to Palm Springs, and birds rarely winter to Los Angeles. In Arizona, Costa hummingbirds as well as some swallows are the first of the spring migrants, sometimes arriving in late January, and frequent by late February and March. However, Costas breed early, and by the end of May they have disappeared from the desert areas, apparently having migrated to the Pacific coast of California and Baja California. They reappear in October, to spend the winter in western Pima County and southern Yuma County (Phillips et al., 1964).

In spite of these limited migratory tendencies, vagrant birds sometimes stray considerable distances. They have appeared on San Clemente, Santa Barbara, and Santa Catalina islands (Grinnell and Miller, 1944), have reached Oregon at least eight times (*American Birds* 34:924), have occasionally reached New Mexico (*American Birds* 26:640; 31:1033) and Texas (*American Birds* 28:824, 29:709; Oberholser, 1974), and have even been reported from Alaska (*American Birds* 49:86). There is also a recent Oregon breeding record (*American Birds* 45:1153).

Foraging Behavior and Floral Ecology

Costa hummingbirds utilize a wide variety of flowering plants for foraging. On the Colorado Desert of California some of the important ornithophilous flowers include the early-blooming ocotillo and chuparosa, which begin to flower shortly after the arrival of Costa hummingbirds in that area (Grant and Grant, 1968). There are several other desert species of importance, including boxthorn (*Lycium* species), desert lavender (*Hyptis emoryi*), desert willow (*Chilopsis*), and sages (*Salvia mellifera* and *S. apiana*) (Grinnell and Miller, 1944).

In a study in Claremont, California, Grant and Grant (1968) observed territorial behavior of male Costa hummingbirds among a population of larkspurs (*Delphinium cardinale*). Males occupied posts in nearby trees, defending their plants from intruders. They fed primarily on plants near their tree posts, but at times flew elsewhere to feed. Single males sometimes held posts for several days; in one instance they defended an area that included some 40 flowering stalks of larkspurs. During 6.5 hours of observation, this male made 42 foraging trips, visiting 1311 flowers, nearly 90 percent of which were located within its territorial limits. Intruder birds made only 12 feeding trips within the male's territory during this period, visiting 192 flowers, indicating the nearly exclusive use of the area by a single male.

According to Woods (1927), during the breeding season females of this species visit flowering plants less often than do males, perhaps because they must spend more time looking for nesting materials. Perhaps also they are then concentrating more strongly on obtaining minute insects and spiders that may be acquired incidentally during such searches. Woods believed that Costa hummingbirds are also less partial to the flowers of the introduced tree-tobacco shrub than are the larger species of California hummingbirds.

Breeding Biology

Nesting in Arizona begins fairly early, with well-developed young reported as early as mid-April, but others are still being fed by their mothers during June (Phillips et al., 1964). On the other hand, in the Santa Barbara area of California the Costa is a relatively late nester; together with the black-chinned hummingbird it begins nesting in middle or late April and raises a single brood, and nearly all the eggs hatch by the end of June. Bent (1940) reported a total period of egg records for California from March 11 to June 29, with a peak from May 12 to June 10; fewer records from Baja California ranged from February 24 to June 5, with a probable peak in late May. Bakus (1962) reported an unusual late January nesting for the San Diego area. Stiles (1972b) suggested that this species may raise one brood in the Borrego desert of southern California in early spring, then move to the chaparral in April to breed a second time.

Territories are established almost as soon as males arrive on their nesting areas. In the Joshua Tree National Monument, males begin active display in early February, as soon as they become evident. Between aerial displays the males perch on lookouts at the tips of the highest available vegetation and periodically perform their diving displays. This usually consists of a very open U-shaped arc, with the male typically circling back to retrace the arc in the same direction rather than retracing it in the opposite direction, according to Miller and Stebbins (1964). These authors also noted a short swinging performance of not quite 1 meter in length, with the bird swinging slowly back and forth. The more typical display occurs throughout the day, and before sunup as well as after sundown. Woods (1927) likewise noted that the long U-shaped swoop may be performed repeatedly from the same direction or at a new angle. The usual sound produced is extremely prolonged and intense, something like the shriek of a glancing bullet; it is apparently of vocal origin, since a very similar sound may be produced by the bird while it is perched (Wells et al., 1978). At times, however, a more "booming" noise is made, and the dives are unusually narrow and steep (Short and Phillips, 1966).

Territories of males are very large, often 1 to 1.5 hectares, with no clearly defined core area. The vegetation is usually low and uniform in height, with scattered higher perching sites. Food is usually widely available, or present in large patches (Stiles, 1972b).

Miller and Stebbins (1964) reported copulation as occurring on a mesquite plant. The two birds were observed 1.5 meters above the ground on open twigs, when the male approached the female from in front and made short 50- to 75-millimeter darting flights toward her at various angles within an arc of about 90°. He then mounted her while she was perching, flew off, and again darted at her from the front. The two birds then flew off in close pursuit.

Nests are constructed in the typical hummingbird manner and average 37.5 millimeters in outside diameter. There is usually a framework of fibers, stems, and other materials attached to the support by cobwebs and a lining of plant down or small feathers. The external surface is lined with a variety of items, including bark, paper, miscellaneous vegetable matter, and similar materials (Woods, 1927). Generally the nests are quite diverse in size, shape, and construction materials. The supporting vegetation is likewise highly variable; Bent (1940) listed 16 different types of trees used, as well as sage, yuccas, *Opuntia* and other cacti, and various weeds.

Nests have been found on vines clinging to rock faces, in citrus trees in open orchards, and in palm trees. The range of nest heights is also variable; Woods (1927) noted that it is usually less than 1 meter to more than 3 meters, but often about 1.5 meters. When a bush or small tree is selected, the nest is often at a height about halfway to the top of the tree; in larger trees the nest is usually on a small twig near the end of a projecting lower limb. Although the birds sometimes build nests near water, they typically nest at considerable distances from surface water. In favorable locations, several nests may be closely situated to one another; there is one instance of six nests within a 40-meter radius in a thicket of dead cockleburs (Bent, 1940). Stiles (1972b) noted that a favorite location is at a break in the chapparal, either along an edge or in particularly tall bushes, providing the female with a clear view of the nesting vicinity.

The first stages of nest construction usually require two or three days, after which a similar period may ensue before the first egg is deposited. Almost invariably two eggs are laid, normally at an interval of about two days. Incubation apparently begins with the first egg, since the young usually hatch a day or so apart. The observed incubation periods range from 15 to 18 days, probably averaging about 16 days (Woods, 1927).

At hatching the young are almost totally devoid of feathers, but by the sixth day pinfeathers begin to appear. The young are fed at roughly half-hour intervals and, according to Woods (1927) require from 20 to 23 days to fledge, based on eight observations. In Woods' experience no instances of successful raising of a second brood were known. Among a total of 29 nests both eggs hatched in 15 cases; in 4 other cases a single egg hatched, for a total hatching success rate of 58.6 percent. Both young fledged in 7 of these instances, and in 5 additional cases a single young was fledged. Thus the fledging success was approximately one-third (32.7 percent) of the total number of eggs that were laid, and the largest single mortality factor was destruction by cultivation or by wind.

Evolutionary and Ecological Relationships
Costa hummingbirds have usually been placed along with Anna in the genus *Calypte*, but various workers such as Mayr and Short (1970) have advocated the merger of *Calypte* with *Archilochus*. These authors believe that the Costa is a close relative of the superspecies *A. colubris*, which includes the ruby-throated and black-chinned hummingbirds. Wild hybridization has been reported with the broad-tailed and calliope hummingbirds (Banks and Johnson, 1961), and more recently with the black-chinned hummingbird (Short and Phillips, 1966).

Short and Phillips have commented on the similarity of the courtship dives of black-chinned and Costa hummingbirds, and the Costa at least at times performs a deep power dive similar to that of the black-chin. These authors implied that the two species are closely related, and they described several natural hybrids between them.

The center of the Costa hummingbird's breeding range is the Colorado Desert, where the ocotillo and chuparosa are among the most prevalent ornithophilous flowers, but several other flower species are commonly utilized, including desert willow and *Lycium* species. However, the Mojave Desert to the north is lacking in ornithophilous flowers, as well as in breeding hummingbirds (Grant and Grant, 1968).

I think that the nearest relative of the Costa is probably the Anna hummingbird; the two are now incompletely isolated from each other by the southern Sierra Nevada and the southern Coast Ranges of California, but they are effectively isolated ecologically and show considerable segregation in breeding seasons (Pitelka, 1951a). However, they are locally sympatric on the Palos Verdes Peninsula of California, where hybrids have been reported several times (Wells et al., 1978). Baltosser and Scott (1996) agree that the Costa and Anna hummingbirds are each other's nearest relatives, with a probable ancestral divergence in the southwestern United States or northwestern Mexico.

Breeding (hatched) and wintering (stippled) ranges, plus migration routes (broken hatching), of the black-chinned hummingbird. Dots represent states or provinces with extralimital sightings. The residential range of the Bahama woodstar is also shown (cross-hatching).

BAHAMA WOODSTAR

Calliphlox evelynae (Bourcier)
(*Nesophlox evelynae* of Ridgway, 1911)

Other Names
None in general English use; God bird (Bahamas).

Range
The Bahama Islands. Accidental in North America (southern Florida).

North American Subspecies
C. e. evelynae (Bourcier). The Bahama Islands except for the southernmost islands (Inagua, Caicos, and Turks).

Measurements
Wing, males 37–40.5 mm (ave. of 17, 38.9 mm), females 41.5–45.5 mm (ave. of 19, 43.3 mm). Exposed culmen, males 15.6–16.5 mm (ave. of 17, 15.9 mm), females 15.5–18 mm (ave. of 18, 16.6 mm) (Ridgway, 1911). Eggs, ave. 12 × 8 mm.

Weights
Two females in the U.S. National Museum weighed 2.2 and 2.6 g. No other weight data are available for the species.

Description (After Ridgway, 1911)
Adult male. Above rather dull metallic green or bronze-green, including middle pair of rectrices; remiges dark brownish slate or dusky, very faintly glossed with purplish; tail (except middle pair of rectrices) purplish black, the second and third rectrices with inner web cinnamon-rufous (except for a narrow space along shaft toward tip), the third (from outside) with basal portion of outer web (extensively) also cinnamon-rufous, the fourth with outer web (sometimes basal portion of inner web also) mostly cinnamon-rufous; a small postocular spot (sometimes a rictal spot also) of dull white; chin and throat brilliant metallic solferino purple, passing into violet or violet-blue posteriorly and laterally, the chin and anterior portion of throat decidedly reddish purple or

purplish red; chest white, passing into light buffy grayish posteriorly; rest of underparts cinnamon-rufous, paler medially, the sides and flanks glossed with metallic bronze or bronze-green; under tail-coverts cinnamon-rufous medially, passing into cinnamon-buff or white laterally; femoral tufts white; bill dull black; iris dark brown; feet grayish brown (in dried skins).

Adult female. Above as in the adult male, but slightly duller metallic bronze-green, especially on pileum, where sometimes dull grayish brown, at least on forehead; three outer rectrices (on each side) extensively light cinnamon-rufous basally, broadly tipped with pale cinnamon-rufous or vinaceous-rufous, and crossed by a very broad subterminal band of purplish black, the latter separated from the rufescent basal portion on outer web by more or less of metallic green; third rectrix (from outside) mostly metallic bronze-green, extensively black terminally, the concealed basal portion light cinnamon-rufous; chin and throat dull grayish white, sometimes tinged with pale cinnamon or cinnamon-buff; chest grayish white; rest of underparts cinnamon-rufous, paler medially, the under tail-coverts paler cinnamon-rufous or cinnamon-buff, sometimes indistinctly whitish along edges; femoral tufts white; bill etc., as in adult male.

Immature male. Similar to the adult female, but lateral rectrices relatively longer and narrower and with the black relatively more extended; in older individuals the throat has a variable number of metallic reddish purple feathers.

Immature female. Similar to the adult female, but feathers of upperparts (especially rump) indistinctly margined with rusty.

Identification
In the hand. Identified by the combination of a fairly short, nearly straight bill (culmen 15–18 mm) and a tail that is either deeply forked (central tail

feathers 13.5–17 mm) in males or rounded in females, with the central rectrices distinctly shorter (23–28.5 mm) than the outer ones. In both sexes the chest is white, becoming cinnamon-rufous posteriorly.

In the field. The strongly forked black and rufous tail of the male, together with the reddish to purplish gorget and crown, plus cinnamon underparts, identify this species. Females are much more difficult, but have a white throat and breast, with the rest of the underparts buffy and the tail buffy, with pale cinnamon tips and a blackish subterminal band. The birds are associated with scrubby woodlands, but in the Bahamas are also common around residential gardens.

Habitats
According to Brudenell-Bruce (1975), this species is common in the Bahamas wherever there are flowers, as in gardens, woods, copses, or open country.

Movements
The Bahama woodstar is apparently sedentary, to the extent that a distinct local race has developed in the Inagua group of islands. However, there are three North American records, warranting inclusion of the species in the current account. The first occurred in Lantana, Florida, between August 26 and October 13, 1971 (*American Birds* 26:52); the second was observed at Homestead, Florida, between April 7 and May 15, 1974 (*American Birds* 28:855); and the first actual specimen record was obtained in Miami, when a dead individual was found on January 31, 1961 (Owre, 1976).

Foraging Behavior and Floral Ecology
Little is known of the floral ecology of this species in the Bahamas, although one early observer commented (*Bonhote*, 1903, p. 292) on its preference for the flowers of sisal (*Agave sisalina*). In Florida it has foraged in a considerable variety of flowering plants including redbird cactus (*Euphorbia pedilanthus*), firecracker plant (*Russellia equisetiformis*), red pentas (*Pentas lanceolata*),

yellow elder (*Stenolobium stans*), and Cape honeysuckle (*Tecomaria capensis*) (*American Birds* 28:855).

Breeding Biology
Northrop (1891) provided an excellent early description of the male's display. The male hovered 15 to 20 centimeters in front of a perched female, with his tail and gorget fully exposed to her view. His wings were rapidly vibrating, and he swung quickly to-and-fro, from side to side, at the same time rising and falling, like a ball suspended on an elastic thread that stretches and contracts as the ball sways. Suddenly the male expanded his tail and threw himself almost violently from side to side, making a rustling sound and uttering a few sharp notes. He then darted toward the female, seemed almost to touch her, and as quickly darted away. This was probably a precopulatory display.

The nests, little more than 25 millimeters in diameter, are made of a soft, wooly material similar to thistledown, camouflaged on the outside with tiny pieces of bark. Many different sites are used, but they are usually in or on the fork of a twig of a bush or tree, from 1 to 4 meters above ground (Brudenell-Bruce, 1975).

Breeding may occur year-round, but Brudenell-Bruce indicated that the main nesting season is in April. Miller (1978) observed nests with eggs or young in December on San Salvador Island of the Watling group; one of them contained three young. James Bond (in litt.) has found nests on Marguana in October, on Providencia in December, on Grand Turk from January through March, on Little Inagua in April, and has observed nest-building on Great Inagua in May.

There is no information on incubation periods, nestling development, or fledging periods.

Evolutionary and Ecological Relationships
The genus *Nesophlox* was erected by Robert Ridgway in 1910, and was envisioned by him to include three species: *evelynae*, *lyrura* (now considered conspecific with *evelynae*), and *bryantae*. He considered this genus most closely related to *Calliphlox*, but distinguishable from it on the basis

of tail and primary shapes. Later, Todd (1942) included *evelynae* within the expanded genus *Calliphlox*. I think there should be a merger of at least these genera, as well as *Doricha* and *Philodice* with *Calothorax*.

No hybrids involving this species are known. Its probable nearest relative is the magenta-throated woodstar, and it is probably slightly less closely related to the purple-throated woodstar.

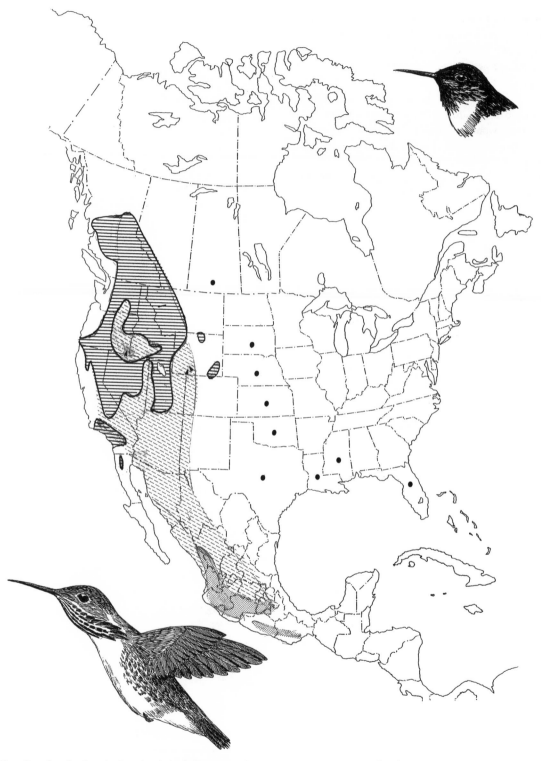

Breeding (hatched) and wintering (stippled) ranges, plus migration routes (broken hatching),
of the calliope hummingbird. Dots represent states or provinces with extralimital sightings.

CALLIOPE HUMMINGBIRD

Archilochus calliope (Gould)
(*Stellula calliope* of AOU, 1983)

Other Names
None in general English use; Chupamirto rafaguitas (Spanish).

Range
Breeds in the mountains of western North America from central British Columbia and southwestern Alberta; south through Washington, Oregon, Nevada, and California; to northern Baja California; east to Utah and western Colorado; locally to eastern Wyoming. Winters in Mexico.

North American Subspecies
A. c. calliope (Gould). Range in North America as indicated above, except for Guerrero, Mexico.

S. c. lowei (Griscom). Endemic to the mountains of Guerrero.

Measurements
Wing, males 37–40 mm (ave. of 10, 38.7 mm), females 40–44 mm (ave. of 10, 42.8 mm). Exposed culmen, males 13.5–15 mm (ave. of 10, 14.3 mm), females 15–16 mm (ave. of 10, 15.6 mm) (Ridgway, 1911). Eggs, ave. 12.1 × 8.3 mm (extremes 10.7–13.0 × 7.4–9.6 mm).

Weights
The average of 46 males was 2.50 g (range 1.9–3.2 g); that of 26 females was 2.83 g (range 2.2–3.2 g) (from various sources, including specimens in the Museum of Vertebrate Zoology). The average of six spring migrant males was 2.6 g (range 2.4–2.9, SD 0.17), whereas 33 breeding males averaged 2.71 g (range 2.3–3.4) and 23 females averaged 2.85 g (range 2.6–3.2) (Calder and Calder, 1994).

Description (After Ridgway, 1911)
Adult male. Above metallic bronze-green, usually rather duller on forehead; middle pair of rectrices subspatulate or with subterminal portion contracted, dull purplish black or dusky, edged basally (more or less distinctly) with cinnamon-rufous, and broadly tipped with dull brownish gray; remaining rectrices similar in coloration, but cinnamomeous basal edgings less distinct (sometimes obsolete) and grayish tip less distinct; remiges dull brownish slate or dusky, very faintly glossed with purplish; feathers of gorget narrow and distinctly outlined, much elongated posteriorly, pure white basally and metallic reddish purple (solferino) terminally, the basal white much exposed, especially on chin and upper throat; foreneck and chest white, or grayish white; rest of underparts more grayish, strongly tinged or suffused with cinnamon-buff laterally, the sides glossed, or overlaid, with metallic bronze-green, the under tail-coverts white, tinged with cinnamon-buff toward anal region; maxilla dull black or dusky; mandible dusky terminally, more flesh-color basally; iris dark brown; feet dusky.

Adult female. Above metallic bronze-green or greenish bronze (usually much more bronzy than in adult male); remiges brownish slate or dusky, very faintly glossed with purplish; middle pair of rectrices dull bronze-green or greenish bronze, sometimes with terminal portion (more or less extensively) purplish dusky or blackish; next pair dull bronze-green with terminal third (more or less) black; the subbasal portion edged (on both webs), more or less distinctly, with cinnamon-buff; next pair similar but with the black relatively more extended and with an apical spot (usually small and wedge-shaped) of white; next similar but white apical spot larger and basal half mostly brownish gray; outermost rectrix like the last but only about the basal third grayish and white apical spot still larger; auricular region light brownish gray; a dusky triangular space in front of eye; chin and throat dull brownish white, usually more or less streaked or flecked with dusky or bronzy brownish; chest pale grayish cinnamon-buff or dull whitish, the median portion of breast and abdomen similar; sides and flanks cinnamon or deep cinnamon-buff; the under tail-coverts similar but paler; femoral tufts and tuft on each side of rump white; bill, etc., as in adult male.

Immature male. Juvenile and subadult males differ from females of the same age in lacking any rufous at the bases of their middle rectrices, and young birds of both sexes have wrinkled upper mandibles, a trait that may persist into early winter. Young birds can be distinguished reliably from all other native hummingbirds and the bumblebee hummingbird by their spatulate middle rectrices (Baltosser, 1994; Calder and Calder, 1994).

Immature female. Similar to the adult female but general color of upperparts more decidedly bronzy, with feathers very narrowly and indistinctly margined terminally with dull brownish or grayish buffy.

Identification

In the hand. Unusually narrow and white-based feathers of the male's gorget are unique among North American hummingbirds; also, in both sexes the inner rectrices are expanded near the tips, rather than being gradually tapering. The very small size (wings 37–44 mm) of these birds is also a helpful criterion.

In the field. Usually found in mountain meadows near coniferous forests. The very small size of both sexes and the red-and-white striped gorget of the males are unique. When feeding the male sometimes utters a soft *tsip*, and also utters very chippy and squeaking noises; the most distinctive sound is a muffled *bzzt* or *pfft* produced near the bottom of each display swoop, which is a shallow U-shaped course. A high-pitched *see-ree* note is produced by the male as it approaches a female. Females closely resemble females of rufous and Allen hummingbirds, but are paler brown on the flanks, and the cinnamon of the tail feathers does not extend to the central feathers. The chip note of the calliope is very high pitched and similar to that of the Costa, but it is relatively silent under most conditions. Females are distinctively small and short-billed, usually with a pale rusty wash on the sides, and rather little rufous in the tail. When hovering the birds hold the tail very still and unusually high above the plane of the body (Stiles, 1971).

Habitats

In California this species breeds in mixed brushland and forest, either of deciduous or coniferous trees or intermingled, in canyon bottoms, valley floors, and forest glades. Males are typically found in open areas, whereas females nest mostly in woods (Grinnell and Miller, 1944). Coniferous forest edges adjacent to meadow areas are also favored breeding locations. In the Grand Teton National Park area, territorial males frequented areas of low willows near grass- and sage-dominated meadows rich in scarlet gilia (*Ipomopsis*) and Indian paintbrush; females nested in nearby groves of lodgepole pines (personal observations). In Canada the birds are associated with openings in woodlands, glades, burntlands, and flowering meadows, and typically nest in open woodlands (Godfrey, 1986).

The species occupies a remarkably broad vertical range during the breeding season. In Washington it has been found nesting as low as 180 meters above sea level in the Columbia River valley; in California it is usually not found below 1200 meters and more commonly nests above 2400 meters. In the Sierra Nevada it often nests nearly to timberline, between 3000 and 3500 meters, and follows the advancing summer season to the limit of flowers. On the other hand, near the northern limit of its range in Montana, it probably nests at elevations of less than 630 meters in the Kootenai Valley to perhaps as high as 2100 meters (Bent, 1940).

On migration, calliopes use a wider array of habitats. In Texas, where it is very rare, the birds are usually found along streams on mesas or in broad valleys where fringes of timber or tall thickets occur (Oberholser, 1974). In New Mexico it occurs from about 1500 to 3400 meters' elevation during the fall migration and is probably most numerous on the high mountain meadows (Bailey, 1928).

Movements

During the winter months this species is largely confined to central Mexico, particularly the states of Michoacan, Mexico, and Guerrero. The species is rare east of Michoacan and Guerrero, and its migratory route is oval-like, similar to (but narrower than) that of the Allen and rufous hummingbirds. In the spring, the species migrates up the western portions of Mexico, arriving in southern California in early March. It apparently is limited in Arizona primarily to the southwestern and central deserts, and skirts most of Arizona, New Mexico, and Utah (Phillips, 1975).

The spring migration apparently passes slowly through California, with the birds arriving in Oregon during early May, in Washington from late April to mid-May, and in central British Columbia at about the same time. As with the rufous hummingbird, Idaho and Montana are reached last, usually not before the middle or latter part of May (Bent, 1940).

The fall migration likewise begins early, with adult males preceding females and immatures by a week or more. The birds are rare but regular in Colorado during late July and August; they sometimes summer in the state, although it has not been proved that they nest there (Bailey and Niedrach, 1965). Late dates of departure are in late August for Washington, British Columbia, and Idaho; early to mid-September for Montana; and early September for Nevada. Similarly, late California dates are for early to mid-September (Bent, 1940). In northern Arizona the fall records are from July 14 to September 29 (Phillips et al., 1964), and the extreme dates in Texas are from July 25 to September 14 (Oberholser, 1974). There are no winter records for the United States, and probably the normal wintering limits are south of the central Mexican transverse volcanic belt (Phillips, 1975). Calliopes seem to wander from their normal migratory routes less than many of the other western hummingbirds, and there are few extralimital records. They sometimes wander out to the foothills and plains of Colorado during late summer (*American Birds* 32:1192), and they have also reached western South Dakota, western Nebraska, and Kansas, as well as extreme eastern New Mexico (Clayton) and adjacent western Texas. Recently, stray birds have been seen in Florida, Mississippi, Louisiana, Oklahoma, and British Columbia.

Foraging Behavior and Floral Ecology

This is the smallest of the hummingbirds that regularly occur in the United States, and as such is probably at a disadvantage in competing with the larger species. Migrant calliope hummingbirds during late May and June in Nevada sometimes have territorial encounters with the larger broad-tailed hummingbirds; yet, in the cases observed,

the calliopes were the victors when broad-tails entered their feeding territories. A male has also been reported diving at male black-chinned hummingbirds, and was defending a larger territory than either rufous or broad-tailed hummingbirds that were in the vicinity (Bailey and Niedrach, 1965).

In most areas of its range, the calliope breeds in company with various other western species, but tends to occur at higher elevations during summer than the others. Grant and Grant (1968) listed seven species of hummingbird-pollinated flowers (three of *Castilleja*, two of *Penstemon, Aquilegia formosa*, and *Ipomopsis aggregata*) that commonly occur in the higher Sierra Nevada and are regularly visited by this species during the breeding season. The birds forage almost to timberline on *Penstemon newberryi*, and occasionally have been seen above timberline (at 3300 m) foraging on an introgressive population of *P. menziesii davidsonii*.

According to Grinnell and Miller (1944), favorite flowers include gooseberries and currants (*Ribes*), manzanitas (*Arctostaphylos*), paintbrush, and penstemons. Bent (1940) indicated that they prefer red columbine and scarlet Indian paintbrush, but also use the yellow monkey flower (*Mimulus implexus*), as well as a lousewort (*Pedicularis semibarbata*) and a snow plant (*Sarcodes sanguinea*). Calliopes are also adept at hawking insects in flight, perhaps most often dipterans, hymenopterans, and coleopterans, but probably including almost any small insect (Bent, 1940).

In contrast to some of the larger hummingbird species, the calliope tends both to perch and to forage rather close to the ground; when foraging with larger species it feeds on flowers near the bottom of the plant, whereas other hummers more often forage on the topmost flowers (Bailey and Neidrach, 1965).

Breeding Biology

Males probably establish territories as soon as they arrive on their breeding grounds. No specific information on territory sizes is available, but there in one instance four males maintained territories within a 23,200-square-meter plot (Bent, 1940). By the first day of Calder's (1971) study on June 13 in the Grand Teton area, the males were already actively engaged in territorial disputes; the territoriality continued until late July, when hatching occurred.

Bent (1940) summarized descriptions of the aerial display of the male. Like the other species, the display consists of a series of swooping flights along a U-shaped course that may be 7 to 9 meters across and begin from a height of 9 to 18 meters. It is perhaps a more shallow course than in some of the other species, and the accompanying sound is a mechanical buzzing, sounding like a *bzzt*. Evidently the wings are able to generate a more explosive and repetitive metallic *tzing* sound as well, which was produced by a male as it was hovering near a perched female (Wyman, 1920).

Observations by Tamm et al. (1989) indicate that dive displays used toward territorial intruders of both sexes. However, intruding males were typically vigorously chased out of the territory, whereas females seemingly were "followed" by resident males in a less vigorous and aggressive manner than when chasing other males. "Hover displays" consisted of the male descending in a series of discontinous alternations of sudden vertical drops and stationary hovering or slow descent, as the male often turned from side to side while facing the object of the display, usually a female. Whenever a female landed within a male's territory, the male would hover in front of and slightly above her, producing a loud buzzing sound (a so-called "buzzing bout"). At times the female would leave her perch and the two would spin about in an aerial "circle dance," the male maintaining his same relative position in front of and above the female. Displays that directly preceded a completed copulation included 19 dives, 4 hover-displays, and 6 buzzing bouts. Copulation occurred in a bush less than a meter from the resident male's main perching site that day.

In one observation of copulation, the female was perched on a dead weed when a male shot past her close to the ground, flew about 7 meters upward along a hillside, then turned and darted back down in a narrow ellipse. On reaching the female he alighted on her and immediately copulated (Wyman, 1920).

Calliope nests are typically located below a larger branch or canopy of foliage, usually on a small branch that has small knots or cones on it. Frequently the tree is a conifer, and the nest is placed among a cluster of old cones in such a way as to resemble one of them extremely closely. When placed in aspens, the nests may mimic mistletoe knots. In height, they may range from 50 centimeters above the ground to as high as 21 meters. Of nine nests in the San Bernardino Mountains of southern California, six were in yellow (ponderosa) or Jeffrey pine, two were in silver fir, and one was in an alder (Bent, 1940). However, in Montana nests are commonly placed in Engelmann spruce, western hemlock, and arborvitae, or in alpine fir at higher elevations. Of 21 nests observed by Weydemeyer (1927), 9 were in Englemann spruce, 6 in arborvitae, 3 in western hemlock, 2 in alpine fir, and 1 in Douglas fir. The nests ranged from 1 to 3 meters above the substrate (averaging a little more than 2 m). Montana nests are often placed above a creek bank or along the edge of the forest, and invariably where there is a low branch hanging free of the rest of the foliage and providing a clear view in all directions (Weydemeyer, 1927).

Calder (1971) reported that in the Grand Teton area the nest is always located under a protective branch. Thus, it does not face the heat-sink of the cold nocturnal sky, and the overhead cover also provides protection from predators and precipitation. In this area the birds often nest at the eastern edge of wooded areas, in sites where the sunlight strikes the nest immediately as it clears the eastern horizon, and thus begins to warm the nest and incubating female.

Like some other hummingbirds, female calliopes often use the same nest site in subsequent years. Series of two-, three-, and even four-story nests have thus been reported, suggesting that at least a few individuals of this species may live as long as five years. Nests of the species are typically fairly small (about 31–44 mm outside diameter), and invariably closely blended to the appearance of the nest substrate, with green to brownish lichens. When built among pine cones the outer surface is covered with bits of tree bark and small cone shreds, and usually rests against the side of a cone or is saddled between two cones. The shell of the nest is made of mosses, needles, bark, small leaves, and various miscellaneous materials, bound and covered with spider webbing and associated camouflage. The nest lining is composed of plant down, frequently the silk of willows or cotton-woods. Second-year additions to a nest are composed mainly of such materials, but at times the rim of the nest will also be heightened (Bent, 1940).

There is no good information on the timing of nest construction, but it is probably fairly rapid. In an Alberta nest, the first egg was deposited well before the sides of the nest were fully completed, and the second egg was deposited 3 days later. Nest construction continued well into the incubation period, which lasted 16 days. Calder (1971) reported a 15-day incubation period in the Grand Teton area, and in two nests noted that the number of foraging trips during incubation ranged from 87 to 111 times per day (about 3.6 hours per day). There was no relationship between duration or frequency of foraging flights and environmental temperatures, but the females did not feed the young after their last trip of the evening. Evidently this food was used for maintaining the body metabolism for the 8- to 9-hour period of nocturnal fasting, when the environmental temperature sometimes dropped nearly to freezing. Minimum night-time body temperatures were from 19.9° to 29.8° C above the corresponding minimum air temperatures, confirming the general view that incubating females do not become torpid at night.

The number of foraging trips after hatching was similar to that prior, but of longer duration; the total hours of absence per day (5–11) was also considerably longer. By the time the young were 8 days old they had attained a midday homeo-thermal condition, and brooding was discontinued after 11 to 12 days (Calder, 1971). In Alberta observations, the nesting period lasted 18 and 21 days in two different years, compared with a 21- to 23-day nesting period observed by Calder in Wyoming. The total nesting period (egg-laying to fledging) is about 37 days in British Columbia, and 34 to 38 days for western North America in general. In British Columbia the egg dates range from mid-May to mid-July, with a peak in mid-June

(Brunton et al., 1979). Likewise, in California the egg dates are from late May to July 30, with a peak between June 10 and June 28, suggesting a very similar nesting cycle there. On the other hand, a few Utah egg records are entirely for July (Bent, 1940).

There is no information on possible second-brooding in this species, nor on relative reproductive success. The longest known lifespans are for two females that each were recaptured after six years; one male has been recaptured five years after banding (Calder and Calder, 1994).

Evolutionary and Ecological Relationships
Mayr and Short (1970) suggested that this species is a close relative of, and probably should be considered congeneric with, the forms currently separated in three other genera (*Selasphorus*, *Calypte*, and *Archilochus*). They did not suggest a possible nearest living relative. However, wild hybrids with the Costa and Anna hummingbirds have been reliably reported (Banks and Johnson, 1961), and one collected specimen (now lost) was a probable hybrid with the rufous hummingbird (Banks and Johnson, 1961). Thus, there are intergeneric hybrids with both *Selasphorus* and *Archilochus* ("*Calypte*"); the coloration and shape of the female's tail put the species somewhat intermediate between these two very poorly defined genera.

This species has a somewhat shorter bill than the other common western hummingbirds, and as a result is perhaps less able to reach the nectar of deeply tubular flowers. However, as Grant and Grant (1968) have noted, the very long tongue of hummingbirds helps to complement actual bill length, and the correspondence between bill length and corolla tube length is only a general one. On its breeding grounds, the calliope hummingbird often seems to be associated with *Ipomopsis aggregata*, which has a fairly short corolla tube, and with various species of *Castilleja*.

BUMBLEBEE HUMMINGBIRD

Selasphorus heloisa (Lesson and DeLattre)
(*Atthis heloisa* of AOU, 1983)

Other Names
Heloise's hummingbird, Morcom's hummingbird; Chupamirto garganta violada (Spanish).

Range
Breeds in the mountains of Mexico from south-western Chihuahua and southeastern Sinaloa, Nuevo Leon, and Tamaulipas, south to Oaxaca and Veracruz (Friedmann et al., 1950). The probably conspecific form *S. ellioti* (the wine-throated hummingbird) occurs from Chiapas south to Honduras.

North American Subspecies
S. h. morcomi (Ridgway). Breeds in the Sierra Madre Occidental of Mexico from southern Chihuahua southward (*Anales del Instituto de Biología, Univ. Mex.*, 32:338–9). Accidental in Arizona.

Measurements
Wing, males 32.5–38 mm (ave. of 9, 34.6 mm), females 35.5–38 mm (ave. of 8, 36 mm). Exposed culmen, males 11.5–13 mm (ave. of 9, 12.1 mm), females 11.5–13 mm (ave. of 8, 12.4 mm) (Ridgway, 1911). Eggs, no information.

Weights
The average of 11 males was 2.13 g (range 1.95–2.7 g); that of 4 females was 2.33 g (range 2.2–2.5 g) (Delaware Museum of Natural History).

Description (After Ridgway, 1911)
Adult male. Above metallic bronze-green, greenish bronze, or golden bronze (sometimes tinged with copper-bronze on back); middle pair of rectrices metallic bronze-green or greenish bronze (some-times dusky at tip), both webs edged for basal half or more with cinnamon-rufous; next pair of

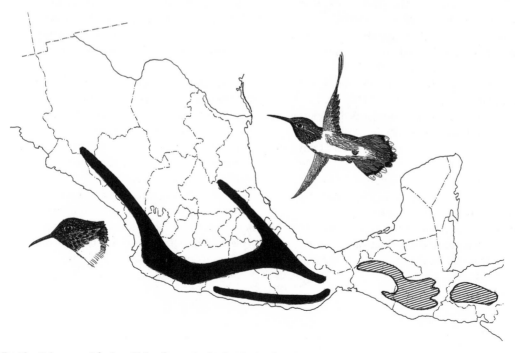

Residential range of the bumblebee hummingbird (inked). The allopatric range of the extralimital wine-throated hummingbird (hatched) is also shown. *(Adapted from Howell and Webb, 1995)*

rectrices with basal half or more cinnamon-rufous, the terminal portion black, this usually separated from the cinnamon-rufous by a space of bronze-green or greenish bronze; other rectrices with basal half or more cinnamon-rufous, the subterminal portion extensively black (usually with more or less of bronze-green or greenish bronze between the black and the cinnamon-rufous portion), the tip broadly white; remiges brownish slate color or dusky, very faintly glossed with violaceous; sides of head brownish gray or grayish brown, passing into dull white (except in *S. ellioti*) on anterior portion of malar region and on postocular region; chin and throat brilliant metallic magenta purple, changing to bluish purple and even, partly, to greenish blue (but generally more reddish in *S. ellioti*), the more posterior feathers of the throat much elongated, especially laterally; chest, sides of neck, breast (medially), abdomen, and under tail-coverts dull white or grayish white; sides and flanks light cinnamon-rufous, overlaid more or less extensively by metallic bronze or bronze-green; bill dull blackish; iris dark brown; feet dusky.

Adult female. Above similar to the adult male, but tail with relatively much more black and less cinnamon-rufous, the latter also duller, especially on lateral rectrices, the middle pair of rectrices without cinnamon-rufous edgings, the white tips also less purely white, those of inner rectrices sometimes cinnamomeous; chin and throat white, conspicuously spotted with metallic bronze; rest of underparts as in the adult male, but sides and flanks more extensively and uniformly cinnamon-rufous; and under tail-coverts more or less strongly tinged with the same.

Immature male. Similar to the adult female, but tail, sides, and flanks as in adult male (the middle pair of rectrices, however, wholly bronze-green or greenish bronze).

Identification

In the hand. Extremely small hummingbird (wing 32–38 mm) with very short bill (exposed culmen 11–13 mm), smaller than any other North American species. The male has an elongated reddish gorget, and both sexes have a rounded tail with cinnamon at the base. The outermost primary is narrower and more or less attenuated (but not

sharply pointed) at the tip. The individual rectrices are normal in shape, and rounded at the tips.

In the field. Usually found in pines, pine-oak woodlands, or cloud forests (in Mexico). Its very small size helps to identify it; it is similar to the calliope hummingbird, but the male's gorget is not streaked with white, and the tail is rounded, with white tips. Females also closely resemble those of the calliope, but tend to have buff-tipped tails. In flight, the wings make little sound, and the movements are relatively sluggish. Males gather in loose groups to sing within earshot of one another; the song lasts 30 to 40 seconds and, although weak, is said to be beautiful. It has been described as a shrill whistle, changing rapidly in pitch, and the closely related (probably conspecific) wine-throated hummingbird of Guatemala and Honduras is said to utter a "sweetly varied outpouring" of sounds lasting for most of a minute (Skutch, 1973).

Habitats

In Mexico this species occurs mainly in the transition zone between 1500 and 2100 meters of the Sierra Occidental, and to as high as 2800 meters in the western portion of the State of Mexico (Friedmann et al., 1950). Its habitats include cloud forest edges, pines, and open pine-oak woodlands. Farther south, *S. ellioti* occurs in similar scrubby and open woodland habitats from the highlands of Chiapas southward to Honduras. In Guatemala it ranges from 900 to 2600 meters' elevation in open pine and oak woodlands, and also in bushy areas (Land, 1970). In El Salvador it has been found in low scrub at 2150 meters in the crater of the Volcan de Santa Ana (Dickey and van Rossem, 1938). In Honduras it occurs in and around cloud forests from 1500 meters upwards, especially in scrubby areas outside the forests that are rich in flowers (Monroe, 1968).

Movements

Bumblebee hummingbirds are probably fairly sedentary throughout most of their range, but perhaps the northwestern race *S. morcomi* is subject to some seasonal movements. This might account for the only records for this species in the United States. Two female specimens were obtained in

Ramsey Canyon of the Huachuca Mountains in Arizona on July 2, 1896; otherwise the species has not been collected within several hundred kilometers of U.S. borders (Phillips et al., 1964). One other presumed early Texas specimen later proved to be a young calliope hummingbird (*Auk* 8:115).

Foraging Behavior and Floral Ecology

Little is known on this subject. Moore (1939a) described the species as one of several foraging on a large shrub in southeastern Sinaloa bearing thousands of grayish-lavender blooms. It also was observed foraging on a maroon-colored species of legume, and on some tawny flowers of a huge *Opuntia* cactus (Bent, 1940). Schaldach (1963) observed foraging on mints and sages (*Salvia*), and Wagner (1946b) reported foraging on *Erythrina americana*.

In a study area in Oaxaca, Lyon (1976) found that bumblebee hummingbirds were nonterritorial, but their small size enabled them to survive on the blossoms of unprotected plants located outside the territories of the larger species of hummingbirds found in the same area. Furthermore, they were able to utilize some bee-pollinated species of flowers (*Penstemon perfoliatus* and *P. gentianoides*) that produce only small amounts of nectar and thus are not exploited by the larger hummingbird. They also used another small species (*Cuphea jorullensis*) that otherwise was exploited only by white-eared hummingbirds. Furthermore, their curious bee-like flight apparently allowed them to forage for long periods in stands of a favored food plant (*Rigidella orthantha*) within white-ear territories without being evicted.

Breeding Biology

Although the behavior of the northern populations remains essentially unknown, A. F. Skutch (in Bent, 1940) has described finding singing assemblies of males of *S. ellioti* in Guatemala between 1800 and 3300 meters' elevation. He noted that these are not common anywhere, but where a single male is heard singing, one or more others are likely to be within earshot. In mid-October, Skutch found one such assembly on a steep brushy slope at about 2700 meters. Each male had a singing site on the exposed twig of a bush or low tree, and was sepa-

rated from the others by 22 to 27 meters. No others were found within 1.6 kilometers of these four, and the birds seemed to be oriented in a line, with those at the extreme ends apparently out of hearing range of one another. The group was located near a highway, but Skutch did not comment on the distribution of food plants in the area.

Each male would sing without pause for 30 to 40 seconds, with a weak but melodious voice that had rising and falling cadences similar to those of a small finch such as a seedeater (*Sporophila* species). As the bird sang it spread its gorget to form a shield-like appearance, and turned its head from side to side, shifting the apparent coloration of the gorget from magenta to velvety black. At times the bird would also vibrate its wings, or suspend itself in midair, while still singing, or it would make a long, looping flight, eventually returning to the perch from which it began, singing the entire way.

The only record of nesting is for the form *S. ellioti*. Baepler (1962) reported that this species was fairly common after late July in scrub oak thickets between 2100 and 2400 meters' elevation in the vicinity of Huehuetanango, Guatemala. On August 12, a nest containing two nestlings was located in a slender oak, about 1 meter above the ground, near the end of a branch. No description of the nest or the eggs was provided.

Evolutionary and Ecological Relationships

The southern form *S. ellioti* is probably no more than subspecifically distinct from *S. heloisa*, as Ridgway (1911) has treated it, but recent tradition has been to maintain the forms as two separate species (without any good reason).

Equally questionable is the practice of maintaining a separate genus (*Atthis*) for these populations. Ridgway (1911) considered that the birds were related to *Selasphorus*, but could be distinguished on the basis of the form and coloration of the tail in the adult male. If one follows the criterion of Brodkorb (in Blair et al., 1968) that a generic trait must not be limited to the condition typical of a single sex, then merger with *Selasphorus* seems reasonable. This had been done by Phillips et al. (1964), although Brodkorb retained the genus, apparently on the basis of relative bill length and tail shape. Brodkorb placed the genus adjacent to *Archilochus* rather than *Selasphorus*. I believe that, together with "*Stellula*," "*Atthis*" occupies an intermediate position between the extremely specialized (in primary and rectrix condition) types of *Archilochus* and *Selasphorus,* and their intermediate state is an argument favoring the merger of all these species into a single large genus.

Unfortunately, too little is known of the ecology of this species to comment on its interspecific relationships with other hummingbirds or with plants, but it seems to occupy a similar ecological niche to that of the black-chinned hummingbird in the United States. The unusual vocal abilities of the male should also receive some attention. Perhaps the relative absence of specialized primaries (the outermost primary of males of the northern population is slightly attenuated) or tail feathers of males has been a source of selection for more elaborate vocalizations.

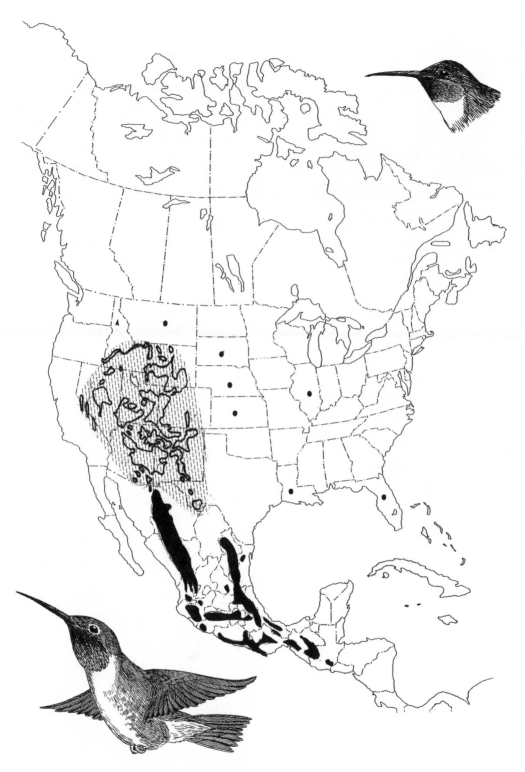

Breeding range (broken hatching, with denser populations outlined with ink and an extra-limital breeding site represented by an arrowhead) and residential range (inked) of the broad-tailed hummingbird. Dots represent states or provinces with extralimital sightings. *(Adapted from Calder and Calder, 1992)*

BROAD-TAILED HUMMINGBIRD

Selasphorus platycercus (Swainson)

Other Names
None in general English use; Chupamirto cola ancha (Spanish).

Range
Breeds in western North America from east-central California, northern Nevada, northern Wyoming, eastern Colorado, New Mexico, and southwestern Texas to southern Mexico and the highlands of Guatemala. Resident from northern Mexico south to Guatemala.

North American Subspecies (After Friedmann et al., 1950)
S. p. platycercus. (Swainson). Breeds in the mountains of western North America from eastern California and Wyoming south through the mountains of Mexico, mainly from 2100 to 3600 meters' elevation, and southward at least to Oaxaca.

Measurements
Wing, males 46.5–50.5 mm (ave. of 19, 48.4 mm), females 45.5–52 mm (ave. of 16, 49.9 mm). Exposed culmen, males 16–19 mm (ave. of 19, 17.6 mm), females 17–20 mm (ave. of 16, 18.6 mm) (Ridgway, 1911). Eggs, ave. 13 × 8.8 mm (extremes 11.9–14.5 × 7.9–10 mm).

Weights
The average of 35 males was 3.16 g (range 2.5–4.1 g); that of 25 females was 3.60 g (range 2.6–3.8 g) (from various sources, including specimens in Museum of Vertebrate Zoology). Sixty-eight territorial males averaged 3.20 g (*SD* 0.188) and 65 females averaged 3.67 g (*SD* 0.237) (Calder and Calder, 1992).

Description (After Ridgway, 1911)
Adult male. Above metallic bronze-green or greenish bronze; middle pair of rectrices metallic bronze-green (sometimes more bluish green); rest of tail dull purplish or bronzy black; the pair next to middle pair usually glossed, more or less, with bronzy green (sometimes mostly of this color), the outer web edged (except terminally) with cinnamon-rufous; the next pair sometimes also narrowly edged with the same; remiges dark brownish slate or dusky, faintly glossed with purplish; chin and throat bright metallic reddish purple (usually more reddish than solferino), the feathers crossed by a broad subterminal (concealed) bar of dull white, their basal portion dusky gray; chest grayish white, passing into very pale gray on breast and abdomen; sides and flanks darker grayish, tinged (especially the flanks) with pale cinnamon, the sides and sides of breast overlaid by metallic bronze-green; femoral tufts white; under tail-coverts white with a central area of pale cinnamon or cinnamon-buff (sometimes partly bronze-green); bill dull black; iris dark brown; feet dusky.

Adult female. Above metallic bronze-green, including middle pair of rectrices, the latter sometimes blackish or dusky terminally; second rectrix (from middle) metallic bronze-green with terminal portion more or less extensively dusky; three outer rectrices, on each side, broadly tipped with white, cinnamon-rufous basally (more or less extensively), the remaining portion purplish or bronzy black with more or less of metallic bronze-green between the blackish subterminal and cinnamon-rufous basal areas; chin and throat dull white, the feathers with small mesial streaks or guttate spots of dusky or dusky bronze; chest dull brownish white or buffy grayish white; the breast and abdomen similar but, usually, more decidedly tinged with buffy; sides and flanks light cinnamon or pale cinnamon-rufous; under tail-coverts pale cinnamon or cinnamon-buff (sometimes partly grayish) centrally, broadly margined with white; femoral tufts white; bill, etc., as in adult male.

Immature male. Similar to the adult female but feathers of upperparts (especially rump and upper tail-coverts) indistinctly margined terminally with pale brownish buff or cinnamon, and lateral rectrices with much less of cinnamomeous on basal portion.

Immature female. Similar to the young male but rectrices as in adult female.

Identification

In the hand. Has a blackish bill about one-third the length of the wing, which in males has the outermost primaries sharply pointed and the outermost slightly excurved. The wing is somewhat longer than in other *Selasphorus* species (minimum 45 mm), and the tail is generally rounded, with the individual feathers fairly wide and pointed (males) or wedge-shaped (females). Cinnamon color on the rectrices is limited to the edges of some in the inner rectrices (in males) or to the basal portions of the outer three rectrices (in females).

In the field. Usually found in mountain meadows associated with coniferous forests. Males are identified by their red throats and the loud buzzing noise made by the wings in flight. The only other sounds are occasional tinny *chip* notes. Where ruby-throated hummingbirds might also occur (as in Texas), the rounded rather than forked tail of the males will help to identify them, and the absence of cinnamon from the tail in males separates them from other *Selasphorus* species. Females closely resemble the other species of *Selasphorus* hummingbirds, and in general are almost impossible to identify in the field. They sometimes may be separated from the Anna, Costa, and black-chinned hummingbirds by their cinnamon at the base of the tail, but are not separable from female Allen or rufous hummingbirds.

There is a shallow 9- to 12-meter dive display by the male, and it produces a musical buzzing sound continously while in flight. The sound is very high pitched and is louder and shriller than most other species, but females produce only the normal humming sound in flight. The chip-note is very similar to that of the Allen and rufous hummingbirds, but slightly higher in pitch. The large size of the female, and its unusually large tail, with extensive black present, and the lack of rufous tinge below, is quite useful in recognition. The throat pattern of adult females consists of faint, dusky streaking or speckling, and young males often have a few magenta feathers (Stiles, 1971).

Habitats

In California, the typical breeding habitat of this species is the woodland belt of pinyon pine, juniper, and mountain mahogany, especially where these trees are in their usual open stands, with interspersed xerophilous shrubbery. Proximity to thickets is particularly favored, such as of willow or silk-tassel bush (*Garrya*) associated with wet or dry stream courses (Grinnell and Miller, 1944). At the eastern edge of its breeding range in Texas, the species favors pine-oak woodlands and scrubby junipers on slopes or in canyons and gulches (Oberholser, 1974).

In Colorado the species is most abundant between 2100 and 2550 meters' elevation in the wooded transition zone. There it often nests along moist canyons in aspens, Douglas firs, or ponderosa pines (Bailey and Niedrach, 1965). After the breeding season the birds move upward at least to timberline and often into the alpine meadows, which then are strewn with flowers. During fall migration the species also ranges well out into open country in the valleys and foothills (Oberholser, 1974).

Movements

This is a highly migratory species, which winters primarily in the highlands of Mexico. In Texas it is scarce to rare in Big Bend National Park from late September to middle or late March; otherwise it is extremely rare during winter. Spring migration there occurs mainly during March and April, with breeding beginning very shortly after arrival (Oberholser, 1974). Likewise in Arizona the species occurs in the southern mountains from the end of February or early March, whereas in the northern part of the state the birds arrive in early April (Phillips et al., 1964). Early arrival dates for New Mexico and Colorado are for April; those for Wyoming and Utah are for May (Bent, 1940). Males typically arrive in advance of females during spring migration.

Fall departures in the northern part of the breeding range are during September in Wyoming and Colorado (Bent, 1940); the latest reported Colorado date is October 24 (Bailey and Niedrach, 1965). In Arizona the birds have been reported as late as the end of September, or perhaps early October, whereas in southern New Mexico (Carlsbad) they have been reported as late as early October. Most of the fall migration in Texas is from early August to middle or late October, with rare observations into November (Oberholser, 1974).

As might be expected with a highly migratory species, extralimital occurrences of broad-tailed hummingbirds are fairly numerous. They have been reported northwest of their breeding range to Washington (*Audubon Field Notes* 15:428) and Oregon (*Audubon Field Notes* 22:569; 23:617), and have reached east as far as Texas (*American Birds* 25:602, 30:864, 31:1021), Arkansas (*American Birds* 33:187), and Louisiana (*American Birds* 33:187).

Foraging Behavior and Floral Ecology
This species forages on a great variety of plants over its wide range, but in the San Francisco Peaks area of Arizona some of the important montane flowers on which it forages during the breeding season and in early fall include *Penstemon barbatus*, *Ipomopsis aggregata*, and *Castilleja miniata* (Grant and Grant, 1968).

In west-central Colorado, broad-tailed hummingbirds interact with four major plant species, including *Delphinium nelsoni*, *D. barbeyi*, *Castilleja miniata*, and *Ipomopsis aggregata* during the breeding season. Nesting begins soon after the onset of blooming by *D. nelsoni* and persists until *Ipomopsis* begins flowering in the same meadows. These two species of plants grow together in dry meadows and flower sequentially. The other two species of food plants are more associated with wetter habitats, and thus tend to be spacially separated from the first two. Studies by Waser (1978) indicate that *D. nelsoni* and *I. aggregata* compete with each other for hummingbird pollination, and during the brief period of flowering overlap some undesirable interspecific pollen transfer occurs. Thus, the two have tended to evolve or maintain sequential flowering periods in this area.

In Texas, this species feeds on a wide variety of plants, including agaves, yuccas, mints (*Salvia*), locusts (*Robinia*), lupines, and others, and on the insects attracted to their blossoms. The birds also glean insects from coniferous foliage, and sometimes hawk them from the air (Oberholser, 1974).

In the Chiricahua Mountains of Arizona, male broad-tailed hummingbirds establish breeding territories in open meadows that support heavy stands of *Iris missouriensis*, which is normally a bee-pollinated species. Most of the hummingbird foraging on this species is "illegitimate" (not resulting in pollination). The birds also forage to some extent on *Robinia neomexicana*, *Penstemon barbatus*, and *Echinocereus triglochidiatus*, the last two of which are more typical hummingbird-adapted species. Lyon (1973) noted that in this area the most important and abundant hummingbird-adapted plant is *P. barbatus*, but it does not begin blooming until the end of June, or near the end of the blooming period of the iris. Thus, the iris provides an adequate early-season interim supply of food until the *Penstemon* is in full bloom, and allows for the establishment of breeding territories in dense stands of this plant. Similarly, in Colorado the breeding season of the species seems to be closely correlated with the flowering periods of the major plant resources (Waser, 1976).

Bent (1940) has summarized information on insect-eating by this species, which includes small flying insects captured in flight, spiders and small insects found in flowers, and probably also the insects that are attracted to the borings of sapsuckers. The flowers of figworts (*Scrophularia*) and ocotillo are especially favored for foraging, as are the insects associated with penstemons, larkspurs, agaves, gilias, gooseberries, and willow catkins.

Breeding Biology
In Texas, breeding occurs from late March or early April until mid-June (Oberholser, 1974). Egg records for Arizona are from May 8 to July 30, with a peak in mid-June to mid-July; those for Colorado are from May 22 to late July, with a peak in the second half of June. A small number of Utah records are from June 6 to July 23 (Bent, 1940).

In Colorado, as many as four nests have been found within an area of 24 to 32 square meters; all were about 7 meters above the ground in ponderosa pines (Bailey and Niedrach, 1965). Near Gothic, Colorado, nesting seasons are closely adjusted to the flowering periods of the major nectar-producing plants. These flowering periods lasted 70 to 80 days in two different years, whereas individual nesting cycles (from egg-laying to fledging) lasted 38 to 41 days. The total nesting season (from first egg laid to last egg hatched) lasted 64 days in both years studied (Waser, 1976).

In a study area in the Chiricahua Mountains of Arizona, Lyon (1973) observed eight male broad-tailed hummingbirds defending breeding territories that averaged about 2040 square meters in area. This is a larger territory than most other estimates, but it may have been a consequence of the birds' defending plants (*Iris*) normally associated with bee pollination and relatively low in nectar production.

By comparison, Barash (1972) observed three males displaying in Colorado while separated from one another by only 7 meters, forming a group that seemed to comprise a lek. On one occasion a female was present for a time in the display area, approaching to within 2 meters of one of the males. While the female was present, the rate of display activity increased (from 4.8 seconds per display circuit to 3.5 seconds per circuit). These small display territories, however, were apparently not centered on feeding areas.

The typical male aerial display consists of a U-shaped dive, often from 9 to 15 meters above ground and passing very close to a perched female. Sometimes females and males will hover nearly bill-to-bill in the air, as the male apparently displays his gorget to the female. Throughout the year the male produces a continuous musical buzzing sound while in flight, except perhaps during molt. Apparently no vocalizations are produced during the display dive, but the overall level of the vibration noises increases as the air passes rapidly through the slotted primaries (Edgarton et al., 1951). The same or similar mechanical wing sounds that occur during display dives also occur during "whisking" flights. This distinctive flight consists of a series of buzzy oscillating movements resembling the motions of a whisk broom or the thrust-like movements of a suspended yo-yo, which may serve to pin down a female sitting in vegetation long enough to allow a copulation attempt by the male (Calder and Calder, 1992).

By cutting off the attenuated tips of their outer pairs of primaries, Miller and Inouye (1983) were able to eliminate selectively the wing whistles of breeding males. Such silenced males were able to penetrate territories of other males more readily than normal males, but also tended to lose their territories to rival males more readily than

unsilenced ones, and were generally less aggressive in their territorial defense behavior.

Hering (1947) reported that during the diving performance the male produces a series of sharp clicking "notes," suggesting that vocalizations may be incorporated into the display. However, Marshall (in Banks and Johnson, 1961) simply noted that three "flups" are produced by the tail at the bottom of the dive. One observed copulation occurred while the female remained perched on a willow branch and the male alighted on top of her (Hering, 1947). This observation calls into question previous assertions that the birds might mate in flight.

According to Wagner (1948), the courtship has two phases: The first, associated with the attraction of the female, consists of the general display flights; the second, concerned with the synchronization of copulatory behavior, consists of the diving flights. Wagner believed that some individuals of this species might breed both on their wintering grounds in Mexico as well as in the United States, but Waser and Inouye (1977) have questioned that suggestion.

Females frequently build their nests in the same location for several consecutive summers. Bailey and Neidrach (1965) mentioned that a female (presumably but not definitely the same bird) nested for six consecutive seasons in a clump of small blue spruces. However, Waser and Inouye (1977) judged that the average life span was probably much shorter than this. One of the males they initially captured as an adult in 1972 was recaptured in 1975, indicating a minimum four-year life span in that individual.

The nests are placed in a wide array of locations, from being saddled on large limbs to being situated on very small twigs. They are usually on low, horizontal branches of willows, alders, cottonwoods, pines, fir, spruce, or aspens, and are generally 1 to 4 meters above the ground. However, they also have been reported in tall sycamores and pines, from 6 to 9 meters above ground. Often they are above water, and they are usually decorated with lichens, shreds of bark, fine leaves, or other plant materials (Bent, 1940).

There does not seem to be any detailed information on the time required to build a nest, but Bailey (1974) mentioned a nest that had been built in

2 days. The incubation period has been reported as 16 to 17 days and the fledging period as 18 days (Bailey and Neidrach, 1965). Calder (1976) reported a 21- to 26-day fledging period for this species. Waser and Inouye (1977) indicated that the mean duration of a nesting cycle (laying to fledging) was 40 days. This would normally preclude second broods in the cool mountain areas inhabited by the species, although Bailey (1974) observed the simultaneous tending of two nests by a female: One nest contained young about 15 days old and a few days prior to fledging; the second nest contained the first egg of the second clutch. This egg was laid 5 days after the nest had been initiated, and the two eggs hatched 16 and 17 days later. Bent (1940) mentioned other earlier reports of second broods.

Calder (1976) described the energy relationships of incubating broad-tailed hummingbirds in the Rockies. He noted that during the nesting period the female initially leaves the nest to feed approximately 11 to 18 minutes before sunrise and makes her last trip at sunset. Throughout the day she may leave the nest about 60 times to feed, and about 30 times for shorter periods. On nights following a day of rain, the bird usually enters a state of hypothermia, and the nest temperature may drop from 32° to 11° C or even lower. This occurred at various times during incubation, hatching, and brooding periods, but did not seem to affect fledging success.

Calder and Calder (1992) summarized the species' breeding biology and reported a fairly high multi-year mean fledging success, with 46 percent of 164 nests fledging at least one young, and a maximum known longevity of 8 years for males versus 12.1 years for females. Mean post-banding life expectancy rates were estimated as 1.6 years for males and 1.9 years for females. The average following-year rate of return of individual birds to their previous breeding areas (philopatry incidence) also was higher for females than for males (36 to 56 percent, vs. 20 to 29 percent) (Waser and Inouye, 1977). Such survival and philopatry estimates collectively suggest higher rates of annual survival than are typical of many larger species of songbirds. Perhaps the birds' elusive flight abilities, their small volume and correspondingly small target size, and also their seemingly unusually high intelligence all allow hummingbirds to live longer lives than biologists familiar with other groups would be likely to predict.

Evolutionary and Ecological Relationships
Mayr and Short (1970) suggested that this species might be more closely related to the ruby-throated superspecies than to any of the other North American species of *Selasphorus,* but they did not explain the basis for this position. Probable hybrids have been found with the black-chinned and the Costa hummingbirds (Banks and Johnson, 1961).

This species overlaps in its breeding range with the calliope hummingbird, but is substantially larger and probably outcompetes it in such areas. However, where broad-tailed hummingbirds occur in the same areas with rufous hummingbirds, the two species are not totally ecologically isolated, and they sometimes compete fiercely for the same food resources (Calder, 1976).

Several flowers interact with broad-tailed hummingbirds in west-central Colorado. At least two of them—*Delphinium nelsoni* and *Ipomopsis aggregata*—have apparently evolved or maintained sequential flowering periods that facilitate their pollination by the hummingbirds without seriously risking cross-pollination.

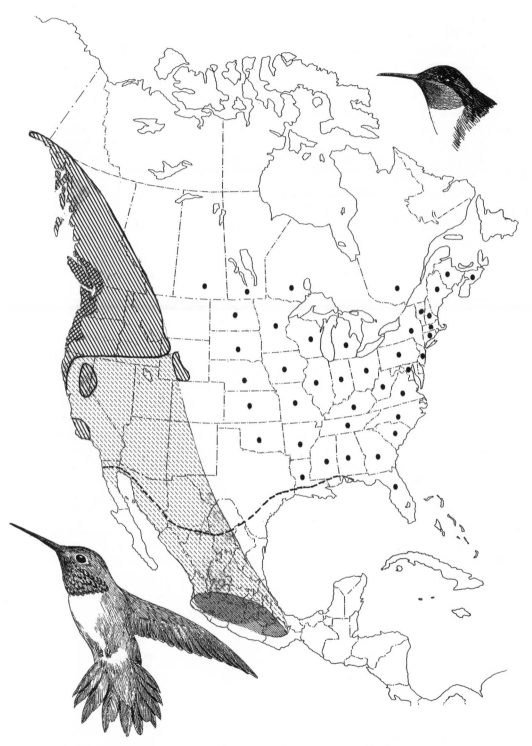

Breeding (hatching) and usual wintering (stippled) ranges, plus migration routes (broken hatching), of the rufous hummingbird. Cross-hatching represents areas of higher breeding densities. Occasional to regular overwintering also may occur north, approximately to the dashed line. Dots represent states or provinces with extralimital sightings. *(Mostly after True, 1993, with additions)*

RUFOUS HUMMINGBIRD

Selasphorus rufus (Gmelin)

Other Names
None in general English use; Chupamirto dorada (Spanish).

Range
Breeds in western North America from southeastern Alaska, southern Yukon, east-central British Columbia, southwestern Alberta, and western Montana south through Washington and Oregon to the Trinity Mountain region of northwestern California and southern Idaho. Winters in Mexico south to Guerrero and northern Oaxaca.

North American Subspecies
None recognized.

Measurements
Wing, males 38–41.5 mm (ave. of 18, 40.3 mm), females 43–45 mm (ave. of 11, 44.4 mm). Exposed culmen 15–17.5 mm (ave. of 18, 16.5 mm), females 17–19 mm (ave. of 11, 18 mm). Eggs, ave. 13.1 × 8.8 mm (extremes 11.4–14.0 × 7.7–10 mm).

Weights
The average of 22 males was 3.22 g (range 2.9–3.9 g); that of 20 females was 3.41 g (range 3.0–3.6 g) (from various sources, including specimens in the Museum of Vertebrate Zoology).

Description (After Ridgway, 1911)
Adult male. Pileum dull metallic bronze or bronze-green; rest of upperparts, including loral, orbital, and auricular regions, sides of occiput, and greater part of tail, plain cinnamon-rufous, the back sometimes glossed with metallic bronze-green; rectrices with a terminal median, more or less fusiform or cuneate, area of purplish or bronzy dusky; remiges dark brownish slate or dusky, faintly glossed with purplish; chin and throat brilliant metallic scarlet, changing to golden green in position *b* (see note at beginning of Part Two for position descriptions); chest white, passing through cinnamon-buff posteriorly into cinnamon-rufous on rest of under-

parts (paler medially); the under tail-coverts whitish basally; femoral tufts white; bill dull black; iris dark brown; feet dusky.

Adult female. Above metallic bronze-green, usually slightly duller on pileum; remiges dark brownish slate or dusky, faintly glossed with purplish; middle pair of rectrices metallic bronze-green (usually more dusky terminally), both webs broadly edged basally with cinnamon-rufous (sometimes with whole basal half or more of this color); next pair with more than basal half cinnamon-rufous, then metallic bronze-green, the terminal portion purplish black; three outer pairs broadly tipped with white, the subterminal portion (extensively) purplish black, the basal half (approximately) cinnamon-rufous, the latter usually separated from the black by more or less of metallic bronze-green; chin, throat, and chest dull white, the throat usually with tips of some of the feathers metallic orange-red or scarlet (changing to golden and greenish) sometimes with a large patch of this color; rest of underparts cinnamon-rufous laterally, fading into dull buffy whitish on breast and abdomen; femoral tufts white; under tail-coverts pale cinnamon-rufous or cinnamon-buff centrally, broadly margined with white or buffy white, the longer ones sometimes with the central area pale grayish or brownish terminally; bill, etc., as in adult male.

Immature male. Similar to the adult female but upper tail-coverts cinnamon-rufous, with a terminal spot of metallic bronze-green; middle pair of rectrices cinnamon-rufous with terminal portion metallic bronze-green (sometimes partly blackish), the lateral rectrices with white tip smaller and cinnamon-rufous deeper and more extensive, and feathers of throat with a terminal mesial spot or streak of dusky metallic bronze or bronze-green.

Immature female. Similar to the adult female, but feathers of upper parts (especially rump and upper tail-coverts) narrowly and indistinctly margined terminally with pale dull cinnamon or buffy, and throat spotted or streaked with dark bronzy, as in young male.

Identification

In the hand. Like other *Selasphorus* species, the outermost primaries are attenuated and sharply pointed in males, and strongly incurved at the tips. The next-to-middle tail feather is also strongly notched in males. There is extensive rufous coloration on the back and rump of adult males, as well as a reddish gorget; in females and immature males the flanks and basal portions of the rectrices are rufous. Females may be distinguished from those of *S. sasin* by their wider fifth or outermost rectrices (at least 2.7 mm wide vs. no more than 2.6 mm in *S. sasin*); they also have slightly longer wings (43–46 mm vs. 39–45 mm in *S. sasin*). Female broad-tailed hummingbirds are slightly larger (wing usually over 45 mm) and less rufous below, with entirely green middle tail feathers; the rufous of the lateral feathers of less extent than the black, and the rufous of the next-to-middle tail feathers limited to the outer edge near the base rather than present on both sides. Although males can normally be separated from the Allen by back color, rufous males may rarely exhibit green backs (*Condor* 77:196).

In the field. Widespread in many western habitats, including forest edges, meadows, woodlands, and chaparral. The rufous back of males sets them apart from all other North American species. Females and immature males are very similar to those of Allen hummingbirds, except perhaps at extremely close range when the difference in the width of the outer tail feathers might be apparent. The calls include various low chipping and buzzy notes and an excited *zee-chupity-chup*. A musical buzz is also produced by flying males. In aerial display the males dive closely toward a female, with the feathers producing a loud whining sound near the bottom of the rather oval course that is followed. The chip-notes of rufous and Allen hummingbirds are almost identical, a fairly sibilant *chip* or *tchup*. In both species the females have white throats speckled with red centrally and greenish laterally; in immatures of both sexes the throat is rather uniformly colored with bronzy markings. Both species also are quite strongly marked with rufous

below and on the tail; while hovering they hold the tail relatively high compared with the plane of the body and move it very little (Stiles, 1971).

Habitats

In California this species occupies all sorts of terrains during migration, from lowland stream bottoms through foothill brushland and heavy chaparral to mountain ridges virtually at timberline (Grinnell and Miller, 1944). Similarly, in Canada it occurs over a great variety of habitats, from seacoast, coastal islands, and valley bottoms to meadows above timberline (Godfrey, 1986). High mountain meadows as well as lowland plains, urban gardens, and other diverse habitats are frequently used by birds on migration. During breeding, coniferous forest habitats are the primary nesting area, but locally the birds occupy forest edges and range out into mountain meadows.

Movements

Because of its very large breeding range and tendency to wander, migration schedules are complicated. Rufous hummingbirds are gone from Alaska by late August, but there are a few September records. There are also some late September and early October records for British Columbia and Washington. Late departure dates for Oregon are in late October, and in California the species is a common migrant during late June, July, and August in the mountains; a few scattered individuals persist into late fall and rarely have been reported (in the Berkeley area) as late as January 1. In Montana the birds become rare in August and are typically gone by the latter half of September. In Idaho, the males begin to disappear in early July; very few immatures or females are left by August; and the latest state records are for early September. From mid-July throughout most of August the species is a common migrant in Colorado, with males appearing seven to ten days ahead of the females and immatures. By late August they are declining, and the latest records are for late September. In New Mexico the species is most abundant in early August, and is mostly gone by early September, although a few linger until the latter part of that month. Likewise in Arizona the few birds seen after August are in female and immature plumages, nearly all of which are gone by October.

In Texas most of the fall migration is from late July to late September, with a few birds persisting until mid-October or (rarely) to the end of November. Many birds are seen in winter months along the coast, from Texas east to Louisiana (where Newfield, 1996, reported that they made up over half of the wintering hummingbirds banded) and Alabama (where Robert Sargent has banded more than 100 birds in a single winter.

During spring the species migrates entirely up the Pacific Coast (Phillips, 1975), avoiding Texas, New Mexico, and Arizona, and passing through California from February to May. It arrives in Oregon by early March, and likewise reaches Washington in late February or early March and British Columbia by early April. By mid-April it has reached Alaska and usually reaches Idaho and Montana by the end of April or early May (Bent, 1940).

Vagrant birds often stray well away from their usual migration routes, especially in fall. Conway and Drennan (1979) summarized available records of rufous hummingbirds for the eastern states, and reported that for Florida alone there are more than 50 published records, involving 50 to 55 birds. According to them, birds have been reported in the eastern states from July 19 to May 1, but more than half of the records are for November and December. The northernmost eastern record so far is from Nova Scotia. It is increasingly apparent that over-wintering in the Gulf Coast region is now a regular phenomenon (Newfield, 1966).

Records for rufous hummingbirds outside their normal breeding and migration range now include all of the midwestern and eastern states, plus six Canadian provinces east of Alberta. Among banded hummingbirds, all five records of the greatest distances attained between banding and recovery points are held by the rufous; the longest is 2788 kilometers for an adult female, and the second-longest is 2211 kilometers for another adult female. The latter averaged 53 kilometers of movement per day between the points of initial banding (Vancouver Island) and recapture (Tijeras, New Mexico). The following summer it was captured again, at its original point of banding (True, 1996). A third juvenile rufous female traveled 1202 kilometers in no more than 15 days (averaging 80.1 km per day) and managed to gain 0.2 gram in the process (Calder and Jones, 1989)! Although more than 70,000 hummingbirds of 12 species have been banded in North America, fewer than 60 individuals have been recovered away from their banding sites, so migration information based on hummingbird banding data is still extremely fragmentary.

Foraging Behavior and Floral Ecology
Like choosing its diverse habitats, the rufous hummingbird uses a wide array of flowers. In California, favorite native species are the goose-berries, currants, and manzanitas; non-native plant species include eucalyptus, tree-tobacco, fuchsia, red-hot-poker, and the flowers of orange and peach (Grinnell and Miller, 1944). In Texas, plant species such as mints, columbines, penstemon, larkspur, bouvardia, tree-tobacco, and agave are all used at various seasons (Oberholser, 1974). In New Mexico the species is associated with currants, gooseberries, ocotillo, fireweed (*Epilobium*), Indian paintbrush, penstemon, and agave. In late summer it has been foraging almost exclusively on a red species of figwort (*Scrophularia*) growing in mountain valleys at about 2250 meters (Bailey, 1928). The birds also have been observed feeding in large numbers on Rocky Mountain bee plant (*Cleome*). Their fall arrival in southwestern New Mexico coincides with the blooming of agave in late July (Ligon, 1961). Likewise, during migration in Oregon and Washington the birds arrive just as crimson currant (*Ribes sanguineum*) is coming into bloom, and the birds feed heavily on this food resource. While in Oregon they also feed preferentially on red columbine and the blossoms of madrone (*Arbutus menziesii*). In Washington they often concentrate around the flowers of salmon-berry (*Rubus spectabilis*), thimble-berry (*R. parviflorus*), and honeysuckles (Bent, 1940), but only the honeysuckle is a typical nectar-producer.

This species defends feeding territories during migration as well as on breeding grounds. Armitage (1955) observed defense of areas of toad-flax (*Linaria vulgaris*) by fall migrant rufous hummingbirds (probably immature males) in Yellowstone National Park; there the territories were very small (2–3 m × 2–6 m).

In the Grizzly Lake area of California, the birds hold territories in late summer that include two primary food sources, *Aquilegia formosa* and *Castilleja miniata*. Of these, the columbine produces about four times the amount of nectar as the paintbrush, but the feeding territories that are held have similar daily caloric productivity, and thus are size-regulated to adjust for differences in floral composition (Gass et al., 1976). In this area the individual bird's abilities to hold territories is dependent on age, sex, and the level of competition from conspecifics. The population is more than half immatures, the adult males typically leaving shortly after breeding is finished, perhaps to release food supplies for the young. The length of time they hold the territories varies greatly, but adult females remained on territories longer than did immatures of either sex (Gass, 1979). In an Arizona study, adult males dominated all other age or sex classes of their own species and all other species, tending to defend more densely flowered areas than did adult females (Kodric-Brown and Brown, 1978).

Breeding Biology

Males probably establish territories as soon as they arrive in their breeding areas, several weeks in advance of most females. The adult male's display flights normally comprise a complete oval, with a slanted axis and broader at the bottom than at the top. In the downswing the bird produces a mechanical sound starting with an exaggerated wing buzz, followed by a staccato whining note, and ending with a rattle. These notes resemble sounds made by the Allen, but there are several differences: One is the absence of a "pendulum" sequence such as that of Allen males; another is the whining note, broken three or four times in the rufous display but continuous in the Allen's; a third is the several oval dives in a sequence by the rufous, as opposed to the Allen's single pendulum sequence and a single dive at most times. Spectrograms show the whining notes to be richer in overtones in the rufous, purer in the Allen, but this difference is not apparent to the untrained ear. Apparently, however, the male's displays are sufficiently species-specific to aid in separation of rufous and Allen immature males as well (F. I. Ortiz-Crespo, personal communication).

Nests are evidently built in a wide variety of locations, from near the ground in low blackberry bushes to some 15 meters above ground in tall firs. In Puget Sound a frequent location is among huckleberry bushes, or at times among alders and blackberry vines. The drooping branches of conifers are also favored locations, with the nest often on the lowest branch having a sharp downward bend (Bent, 1940). According to Kobbe (1900), most nests are placed between 2 and 5 meters above ground, with extremes of 75 centimeters and 9 meters, and are usually well hidden in evergreen foliage. Nearly all nests are placed above paths or gullies. All but 2 of 20 nests found by Kobbe were in spruce trees, and nearly all were decorated with lichens and lined with willow down.

Renesting in subsequent years on the old nest is apparently frequent, and occasionally a group of three nests stacked on top of one another has been found. Kobbe stated that the birds regularly renest if the first clutch is destroyed; usually the second nest is placed in a less exposed site and often is higher in the tree.

In coastal areas of Washington and Oregon, overhead protection from rain is apparently also a prominent characteristic of most nest sites. Vines that overhang embankments, especially those having a southern exposure, or dry clusters of roots from upturned trees are also frequent nest sites in western Oregon (Bent, 1940).

Studies in British Columbia by Horvath (1964) indicate that there are seasonal differences in heights of nests, which are related to nest microclimate. Early nests are typically in conifers at low levels, whereas in summer they are in the crowns of deciduous trees. Low nests among conifers in the spring are protected from extremes of temperature better than higher ones, but by summer the nests in the tops of deciduous trees benefit from the temperature-reducing effects of evapotranspiration from the trees and avoid overheating the young.

Although the species has not been studied well in Alaska, the nests found there have similar interior temperatures to those of broad-tailed hummingbirds in the Rocky Mountains. Likewise food supplies during the breeding season in Alaska need intensive study, since the birds must at that time feed primarily on flowers adapted for pollination by insects, such as blueberries, salmonberries, and *Menziesia*. Only near the end of the nesting season is there a significant bloom of flower species adapted for hummingbird pollination (Calder, 1976).

In Washington, eggs have been found from April to July, indicating a fairly protected breeding season, but probably the month of May represents the peak of the season. There is no good information on the length of time required to build the nest, but Kobbe (1900) observed one nest that had a 3-day interval between the laying of the two eggs. There are likewise no good estimates of the incubation period; several early estimates of 12 to 14 days are clearly erroneous.

The total nestling period is approximately 20 days, and Dubois (1938) has carefully described the development of the young for the first 12 days. He noted that at 6 days the pinfeathers were becoming visible along the sides and edge of the wings, the eye-slit was noticeable, and slight ticking sounds were first heard. By the next day the pinfeathers had emerged on the sides, and at 8 days the eyes were partly open and there were filaments at the top of the head. On the 9th day there were prominent feather tracts visible along the sides of the belly, and the voice was becoming stronger. On the 11th day the young was well covered with pinfeathers; the eyes were opened to an ellipse; and the feathers of the wings and tail were well sprouted. The eyes were well opened by the 12th day following hatching.

Although it must be regarded as extremely aberrant behavior, there is a single record of a male incubating eggs (Bailey, 1927). The female incubates almost constantly, and is usually off the nest for no more than about 20 minutes. However, by the time the young were 4 days old, the average interval was about 44 minutes in Dubois' (1938) study. After 7 days the female was absent for much of the morning, but would return in the afternoon to shield the nestling from the hot sun.

Evolutionary and Ecological Relationships

The rufous and the Allen hummingbirds apparently constitute a superspecies (Mayr and Short, 1970); probably only their allopatric breeding distributions prevent them from hybridization. The only reported hybrid combination involving the rufous is with the calliope hummingbird (Banks and Johnson, 1961).

Rufous hummingbirds have certainly evolved in association with a variety of plant species throughout their broad breeding range, and probably no one plant can be singled out as crucial to the success of this species. However, Grant (1952) described a remarkable instance of divergent evolution between two Sierra Nevada species of columbine (*Aquilegia formosa* and *A. pubescens*), associated with differing adaptations for pollination. In *A. formosa*, which is often pollinated by rufous hummingbirds, the flower is red and yellow, nodding, and has a relatively short spur (under 20 mm). These traits facilitate feeding by hummingbirds but not by hawkmoths, which are less able to feed easily on nodding blossoms or to see the flowers in semidarkness. On the other hand, *A. pubescens* has yellow-to-white blossoms, erect flowers, and a much longer spur (at least 29 mm). In this case, hummingbirds are unable to reach the nectar, which helps to avoid cross-pollination and reduces hybridization between these two closely related species.

In Alaska, only five plant species (*Aquilegia formosa* and four species of *Castilleja*) have floral characteristics associated with hummingbird pollination, and all are confined to southern Alaska (Grant and Grant, 1968). Probably the rufous hummingbird invaded Alaska fairly recently, and the spread of hummingbird-adapted flowers to this region has lagged behind that of the birds. Thus, the birds must get much of their food there from flowers usually adapted to bee pollination (Calder, 1976).

Breeding (hatching) and wintering (stippled) ranges, plus migration routes (broken hatching), of the Allen hummingbird. Arrowheads indicate the residential Channel Islands population. Dots represent states or provinces with extralimital sightings, and the dashed line represents the limit of occasional overwintering.

ALLEN HUMMINGBIRD

Selasphorus sasin (Lesson)

Other Names
Red-backed hummingbird; Chupamirto petirrojo (Spanish).

Range
Breeds in coastal California from about the Oregon line southward to Santa Barbara County, and the islands off the coast of southern California. Winters south to the Valley of Mexico and Morelos (Phillips, 1975).

North American Subspecies
S. s. sasin (Lesson). Breeds on the mainland of coastal California; winters south to the Valley of Mexico and Morelos.

S. s. sedentarius (Grinnell). Resident on San Clemente and Santa Catalina islands of the Santa Barbara group off coastal California, and on the adjoining California mainland (*Western Birds* 10:83–85, 11:1–24).

Measurements
Of *S. s. sasin:* Wing, males 36.5–38.5 mm (ave. of 10, 37.8 mm), females 41–42 mm (ave. of 9, 41.6 mm). Exposed culmen, males 15–16.5 mm (ave. of 10, 15.9 mm), females 17–18.5 mm (ave. of 9, 17.8 mm) (Ridgway, 1911). Eggs ave. 12.7 × 8.6 mm (extremes 11.7–14 × 7.6–10 mm). Measurements of *S. s. sedentarius* average larger (Stiles, 1972b).

Weights
The average of 38 males of *S. s. sasin* was 3.13 g (range 2.5–3.8 g); that of 18 females was 3.24 g (range 2.8–3.5 g) (from various sources). Averages of 19 males and 26 females of *S. s. sedentarius* were 3.52 and 3.73 g, respectively (Stiles, 1971).

Description (After Ridgway, 1911)
Adult male. Above metallic bronze-green, the feathers of rump with basal portion (mostly concealed) deep cinnamon-rufous; upper tail-coverts and tail deep cinnamon-rufous, the

rectrices with a terminal, more or less fusiform streak of purplish black or dusky, the lateral ones with this dusky confined mostly to outer web; remiges dark brownish slate or dusky, faintly glossed with purplish; loral, orbital, auricular, and postocular regions deep cinnamon-rufous, sometimes brokenly extending across nape; chin and throat brilliant metallic scarlet or orange-red, changing in position *b* to golden and greenish, the latero-posterior feathers of the gorget elongated (see note at beginning of Part Two for position descriptions); chest white, passing gradually into pale cinnamon-rufous or cinnamon-buff on breast and abdomen, this into deep cinnamon-rufous on sides and flanks; femoral tufts white; under tail-coverts cinnamon-rufous, paler basally; bill dull black; iris dark brown; feet dusky.

Adult female. Above metallic bronze-green, the upper tail-coverts with basal portion light cinnamon-rufous (partly exposed); middle pair of rectrices with basal half (laterally, at least) cinnamon-rufous, the terminal half (more or less) metallic bronze-green; next pair similar, but terminal portion (extensively) black, the tip of inner web sometimes with a small spot of white; three outer rectrices (on each side) broadly tipped with white, crossed by a broad subterminal area of black, the basal portion cinnamon or dull light cinnamon-rufous, this separated from the subterminal black (at least on third rectrix) by more or less of metallic greenish; remiges dark brownish slate or dusky, faintly glossed with purplish; underparts dull white (sometimes slightly tinged with pale cinnamon-buffy), passing into light cinnamon-rufous on sides, flanks, and under tail-coverts; the throat usually spotted, more or less, with metallic orange-red or scarlet; bill, etc., as in adult male.

Immature male. Similar to the adult female, but upper tail-coverts mostly (sometimes wholly) cinnamon-rufous, rectrices more extensively cinnamon-rufous, and throat strongly tinged with cinnamon-rufous and spotted or speckled with dark bronzy.

Identification

In the hand. A typical *Selasphorus* species in that males have the outermost primary attenuated and sharply pointed as well as strongly incurved toward the tip. Adult males differ from *S. rufus* in that the rufous coloration does not include the back, but rather only the tail and rump, while the back is green. Females closely resemble those of *S. rufus,* but their outer rectrices are narrower (less than 2.7 mm vs. more than 2.7 mm in *rufus*), and the tail is usually less than 25 millimeters (23–26 mm) in contrast to the usual 26 millimeters or more (25–29 mm) in *S. rufus.*

In the field. Adult males can be readily recognized by their rufous tail and rump, interrupted by a greenish back and crown. Females and immature males cannot by distinguishing safely from those of the rufous hummingbird, and both species have a similar call, a sharp *tchup.* However, the courting flight of the male is quite different: The male produces a series of pendulum-like arcs above the female, with each arc about 6 to 9 meters across; at the bottom of the arc the tail is bobbed to produce an interrupted buzzing, after which there are one or two final faster swoops from much greater height, with a veering follow-through. In this final phase it produces a ripping *vrrrrp* sound, probably mechanical in nature.

The wing-noise of the Allen hummingbird is very high-pitched and similar to that of the rufous, and females of both species tend to be more strongly marked with rufous below and on the tail than do other western hummingbirds. Their dorsal color is also quite golden in hue, and the throat of adult females is essentially white, with red specks centrally and bronze-green laterally. Immature birds have the entire throat rather uniformly marked with dusky to greenish spotting, and juveniles have cinnamon-buff feather edging dorsally. In flight both species hold the tail quite high, with little movement (Stiles, 1971).

Habitats

This species is essentially limited during the breeding season to coastal California, in humid ravines or canyons essentially within the summer fog belt. During this period the males occur in territories overlooking "soft chaparral"; the females are in willows, blackberry tangles, or beds of brakes along the bottoms of these slopes. By midsummer the birds sometimes visit the Sierras (Kings Canyon and Sequoia National Parks), and the migration route southward occurs along these mountain slopes (Grinnell and Miller, 1944). In the vicinity of Berkeley, male territories typically occur along canyon bottoms or stream courses where there are areas of massive, deciduous vegetation, including willows, poison oak, or dogwood. Areas between territories often consist of trees or open grassy slopes. Females typically occupy groves or more or less continuous woodland during the nesting period, frequently nesting in live oaks (Pitelka, 1951b). The nonmigratory race on the islands off the California coast is also found in canyons and ravines where there is a heavy growth of brush (Grinnell and Miller, 1944).

Movements

Allen hummingbirds are migratory only at the northern end of their range. Northward migration begins very early; in the San Diego district the birds begin moving northward by January, and spring migration is over by April (Grinnell and Miller, 1944). In the Berkeley area, males take up territories between mid-February and mid-March, and even at the northern limit of their breeding range probably arrive no later than late February (Pitelka, 1951b). The post-breeding southward movement occurs chiefly during July and August. However, in the Santa Monica Mountains of California, post-breeding birds begin to arrive in late May or June, with the earliest arrivals high in adult males; by late June and July most of the birds are juveniles. By early October Allen hummingbirds (and rufous hummingbirds) have begun to leave the area, and all *Selasphorus* hummingbirds are usually gone by the end of the month (Stiles, 1972b). However, on the Palo Verdes Peninsula, the race *S. s. sedentarius* breeds throughout the year (Wells and Baptista, 1979).

In Arizona the species likewise is a regular early fall transient. In July and August the birds pass through the mountains of central southern Arizona, apparently on their way to the Valley of Mexico (Phillips et al., 1964).

According to Phillips (1975), Allen hummingbirds migrate southward along an elliptical route that takes them all the way to the Valley of Mexico by August. Adult males precede the females, but young males linger in California for about a month after the last adult males have gone, and half a month after the last females have departed. In Phillips' view, there are only two local wintering concentrations of Allen hummingbirds: one in the Valley of Mexico during August and the other in Morelos in December. The return route in spring is apparently a more westerly one, resulting in an oval migration route for the entire year.

Extralimital records of this species are not numerous, but it has been reported from as far north as Victoria, British Columbia (*American Birds* 25:792) and as far east as Alabama. It also has been reported from unlikely locations such as Massachusetts, Mississippi, Louisiana, Kansas, and Texas.

Although there have been a considerable number of summer sightings throughout Oregon, the species apparently breeds only in Curry County (AOU, 1957). It is a rare visitor in western Washington.

Foraging Behavior and Floral Ecology
Little of a specific nature has been written on the food and foraging behavior of Allen humming-birds. Bent (1940) mentioned that among the popular food plants are tree-tobacco (*Nicotiana*), the blossoms of California lilac (*Ceanothus*), madrone (*Arbutus menziesii*), and the flowering stalks of the century plant (*Agave*). Also attractive are scarlet sage, mints, monkey-flowers (*Mimulus langsdorfi* and *M. cardinalis*), columbines, hedge-nettles (*Stachys albens*), and Indian paintbrush (*Castilleja grinnelli*). In late summer, the bush or sticky monkey-flower (*Diplacus*) is perhaps used more heavily than any other native plant, and during fall migration a great variety of flowering herbs, shrubs, and trees are utilized (Grinnell and Miller, 1944). Along the coast of south-central California this species frequently visits honeysuckles (*Lonicera involucrata ledebourii*) during the breeding season, and in the mountains of southern California an important late summer or fall plant is California fuchsia (*Zauschneria californica latifolia*) (Grant and Grant, 1968).

Legg and Pitelka (1956) listed a variety of plants used by Allen hummingbirds near Santa Cruz, and

also observed them hawking insects in early morning and late evening hours. Two other observations noted probable foraging on ants on the ground, although the birds may actually have been obtaining grit.

Breeding Biology
In the Berkeley area, male territories are established between mid-February and late March, and typically are situated in shrubby areas along canyon bottoms or beside stream courses. Males guard their areas of shrubs, chasing intruders away, displaying above them, and performing circuit and advertisement flights around them. Yet, they defend only the shrubby areas and the airspaces 3 to 14 meters above them. Thus, territorial encounters are not so frequent in this species as in the Anna hummingbird, and there is no definite buffer zone between adjoining territorial core areas. There also are no challenge flights or patrolling of buffer areas, and boundaries of adjacent males seem to be recognized and respected by adjacent territorial holders (Pitelka, 1951b). There are often two types of territories: feeding territories, which are small and numerous and often contested; and mating territories, which are larger and more formally defined (Legg and Pitelka, 1956).

The aerial display of the male usually consists of a series of arcing pendulum-like flights some 6 to 9 meters across, typically with a pause near the end of each arc. At the bottom of each arc the male spreads the tail and bobs it, producing an interrupted buzzing and shaking movement of the body. This flight is also accompanied by mouse-like squeaking vocalizations. After several arcs the male climbs to a higher elevation and performs a power dive at great speed. A ripping sound, something like that made by rapidly drawing a fine-grained file over the edge of a steel sheet, is produced at the bottom of the dive and lasts for a second or longer (Bent, 1940).

In a photographic analysis, Pearson (1960) determined that at the middle of the power dive the male descends at an angle of about 45° from the horizontal, at between 84 and 102 kilometers per hour. The entire dive, from a height of about 18 meters, requires slightly more than a second and has an average speed of about 64 kilometers per hour. As the bird levels off at the bottom of the dive, when the tail feathers produce the loud sound, it is probably traveling between 54 and 72 kilometers per hour.

Females nest in areas other than those used by males for display. Males occasionally have been observed in such areas, sometimes displaying, but they remain only a short period and do not hold territorial posts. The nests are often located in oaks; 29 of 30 nests in Pitelka's (1951a) study were found in live oaks, and were usually located between 2 and 4 meters above ground (12 of 22 nests), with extremes of 60 centimeters and 7.5 meters. In another study, 21 nests ranged in height from 45 centimeters to 12 meters, averaging 5.5 meters, and were mostly in eucalyptus trees (Legg and Pitelka, 1956).

According to Aldrich (1945), nesting sites are usually provide many separate supports for the first nesting materials; thus females place the nest where lateral supports are present. Dense tangles of vines are favorite sites, and in eucalyptus trees the nests are usually near the tips of drooping, incurved branches where the nest can be saddled between two fruits or among the petioles of leaves. In such tree locations the nests may be up to 15 meters from the ground, typically on limbs less than 25 millimeters in diameter. At times the nest is built between pieces of loose bark on the main eucalyptus trunk. Cypress trees, when nearby, are favored over eucalyptus, apparently because of the rough surface provided by their branches and twigs. When nests are in vines, they are frequently among wild blackberries and ferns, shaded by live oaks. Then they are often at the intersection of several leafless stems, whereas on ferns they are usually supported by both the leaves and stems. In such locations they may be as close as 15 centimeters to the ground. Overhead shade is apparently an important part of nestsite selection, and patchy shade may be favored over unbroken shade.

The length of time required to build a nest varies, but it may range from 8 to 11 days, with shorter periods associated with nests at old nest sites. Most of the materials are gathered within 22 meters of the nest and invariably include spider webs. Bits of shredded leaves, bark, and grass are also used, and the down from willow trees or composite seeds is used for nest lining, as are downy feathers. Hair is sometimes also used for lining,

and lichens are invariably used for the outer nest surface. Typically, moss makes up the majority of the outer layer, giving the nest a distinctive greenish color, whereas the inner layer makes up the greater bulk and always consists of white downy materials. The first material to be deposited is usually down, and as a rim is gradually developed the two layers become distinct. Lichens are placed with the light green side outward and are attached with spider webs. Frequently lichens are added after the eggs are laid, and lining may also be added throughout the incubation period (Aldrich, 1945).

Almost invariably there are two eggs, deposited on alternate days and rarely laid on successive days; two freshly laid eggs weighed 0.323 and 0.388 grams (Aldrich, 1945). Incubation sometimes begins after the laying of the first egg, but it becomes more intense after the second egg is deposited, so that the eggs typically hatch a day apart. By 3 days after the laying of the second egg, incubation reaches a constant level, with the bird on the nest about 80 percent of the daytime hours. The longest periods of absence are in midday, when the temperatures are highest, and the last departure is apparently dictated by daylight levels rather than temperatures.

When incubating, the bird keeps its back directed toward the source of light, and temperature strongly affects the posture of the bird in the nest. Incubation lasts from 17 to 22 days, during which time females pay little attention to courting males (Aldrich, 1945).

During the first few days after the young hatch, the female probably spends nearly as much time brooding on the nest as she had previously spent incubating. However, brooding nearly ceases by the time the young are 12 days old. In one observed nest, the young were very darkly pigmented when 6 days old, but their eyes were still closed and juvenile feathers were not evident. On the 7th day a few feathers appeared along the spine, and by the 8th day the first tail feathers had emerged. By the 11th day all the feathers of the dorsal tracts had emerged, and on the 12th day the eyes of one nestling were opening. The flight feathers emerged on the 13th day; wing fanning was observed on the 19th day; and fledging of one young occurred on the 22nd day after hatching.

The second-hatched bird left the nest 25 days after hatching (Orr, 1939).

In the Berkeley hills, it is typical for two nestings to occur per season, but almost all breeding activity is over by mid-July. Males abandon their territories by mid-June and gradually move out of the area (Pitelka, 1951b).

Evolutionary and Ecological Relationships

Mayr and Short (1970) indicate that the Allen and rufous hummingbirds have apparently mutually exclusive ranges and in their opinion constitute a superspecies. Apparently the only recorded wild hybrids are with the black-chinned hummingbird (Lynch and Ames, 1970) and the Anna hummingbird (Pitelka, 1951b; Wells and Baptista, 1979).

Williamson (1957) has reviewed the criteria for the generic separation of the Anna and Allen hummingbirds, as well as their hybrid traits, and has suggested that the hybrids might be fertile. He thus questioned whether the two species should be retained in separate genera.

The Allen interacts with the Anna hummingbird during the breeding season and competes with it for resources. The territories of Allen hummingbirds are often peripheral to or within the upland territories of Anna hummingbirds, and areas of overlap favor the Anna. In such areas a mutual but unbalanced population depression of males was found, with about 80 percent occupancy of suitable sites by Annas and no more than 48 percent occupancy by Allens. The larger size of the male Anna probably helps to account for their dominance, but there have been several records of apparent displacements of Anna males by Allen males (Legg and Pitelka, 1956; Pitelka, 1951b).

APPENDIXES

Key to Identification of North American Hummingbirds

A.   Larger, wing at least 60 mm
  B.   Violet ear-patch present, tail with a black band near tip (Green violet-ear)
  BB. No violet ear-patch or black band near tip of tail
    C.   Bill very long (culmen 33–36 mm), white rump patch present (Plain-capped starthroat)
    CC. Bill shorter (culmen less than 32 mm), no white rump patch
      D.   Tail bluish or black with white tip, that of male slighty forked (Magnificent hummingbird, Figure 24B)
      DD. Tail green, with grayish tip, that of male somewhat rounded (Blue-throated hummingbird, Figure 24A)
AA. Smaller, wing no more than 56 mm
  B.   Bill reddish, at least on lower mandible, nasal operculum at least partly exposed
    C.   Nasal operculum wholly exposed, a white eye-stripe present above a blackish ear-patch
      D.   Ear-patch black, with a long white eye-stripe above (White-eared hummingbird, Figure 24D)
      DD. Ear-patch grayish, bordered above with a short, dull whitish eye-stripe (Broad-billed hummingbird, Figure 24C)
    CC. Nasal operculum partially concealed, not with combination of a white eye-stripe above a blackish earpatch
      D.   Tail forked and dusky violet to blackish, a small white spot present behind eye (Cuban emerald)
      DD. Tail square or only notched, brownish; no white spot behind eye
        E.   Bill only slightly widened basally, only lower mandible reddish basally (Berylline hummingbird)
        EE. Bill abruptly widened near its base, both upper and lower mandibles reddish
          F.   Chin and throat white (Violet-crowned hummingbird)
          FF. Chin and throat metallic green

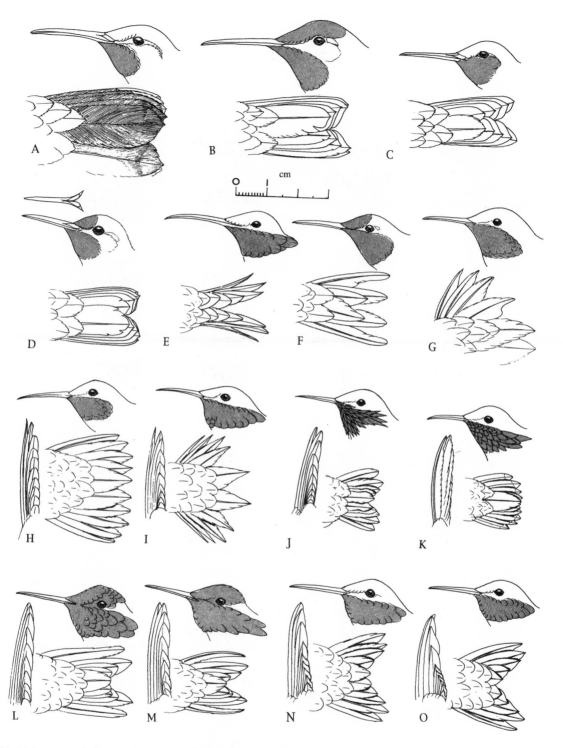

24. Male structural characteristics of blue-throated hummingbird (A), magnificent hummingbird (B), broad-billed hummingbird (C), white-eared hummingbird (D), lucifer hummingbird (E), Bahama woodstar (F), rufous hummingbird (G), broad-tailed hummingbird (H), Allen hummingbird (I), calliope hummingbird (J), bumblebee hummingbird (K), Anna hummingbird (L), Costa hummingbird (M), black-chinned hummingbird (N), and ruby-throated hummingbird (O). Stippling indicates areas of high iridescence. *(Drawn to scale, after Ridgway, 1911)*

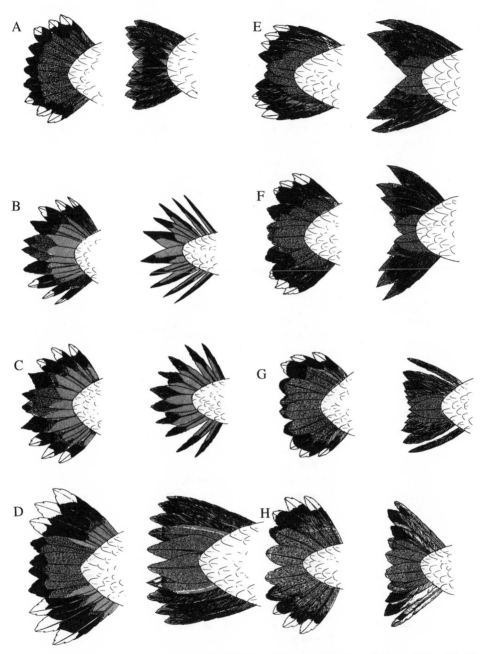

25. Tail shapes and color patterns of female (left) and male (right) hummingbirds, including calliope (A), Allen (B), rufous (C), broad-tailed (D), ruby-throated (E), black-chinned (F), Costa (G), and Anna (H). Dark stippling indicates green; light stippling indicates rufous to cinnamon. *(Drawn to scale, after various sources)*

        G. Central tail feathers brownish, abdomen brownish gray (Rufous-tailed hummingbird)

        GG. Central tail feathers greenish, abdomen buffy (Buff-bellied hummingbird)

BB. Bill blackish, not broader than deep at base, nasal operculum concealed by feathers

    C. Bill long and decurved, tail deeply forked in males (Lucifer hummingbird, Figure 24E)

    CC. Bill straight or only slightly decurved, one-fourth to one-half as long as wing

        D. Plumage usually with considerable rufous, at least on tail, which is usually rounded in both sexes; outermost primary of male sometimes sharply pointed

            E. Abdomen pale rufous, bounded by a whitish breast, central tail feathers distinctly shorter than the rest and the tail forked in males (Bahama woodstar, Figure 24F)

            EE. Abdomen and breast both whitish, with darker flanks, central tail feathers as long as the others

                F. Bill short (exposed culmen 10–13 mm), wing usually less than 36 mm (Bumblebee hummingbird, Figure 24K)

                FF. Bill longer (at least 13.5 mm), wing usually at least 37 mm

                    G. Lateral rectrices mostly blackish, wing 45–52 mm (Broad-tailed hummingbird, Figures 24H and 25D)

                    GG. Lateral rectrices mostly rufous, wing 36–45 mm

                        H. Throat metallic red or purple (males)

                            I. Back metallic green (Allen hummingbird, Figure 24I)

                            II. Back cinnamon-rufous (Rufous hummingbird, Figure 24G)

                      HH. Throat whitish, usually flecked with dusky (females)

                            I. Outermost rectrix no more than 2.7 mm wide (Allen hummingbird, Figure 25B)

                            II. Outermost rectrix more than 3 mm wide (Rufous hummingbird, Figure 25C)

        DD. Plumage lacking any rufous, tail square or slightly forked in males and outermost primary never sharply pointed

            E. Inner rectrices broadening subterminally, throat feathers of adult males very narrow and pure white basally (Calliope hummingbird, Figure 24J)

            EE. All rectrices more or less tapering toward tips

                F. Inner primaries with small notch near tip of inner web, lateral rectrices pointed

                    G. Throat metallic purplish or violet (males)

    H. Throat purplish red (Ruby-throated
      hummingbird, Figure 24O)
    HH.Throat black and violet (Black-chinned
      hummingbird, Figure 24N)
   GG. Throat dull white, tail double-rounded or
    rounded (females)
     H. Exposed culmen 17–19.5 mm, tail double-
      rounded (middle rectrices shorter than
      outer ones) (Ruby-throated hummingbird,
      Figure 25)
     HH.Exposed culmen 19.5–22 mm, tail rounded
      (Black-chinned hummingbird, Figure 25F)
  FF. Inner primaries not notched near tip of inner web,
   lateral rectrices mostly rounded
    G. Gorget and crown metallic-colored (males)
     H. Gorget and crown purplish red, outermost
      rectrices normal in width (Anna humming-
      bird, Figure 24L)
     HH.Gorget and crown violet, outermost rectrices
      distinctly narrowed (Costa hummingbird,
      Figure 24M)
    GG. Throat pale gray or dull whitish, no gorget
     (females)
     H. Wing 48–51 mm, pale grayish underparts
      and tail corners (Anna hummingbird,
      Figure 25H)
     HH.Wing 43.5–46 mm, whitish underparts
      and tail corners (Costa hummingbird,
      Figure 25G)

Key to Identification of Mexican Hummingbirds

A. Outer rectrices extensively white
  B. Larger (wing minimum 73 mm), bill distinctly decurved (Violet sabrewing, Figure 27A)
  BB. Smaller (wing maximum 70 mm), bill slender, nearly straight
    C. A white band across hindneck; outer tail feathers almost entirely white (White-necked jacobin, Figure 26C)
    CC. No white band on hindneck, wing-coverts bright cinnamon-rufous (Stripe-tailed hummingbird, Figure 26D)
AA. Outer rectrices without white or white only at base or tips
  B. Bill more than half as long as wing
    C. Middle tail feathers distinctly elongated
      D. Tips of middle tail feathers not whitish (Wedge-tailed sabrewing, Figure 26B)
      DD. Tips of middle tail feathers whitish
        E. Wing no more than 40 mm (Little hermit, Figure 27B)
        EE. Wing over 50 mm (Long-tailed hermit, Figure 26A)
    CC. Middle tail feathers not markedly elongated
      D. Larger (wing at least 60 mm), bill nearly straight (Long-billed starthroat, Figure 26L)
      DD. Smaller (wing no more than 40 mm), bill distinctly decurved
        E. Bill shorter (under 20 mm) (Beautiful hummingbird, Figure 27G)
        EE. Bill longer (over 20 mm) (Mexican sheartail, Figure 26P)
  BB. Bill less than half as long as wing
    C. Bill very short (maximum 15 mm) and straight
      D. Tail slate-colored; no white on upper rump area (Emerald-chinned hummingbird, Figure 26M)
      DD. Tail mostly brown; a conspicuous white patch or buffy band in upper rump area
        E. White rump band continuous over middle of rump

Note: Hummingbird species not in this key are identifiable with the North American key.

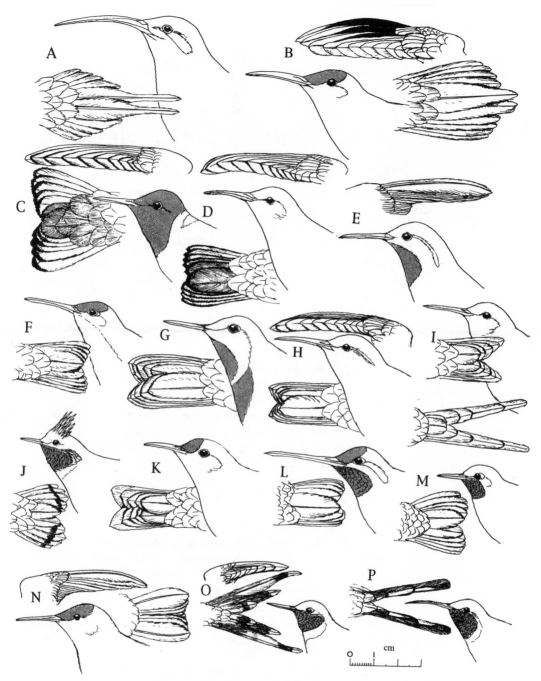

26. Adult male traits of 16 Mexican hummingbirds, including long-tailed hermit (A), wedge-tailed sabrewing (B), white-necked jacobin (C), stripe-tailed hummingbird (D), amethyst-throated hummingbird (E), azure-crowned hummingbird (F), garnet-throated hummingbird (G), dusky hummingbird (H), fork-tailed emerald (showing extreme tail profiles) (I), rufous-crowned (= short-crested) coquette (J), crowned woodnymph (K), long-billed starthroat (L), emerald-chinned hummingbird (M), green-fronted hummingbird (N), sparkling-tailed hummingbird (O), and Mexican sheartail (P). *(Mainly after Ridgway, 1911)*

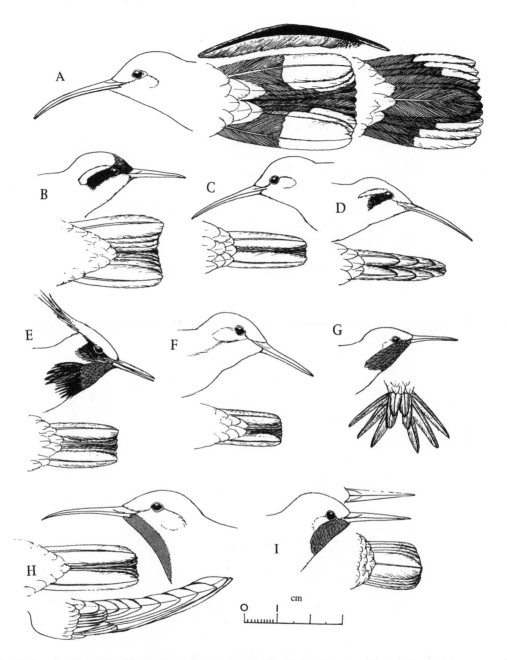

27. Adult male traits of nine Mexican hummingbirds, including violet sabrewing (A), little hermit (B), cinnamon hummingbird (C), Xantus hummingbird (D), black-crested coquette (E), white-bellied emerald (F), beautiful hummingbird (G), green-breasted mango (H), and blue-throated goldentail (I). *(Drawn from specimens)*

F.   Area around and behind eye green to buff; tail
banded black (Rufous-crested coquette, Figure 26J)

FF. Area around and behind eye black; tail not banded
with black (Black-crested coquette, Figure 27E)

EE. Middle of rump green, separating two white lateral
patches (Sparkling-tailed hummingbird, Figure 26O)

CC. Bill longer (minimum 15 mm), straight or slightly decurved

D.   Larger species (wing at least 65 mm) with almost straight,
blackish bills

E.   Wings blackish; bill slightly decurved (Amethyst-throated
hummingbird, Figure 26E)

EE. Wing coverts bright rufous; bill straight (Garnet-throated
hummingbird, Figure 26G)

DD. Smaller species (wing maximum 60 mm)

E.   With a blackish, somewhat decurved bill, and either
mostly deep green, with a purple tail (males), or green
with the middle of the green underparts separated from
the sides by a narrow white line (females) (Green-breasted
mango, Figure 27H)

EE. Not like the above; bill usually reddish basally, but if black
then not decurved

F.   Entire underparts from chin to vent white

G.   Crown blue-tinted, sides of neck metallic green
(Azure-crowned hummingbird, Figure 26F)

GG. Crown green-tinted, sides of neck white

H.   Tail uniform coppery or bronzy; wing
minimum 55 mm (Green-fronted hummingbird, Figure 26N)

HH. Tail metallic green, with a paler tip; wing
under 55 mm (White-bellied emerald,
Figure 27F)

FF. Underparts not entirely white

G.   Rectrices variably cinnamon-colored

H.   Underparts uniformly rich cinnamon-toned
(Cinnamon hummingbird, Figure 27C)

HH. Underparts not uniformly rich cinnamon;
with black postocular and white superciliary
stripes (Xantus hummingbird, Figure 26D)

GG. Rectrices not cinnamon-colored near shafts

H.   Rectrices golden-bronze; gorget purplish,
bill straight and broadened basally (Blue-
throated goldentail, Figure 27I)

HH. Rectrices, gorget and bill not as described

I.   Bill and tail both blackish (Crowned
woodnymph, Figure 26K)

II. Bill red, with a darker tip; tail variable in
shape and color

J. Rather uniformly bronzy green on upperparts, and grayish below; bill at least 25 mm (Dusky hummingbird, Figure 26H)

JJ. Upperparts brilliant green; underparts green or grayish white; bill under 20 mm (Fork-tailed emerald complex, Figure 26I)

K. Endemic to Cozumel Island, tail deeply forked in males (Cozumel emerald)

KK. Mainland species; tail sometimes deeply forked

L. Western slope of Mexico; tail deeply forked in males (Golden-crowned emerald)

LL. Eastern Mexico and Isthmus; tail of males somewhat forked (Canivet's emerald)

Glossary

Adaptive peak. A genetic combination that approximates the maximum fitness for a specific population in a particular place and time.

Advertisement behavior. Communication, especially by various displays, evolved to increase the conspicuousness of an individual; frequently associated with territoriality.

Aggregation. A grouping of individuals attracted to some commonly exploited environmental resource.

Agonistic. Referring to aggressive interactions between individuals, ranging from attack to escape behavior.

Altricial. Referring to the condition of being hatched or born in a relatively undeveloped state, usually blind and without locomotory abilities.

Altruism. Behavior that increases the fitness of another individual at the expense of the individual performing it, including both nurturing behavior (usually between parent and offspring) and succoring behavior (nonparental assistance given to other, sometimes unrelated, individuals).

Angulated. Forming a distinct angle.

Anthesis. The period or time of blossoming in plants.

Apical. At the apex or tip.

Apodiformes. The avian order that includes swifts (suborder Apodi) and hummingbirds (family Trochilidae).

Arroyo. A gulley, creek, or dry creek bed (Spanish).

Attenuated. Tapering to a narrow tip.

Allopatric. Having geographically separated distributions, at least during the breeding season.

Auriculars. Feathers surrounding the ear opening of birds.

Axillars. Feathers located at the base of the underside of the wing.

Barbicels. Hooklet-like outgrowths toward the tip or pennulum of a barbule; also called *hamuli.*

Barbs. The individual subunits of a feather, which extend diagonally outward in pairs from the central shaft and contain the feather pigments responsible for its colors.

Barbules. The individual subunits of a barb, which extend diagonally outward in pairs from the main axis of each barb.

Barranca. A cliff or gorge (Spanish).

Boreal. Northern or northerly.

Chapparal. A vegetation type dominated by ever-green shrubs, usually growing in dense thickets.

Choripetalous. A blossom condition in which the corolla is made up of separate petals.

Cinnamomeous. Yellowish or reddish brown, the color of cinnamon.

Cloud forest. Tropical montane forest at elevation where clouds or mist are frequent, usually rich in epiphytic plants and tree ferns.

Coevolution. The development of genetic traits in two species that help to facilitate interactions, usually to their mutual advantage.

Coleopteran. Referring to the insect order of beetles.

Communication. Behavior patterns of an individual that alter the probability of behavior in another individual in an adaptive manner.

Community. The organisms that occupy a particular habitat.

Congregation. A grouping of individuals attracted to one another on the basis of social interactions among them.

Conspecific. Belonging to the same species; likewise applied to genera (*congeneric*).

Contour feathers. The body feathers in general, exclusive of more specialized types.

Convergent evolution. The evolution of structural similarities in organisms that are not closely related, often because of similar ecological adaptations.

Corolla. Collectively, the petals of a flower.

Courtship. Communication between individuals of opposite sexes that facilitates pair-bonding or fertilization; see also *display*.

Culmen. The distance from the tip of a bird's beak to its base, or to the limit of feathering on its forehead (*exposed culmen*).

Cuneate. Wedge-shaped.

Dendrogram. A diagram of phyletic relationships, drawn in the form of a branching tree.

Dicotyledons. Plants having two seed leaves, usually broad-leaved and net-veined.

Dipteran. Referring to the insect order of flies, mosquitoes, and gnats, all having two functional wings.

Display. Behavior patterns ("signals") that have been evolved ("ritualized") to provide communication functions for an organism, usually through stereotypic performance and exaggeration.

Distal. Away from the main body axis, as opposed to proximal.

Diurnal. Referring to the hours of daylight; diurnal segregation is a special type of habitat segregation (*q.v.*).

Double-rounded. A tail condition in which the central and outermost pairs of tail feathers are shorter than the intermediate ones, producing a double-convex profile.

Dusky. Blackish, or dark brownish black.

Ecotone. A transitional area between two community types.

Egocentric. Behavioral adaptations that are directed toward the survival and health of the individual.

Endemic. Referring to species or groups that are native to and restricted to a particular area; see also *relict*.

Entomophilous. Referring to plants adapted for pollination by insects.

Epigamic. Referring to social interactions that lead toward or facilitate mating; see also *courtship*.

Epiphyte. A plant that depends on other plants for its support but not its nutrition, usually located above ground level.

Femoral. Referring to the region of the upper leg.

Fitness. The relative genetic contribution of an individual toward future generations of its species.

Fledging period. The period between hatching and initial flight in birds; in hummingbirds essentially synonymous with nestling period (*q.v.*).

Form. A taxonomically neutral term for a species or some subdivision of a species.

Fusiform. Spindle shaped, tapering at both ends.

Generalist. A species with broad foraging or habitat requirements, a "jack-of-all-trades" not dependent on a single environmental resource.

Genus (plural, *genera*). A taxonomic category within a family representing a grouping of related species.

Gleaning. Foraging for insects or similar food from leafy or bark surfaces.

Gorget. A patch of iridescent color in the throat region of some hummingbirds.

Graduated. Referring to tail shape in which the rectrices shorten progressively from the central pair outwardly.

Guild. A group of species that exploits the same class of environmental characteristics in a similar way, whether or not they are closely related.

Guttate. Shaped like a teardrop.

Habitat. The physical, chemical, and biotic characteristics of a specific environment.

Habitat segregation. A physical division of environmental resources by two or more of the resource users, such as the two sexes of a species, by age classes or by different species having similar niche adaptations; the resources may be subdivided horizontally, vertically, by microhabitats, or by usage time periods (diurnal, seasonal).

Hawking. Capturing insects while in flight.

Herbaceous. Referring to a nonwoody plant (herb), including both grasslike and nongrasslike plants (forbs).

Homeothermal. Warm-blooded, capable of maintaining a constant body temperature; also called *endothermal.*

Hybridization. The production of progeny of unlike genetic lines, such as between subspecies (*intraspecific hybridization*), separate species (*interspecific*), and more rarely between genera (*intergeneric*).

Hymenopteran. Referring to the insect order of bees, wasps, and ants.

Hyoid apparatus. The paired cartilaginous elements that support the tongue and that in hummingbirds are greatly elongated, allowing for unusual extension.

Hypothermia. A condition of lower-than-normal body temperature and a sleeplike state (*torpidity, q.v.*).

Inflorescence. A flower cluster.

Innate. Inherited, such as instinctive patterns of behavior.

Instinctive behavior. Innate responses that typically are more complex than simple reflexes or taxes, and that are dependent upon specific external stimuli ("releasers") as well as variable and specific internal states ("drives" or "tendencies") for their expression.

Interference. The optical process responsible for iridescent color in feathers, brought about by the reciprocal action of waves of reflected light, which may either reinforce or cancel each other.

Intergeneric. Between separate genera; likewise applied to species (*interspecific*).

Iridescence. A spectral color formed by selective interference and reflection of specific wavelengths of light, rather than by differential pigment absorption and reflection, as is the case with simple feather coloration; see also *interference.*

Isabella. Brown, tinged with reddish yellow.

Isolating mechanism. An innate property of an individual that prevents successful mating and subsequent offspring with individuals of genetically unlike populations; mechanisms that prevent successful mating, rather than control the subsequent development of offspring, called *premating mechanisms*, include heritable ecological, morphological, behavioral, and temporal differences between populations.

Isotherm. A line on a map connecting points having the same average temperature.

Jugulum. The lower portion of the throat.

Juvenal. Referring to the plumage stage of juvenile birds during which fledging normally occurs, and immediately following any downy stage.

Lek. An area occupied by males of species that display in groups in relatively small, contiguous territories; in hummingbirds usually called a *singing assembly.* Often also used as an adjective (*lek behavior*, synonymous with *arena behavior*) and sometimes as a verb (*lekking*).

Loral. Referring to the area (lore) between the eye and the base of the bill in birds.

Lumbar. Referring to the lower back region.

Malar. Referring to the area extending from the base of the mandible backward below the eye, sometimes forming a moustache-like streak.

Mandible. The lower component of the beak or bill in birds.

Mating system. Patterns of mating within a population, including length and strength of pair bond, the number of mates, degree of inbreeding, and the like.

Maxilla. The upper component of the beak or bill in birds.

Melanin. A pigment of hair and feathers that is insoluble in water and organic solvents and is responsible for blackish, brown, brownish yellow, and reddish brown coloration.

Mesial. Toward the middle, medial.

Micron. A micromillimeter (0.001 mm).

Monocotyledons. Plants having a single seed leaf, typically narrow-leaved and parallel-veined.

Monotypic. Referring to a taxonomic category that has only one unit in the category immediately subordinate to it, such as a genus with only one species.

Nectarivorous. Adapted for foraging on nectar.

Neotropical region. The zoogeographic region that includes all of South America, Middle America, and Mexico north to the central highlands, as well as the West Indies.

Nestling period. The period between hatching and leaving the nest; in hummingbirds essentially synonymous with the *fledging period.*

Niche. Structural, physiological, and behavioral adaptations of a species to its environment.

Nidicolous. The condition in which the young are raised in the nest until fledging or nearly so.

Nocturnal. Referring to the hours of darkness.

Nuptial plumage. Plumage associated with the breeding season.

Obsolete. Rudimentary, nearly lacking.

Occiput. The region of the head formed by the back of the skull.

Oligocene. An early geological epoch in the Cenozoic Period, extending from about 25 to 40 million years before the present.

Operculum. A shelf-like structure partially covering the nasal opening.

Orbital. The head region associated with the eyes; *suborbital* and *postorbital* areas are those below and behind the eyes, respectively.

Ornithophilous. Referring to plants adapted for pollination by birds.

Pair-bonding. Individualized association between members of the opposite sex lasting for variable periods, but longer than that necessary only to attain fertilization.

Páramo. The moist vegetational zone above timberline in the Andes.

Parapatric. Descriptive of two or more populations whose ranges are in geographic contact but do not overlap.

Phyletic. Referring to an evolutionary lineage; synonymous with *phylogenetic.*

Pileum. The crown of the head, from the forehead to the occiput.

Pollination. The transfer of pollen from the anther to the stigma of the same flower (*self-pollination*) or of another flower (*cross-pollination*).

Polygamy. A mating system in which individuals of one sex maintain simultaneous or serial individualized contacts (*pair-bonds*) with two or more members of the opposite sex, to attain multiple fertilization; includes both *polyandry* (pairing with multiple males) and *polygyny* (pairing with multiple females); see also *promiscuity.*

Promiscuity. A mating system in which no pair-bonds exist, and individual contacts between the sexes occur only at the time of fertilization, typically involving multiple matings by one sex or the other; see also *polygamy.*

Proximal. Toward the main body axis, as opposed to distal.

Protandry. The asynchronous development of sexual parts in a flower, with male structures developing earlier than female structures; in animals refers to the change of sex from male to female during development.

Psilopaedic. Hatched naked or nearly so.

Quasi-social. Referring to behavioral adaptations that facilitate aggregation behavior.

Racquet-like. Formed in the shape of a racket.

Rectrix (plural, *rectrices*). The individual tail feathers of birds.

Reflectance. A measure of the relative amount of light reflected back from a surface, rather than absorbed by it or transmitted through it; when two light rays are reflected from the outer and inner surface of a transparent substance, they may either reinforce or cancel specific wavelengths of light; see also *interference* and *iridescence.*

Refractive index. The ratio of the speed of light in air to its speed in any other transparent substance, resulting in differential bending of the light rays (refraction) in that substance.

Relict. An organism belonging to an earlier time than the other members of the contemporary population in the region where the organism now occurs.

Remix (plural, *remiges*). The individual flight feathers (primaries and secondaries) of birds.

Reproductive isolation. The prevention of successful interbreeding between individuals of different populations, attained through various genetically determined *isolating mechanisms (q.v.).*

Resaca. A surfline or coastal edge (Spanish).

Resource. A feature of the environment, the control of which contributes to an organism's fitness; *resource partitioning* is a division of such a feature, within or between two species, to reduce the intensity of competition over its control; see also *fitness* and *habitat segregation.*

Rictal. Referring to the point where the mandible and maxilla join at the base of the bill.

Riparian. Associated with riverbanks or lake shorelines.

Ritualization. The evolutionary process by which behaviors gradually acquire communication function (become "signals"); the development of signaling behavior and associated signaling devices.

Rufous. Brownish red to reddish.

Rufescent. Somewhat tinged with reddish.

Serrate. Toothed, saw-like.

Sexual dimorphism. Here used to refer to species in which the sexes differ markedly as adults, and including both *dichromatism* (the sexes differ in color) and *diethism* (they differ in behavior).

Sexual selection. Selection favoring divergence between the sexes of a species (*sexual dimorphism*), resulting from differential mating success produced by individual differences in heterosexual attraction or intrasexual dominance.

Singing assembly. A group of two or more territorial males localized in a traditional area (*lek* or *arena*) for courtship and singing; see *lek.*

Social behavior. Behavioral adaptations that result in congregating behavior and facilitate social interactions.

Solferino. Bluish-red in color.

Specialist. A species with narrow foraging or habitat adaptations, or both; a "master-of-one-trade" specialized for exploiting a single resource.

Speciation. The multiplication of species through the development of reproductive isolation in a population or group of populations.

Species. A "kind" of organism or, more technically, a group or groups of actually or potentially interbreeding populations that are reproductively isolated from all other populations; the term remains unchanged in the plural, and also refers to the taxonomic category below that of

the genus and above that of the subspecies (abbreviated *sp.,* plural *spp.*).

Species group. A group of closely related species, usually with partially overlapping ranges.

Squamate. Scale-like.

Stamen. The part of the flower that produces pollen, including the anthers and their supporting filaments.

Standard metabolic rate. The heat production of an organism per unit of time, under controlled conditions; also called *basal metabolic rate.*

Stigma. The tip of the pistil in flowers, which receives the pollen.

Strategy. The evolved niche adaptations of a population that are associated with fitness in a particular environment, including mating strategies, foraging strategies, life history strategies, etc.; *optimal strategies* are those that maximize fitness while minimizing wastage of time, energy, gametes, or other limited attributes.

Style. The portion of the pistil that lies above the ovary and supports the stigma in flowers.

Subbasal. Toward or near the base.

Subgenus. One or more species of a genus that differ from the rest of the genus as to be recognized taxonomically, but are not sufficiently distinctive to be considered a full genus.

Subspatulate. Slightly spoon-shaped in outline.

Subspecies. A group of local populations of a species that occupies part of the species' range and differs taxonomically from other local populations; a geographic race (abbreviated *ssp.,* plural *sspp.*).

Subterminal. Toward or near the tip.

*Superspecies*Two or more species with largely or entirely non-overlapping ranges, which are clearly derived from a common ancestor, but are too distinct to be considered a single species.

Sympatric. Having geographically overlapping distributions, at least during the breeding season.

Sympetalous. A blossom condition in which the corolla is partly to entirely fused, often forming a tube.

Syrinx. The structure in birds responsible for producing vocalizations, analogous to the larynx in mammals.

Taxon (plural, *taxa*). A group of organisms belonging to a formally recognized taxonomic unit, such as a species or genus; taxonomy is the procedure by which taxa are identified, described, and classified.

Territory. An area having resources that are controlled or defended by an animal against others of its species (*intraspecific territories*), or less often against individuals of other species (*interspecific territories*). Hummingbird territories usually control food supplies (feeding territories) and/or facilitate fertilization (mating station territories), but in other groups often include nesting or roosting sites as well.

Territoriality. The advertisement and agonistic behavior associated with territorial establishment and defense.

Torpidity. A sleep-like state of reduced body temperature (*hypothermia*) and metabolism, entered into during the night (*noctivation*) or during colder periods of the year (*hibernation*).

Transition Zone. A "life zone" that occurs between the relatively cooler Canadian Zone and the relatively hotter Austral Zone, and generally conforming to the lower coniferous forest belt in western mountains.

Traplining. Behavior of hummingbirds that do not defend a definite feeding territory, but instead visit a variety of flowers over a regularly traveled route.

Trochilidae. The family of birds that includes all the hummingbirds.

Vicariad. One of two or more geographically isolated but closely related populations having similar ecologies and morphologies.

Vinaceous. Referring to the color of wine or grapes; reddish.

Wing disc loading. The ratio of body mass to the area swept out by the wings during flight.

Xeric. Referring to a dry, desert-like environment.

Xerophilous. Referring to desert- or drought-adapted plants.

Zygomorphic. Blossoms in which the corolla exhibits bilateral or other nonradial symmetry, such as bilabiate blossoms.

Origins of Latin Names of North American Hummingbirds

Abeillia. In honor of M. Abeillé and his wife Felice, mid-19th-century French naturalists.

Amazilia. Evidently not named for the Amazon region, but rather for Amazili, a native heroine in the 1777 novel *Les Incas, ou la Destruction de Empire du Pérou,* by J. F. Marmontel.

    beryllina—the color of beryl, bluish green.

    candida—from the Latin *candidus,* shining white.

    cyanocephala—from the Green *kuanos,* dark blue, and *kephalos,* headed.

    rutila—from the Latin *rutilus,* red or golden.

    tzacatl—probably from Aztec, grass (green). A man named Rieffer collected the first specimens of this species, which is sometimes called Rieffer's hummingbird.

    violiceps—from the Latin *viola,* violet; and *ceps,* head.

    yucatanensis—of Yucatan.

    viridifrons—from the Latin *viridis,* green, and *frons,* forehead.

Anthracothorax. From the Greek *anthrax,* coal; and *thorax,* the chest.

    prevostii—After F. Prevost, French museum artist and author of the mid-19th century.

Archilochus. Apparently after Archilochus of Paros, 7th-century Greek poet; or from the Greek prefix *arch-,* chief; and *lochos,* a body of individuals: thus, first among the birds.

    alexandri—after M. Alexandre, who discovered the species.

    anna—for Anne de Belle Massena (c. 1806–1896), the wife of François Victor (see *Heliodoxa fulgens*).

    calliope—from the Greek *kalliope,* beautiful-voiced; Calliope was the muse of eloquence; or from the Greek *kalle,* beautiful; and *ops,* face.

    colubris—seemingly from the Latin *colubris,* a serpent; but more likely a Latinized form of *colibri,* a South American Indian name. The French *colibre* and the German *Kolibris* both refer to hummingbirds and are of the same origin.

    costae—for Louis Marie Pantaleon Costa (1806–1864), Marquis de Beau-Regard, French nobleman who specialized in collecting hummingbirds.

Atthis. Greek for an Athenian, possibly for Philomena, who was transformed into a bird; or after a mythical priest of the goddess Cybele.

> *heloisa* (transfer from *Selasphorus*)—probably for Heloise (1101–1164), niece of the Canon of Notre Dame, who had a famous but tragic love affair with her mentor Pierre Abelard (1079–1142).

Basilinna. From Greek, a queen.

Calliphlox. From Greek *kalos*, beautiful; and *thoras*, back.

> *lucifer*–from Latin, light-bringing.

Calypte. From the Greek *kalyptos*, covered or hidden.

Campylopterus. From the Greek *kampulos*, bent or curved; and *pteros*, the wing.

> *curvipennis*—from the Latin *curvus*, curved; and *pennis*, winged.
>
> *excellens*—from Latin, excellent or remarkable.
>
> *hemileucurus*—from the Greek *hemi*, half; and *leukouros*, white-tailed.

Chlorostilbon. From the Greek *khloros*, green; and *stilbon*, glistening.

> *auriceps*—from the Latin *aurum*, gold; and *ceps*, capped or headed.
>
> *canivetii*—in honor of E. Canivet de Carentan, French ornithologist and collector of the late 19th century.
>
> *forficatus*—from Latin, scissor-like, or forked.
>
> *ricordii*—after Philippe Ricord (1800–1889), French surgeon.

Colibri. A term derived from the Carib language, meaning "a resplendent area."

> *thalassinus*—from the Greek *thalassios*, marine.

Cynanthus. From the Greek *kuanos*, dark blue; and *anthos*, flower or blossom.

> *latirostris*—from the Latin *latus*, broad; and *rostris*, bill.
>
> *leucotis*—from the Greek *leukos*, white; and *otos*, ear.
>
> *sordidus*—from Latin, shabby or poor.

Doricha. After Doricha or Dorikha, a Greek courtesan (hetaira) of the 7th century B.C.

> *eliza*—after Eliza Lefevre, wife of Amadee Lefevre, French zoologist of the early 19th century.

Eugenes. from the Greek prefix *eu-*, good; and *genos*, birth: thus, well-born.

Eupherusa. From the Greek prefix *eu-*, good; and *pherousa*, bearing.

> *cyanophrys*—from the Greek *kuanos*, dark blue; and *ophrus*, browed.
>
> *eximis*—from Latin, exceptional or distinguished.
>
> *poliocerca*—from the Greek *polios*, gray; and *kerkos*, the tail.

Florisuga. From the Latin *floris*, a flower; and *sugere*, to suck.

> *mellivora*—from the Latin *mel*, honey; and *vorus*, eating.

Heliodoxa. From the Greek *helios*, the sun; and *doxa*, glory.

> *fulgens*—from the Latin *fulgeo*, to shine or glitter. The vernacular name is after François Victor Massena (1798–1863), third Prince of d'Essling and Duc di Rivoli, a patron of natural history (Gruson, 1972).

Heliomaster. From the Greek *helio*, the sun; and *master*, a searcher.

> *constantii*—in honor of C. Constant, French taxidermist of the late 19th century.
>
> *longirostris*—from the Latin *longus*, long; and *rostris*, billed.

Hylocharis. From the Greek *hyle*, a wood or forest; and *charis*, delight or beauty.

> *eliciae*—after Elicia Alain, of the mid-19th century, but about whom nothing is known.
>
> *xantusi*—in honor of A. J. Xantus de Vesey, Hungarian adventurer and collector of the late nineteenth century.

Lampornis. From the Greek *lampros*, shining or beautiful; and *ornis*, bird.

> *amethystinus*—from Latin, amethyst-colored.
>
> *clemenciae*—for the wife of R. P. Lesson (1794–1849), French naturalist.

Lamprolaima. From the Greek *lampros*, shining or beautiful; and *laimus*, the throat.

> *rhami*—in honor of H. C. de Rham, American naturalist and collector of the mid-19th century.

Lophornis. From the Greek *lophos*, a crest; and *ornis*, a bird.

> *brachylopha*—from the Greek *brakhus*, short; and *lophos*, a crest.
>
> *helenae*—in honor of Princesse Helène d'Orleans, wife of Duc d'Orleans of the early to middle 19th century.

delattrei—in honor of H. de Lattre, French naturalist of the mid-19th century.

Nesophlox. From the Greek *nesos,* island; and *phlox,* flame.

Phaethornis. From the Greek *phaethon,* the sun god; and *ornis,* a bird. The vernacular name "hermit" refers to a recluse or monk.

> *longuemareus*—in honor of G. Longuemare, French collector of the mid-nineteenth century.
>
> *mexicanus*—from Mexico.
>
> *superciliosus*—from Latin, supercilious.

Philodice. From Greek mythology, the mother of Hilara and Phoebe.

Selasphorus. From the Greek *selas,* light; and *phoros,* bearing.

> *heloisa*—named by R. P. Lesson and A. De Lattre for an unspecified woman, possibly the wife of De Lattre (Gruson, 1972).
>
> *platycercus*—from the Greek *platys,* flat; and *kerkos,* tail.

rufus—from the Latin, reddish.

sasin—probably from the Nootka Indian name for this species. C. A. Allen (1841–1930) collected the species, and it was named *alleni* for him by H. W. Henshaw, but was later found to have been already described. The vernacular name Allen hummingbird nonetheless has been retained (Gruson, 1972).

Stellula. From the Latin, a little star.

Thalurania. From the Greek *thalia,* abundance or wealth; and *ouranios,* heavenly.

> *colombica*—from Colombia, South America.
>
> *furcata*—from Latin, forked.
>
> *ridgwayi*—in honor of R. Ridgway (1850–1929), premier American ornithologist of his day.

Tilmatura. From the Greek *tilma,* plucked or shredded; and *oura,* the tail.

> *dupontii*—after a Monsieur Dupont, French dealer in natural history during the mid-19th century.

Bibliography

General References and Regional Identification Guides

Austin, O. L., Jr. 1961. *Birds of the World.* Golden Press, New York.

Bond, J. 1971. *Birds of the West Indies,* 2nd ed. Houghton Mifflin, Boston.

Brisson, M.-J. 1760. *Ornithologie.* 6 vols. Bauch, Paris.

Davis, L. I. 1972. *A Field Guide to the Birds of Mexico and Central America.* University of Texas Press, Austin.

de Schauensee, R. M. 1966. *The Species of Birds of South America.* Livingston Publishing, Narberth, Pa.

de Schauensee, R. M., and W. H. Phelps. 1978. *A Guide to the Birds of Venezuela.* Princeton University Press, Princeton.

Elliot, D. G. 1878. A classification and synopsis of the Trochilidae. *Smithsonian Institution, Contributions to Knowledge* 317:1–277.

ffrench, R. 1991. *A Guide to the Birds of Trinidad and Tobago,* 2nd ed. Livingston Publishing, Wynnewood, Pa.

Godfrey, W. E. 1986. *The Birds of Canada,* 2nd ed. National Museums of Canada, Ottawa.

Gould, J. 1849–61. *A Monograph of the Trochilidae, or Family of Hummingbirds.* Taylor and Francis, London. (Second ed. and supplements 1880–85). Reprinted (1990) as a single volume and in slightly reduced format, as *Hummingbirds,* Wellfleet Press, Secaucus, N.J.

Gray, G. R. 1861–71. *Hand-list of Genera and Species of Birds, Distinguishing Those Contained in the British Museum.* Richard and John Taylor, London.

Greenewalt, C. H. 1960a. *Hummingbirds.* Doubleday and the American Museum of Natural History, Garden City.

Grzimek, B., ed. 1972. *Grzimek's Animal Life Encyclopedia,* vol. 8. Van Nostrand Reinhold, New York.

Haverschmidt, F. 1968. *Birds of Surinam.* Oliver & Boyd, Edinburgh.

Howell, S.N.G., and S. Webb. 1995. *A Guide to the Birds of Mexico and Northern Central America.* Oxford University Press, New York.

Land, H. 1970. *Birds of Guatemala.* Livingston Publishing, Narberth, Pa.

Peters, J. L. 1945. *Check-List of Birds of the World,* vol. 5. Harvard University Press, Cambridge.

Peterson, R. T., and E. L. Chalif. 1973. *A Field Guide to Mexican Birds.* Houghton, Mifflin, Boston.

Ridgley, R. S., and J. W. Gwynne, Jr. 1989. *A Guide to the Birds of Panama,* rev. ed. Princeton University Press, Princeton.

Ridgway, R. 1890. The hummingbirds. *Report of the U.S. National Museum for 1890,* pp. 253–383.

Ridgway, R. 1911. The birds of North and Middle America. Part V. *U.S. Natl. Mus. Bull.* 50:1–859.

Rutgers, A. 1972. *Birds of South America. Illustrations from the Lithographs of John Gould.* St. Martin's Press, New York. (Includes reproductions of 80 hummingbird plates.)

Scheithauer, W. 1967. *Hummingbirds.* Thomas Y. Crowell, New York.

Sick, H. 1993. *Birds in Brazil.* Princeton University Press, Princeton.

Skutch, A. F. 1973. *The Life of the Hummingbird.* Crown, New York.

Smithe, F. B., and R. A. Paynter. 1963. *Birds of Tikal.* Natural History Press, Garden City, N.Y.

Stiles, F. G., and A. F. Skutch. 1989. *A Guide to the Birds of Costa Rica.* Cornell University Press, Ithaca.

Thompson, A. L. 1974. *A New Dictionary of Birds.* McGraw-Hill, New York.

Other Relevant Literature

Aldrich, E. C. 1945. Nesting of the Allen hummingbird. *Condor* 47:137–148.

———. 1956. Pterylography and molt of the Allen hummingbird. *Condor* 58:121–133.

American Ornithologists' Union (AOU). 1957. *Check-List of North American Birds,* 5th ed. Lord Baltimore Press, Baltimore.

———. 1983. *Check-list of North American Birds,* 6th ed. Allen Press, Lawrence, Kan.

———. 1995. 40th supplement to the American Ornithologists' Union *Check-list of North American birds, Auk* 112:819–830.

Anderson, J. O., and G. Monson. 1981. Berylline hummingbirds nest in Arizona. *Continental Birdlife* 2:56–61.

Arizmendi, M. del C., and J. F. Ornelas. 1990. Hummingbirds and their floral resources in a dry tropical forest in Mexico. *Biotropica* 22:172–180.

Armitage, K. B. 1955. Territorial behavior in fall migrant rufous hummingbirds. *Condor* 57:239–240.

Arriaga, L., R. Rodriguez-Estrella, and A. Ortega-Rubio. 1990. Endemic hummingbirds and madrones of Baja: are they mutually dependent? *Southwest. Nat.* 35:76–79.

Austin, D. V. 1975. Bird flowers in the eastern United States. *Florida Sci.* 38:1–12.

Austin, G. T. 1970. Interspecific territoriality of migrant calliope and resident broad-tailed hummingbirds. *Condor* 72:234.

Baepler, D. H. 1962. The avifauna of the Soloma region in Huehuetenango, Guatemala. *Condor* 64:140–153.

Bailey, A. M. 1927. Notes on the birds of southeastern Alaska. *Auk* 44:351–367.

———. 1974. Second nesting of broad-tailed hummingbirds. *Condor* 76:350.

Bailey, A. M., and R. J. Niedrach. 1965. *Birds of Colorado.* 2 vols. Denver Museum of Natural History, Denver.

Bailey, F. M. 1928. *Birds of New Mexico.* New Mexico Department of Fish and Game, Santa Fe.

Bakus, G. J. 1962. Early nesting of the Costa hummingbird in southern California. *Condor* 64:438–439.

Baldridge, F. A. 1983. Plumage characteristics of juvenal black-chinned hummingbirds. *Condor* 85:102–105.

Baltosser, W. H. 1986. Nesting success and productivity of hummingbirds in southeastern New Mexico and southeastern Arizona. *Wils. Bull.* 98:353–367.

———. 1987. Age, species and sex determination of four North American hummingbirds. *N. Am. Bird Bander* 12:151–166.

———. 1989a. Costa's hummingbird: its distribution and status. *West. Birds* 20:41–62.

———. 1989b. Nectar availability and habitat selection by hummingbirds in Guadeloupe Canyon. *Wils. Bull.* 101:559–578.

———. 1994. Age and sex determination in the calliope hummingbird. *West. Birds* 25:104–109.

——— and P. Scott. 1996. Costa's hummingbird (*Calypte costae*). In *The Birds of North America,* No. 25 (A. Pool, P. Stettenheim, and F. Gill, eds.). Academy of Natural Sciences, Philadelphia, and American Ornithologists' Union, Washington, D.C.

Banks, R. C. 1990. Taxonomic status of the coquette hummingbird of Guerrero, Mexico. *Auk* 107:191–192.

Banks, R. C., and N. K. Johnson. 1961. A review of North American hybrid hummingbirds. *Condor* 63:3–28.

Baptista, L. F., and M. Matsui. 1979. The source of the dive-noise of the Anna's hummingbird. *Condor* 81:87–89.

Barash, D. P. 1972. Lek behavior in the broad-tailed hummingbird. *Wilson Bull.* 82:202–203.

Barbour, R. 1943. Cuban ornithology. *Mem. Nuttall Ornithol. Club* 9:1–129.

Baumgartner, F. M. 1981. To hold a hummingbird. *Bird Watcher's Digest* 3(3):18–21.

——— and A. M. Baumgartner. 1992. *Oklahoma Bird Life*, Oklahoma University Press, Norman.

Behnke-Pedersen, M. 1972. *Kolibrien*, vol. 1. Aage Poulsen, Skibby, Denmark.

Bendire, C. E. 1895. Life histories of North American birds. *U.S. Natl. Mus. Spec. Bull.* 3.

Bené, F. 1947. The feeding and related behavior of hummingbirds with special reference to the black-chin, Archilochus alexandri (Bourcier and Mulsant). *Mem. Boston Soc. Nat. Hist.* 9:395–478.

Bent, A. C. 1940. Life histories of North American cuckoos, goatsuckers, hummingbirds and their allies. *U.S. Natl. Mus. Bull.* 176:1–506.

Berlioz, J. 1932. Notes critiques sur quelques Trochilidés du British Museum. *Oiseau* (N.S.) 2:530–534.

———. 1944. *La Vie des Colibris.* Histoires Naturelles, Paris.

———. 1965. Note critique sur les Trochilidés des genres *Timolia et Augasma*. *Oiseau* 35:1–8.

Binford, L. C. 1985. Re-evaluation of the "hybrid" hummingbird *Cynanthus sordidus* × *C. latirostris* from Mexico. *Condor* 87:148–150.

———. 1989. A distributional survey of the birds of the Mexican state of Oaxaca. *Ornith. Monogr.* No. 43.

Blair, F., A. Blair, P. Brodkorp, F. Cagle, and G. Moore. 1968. *Vertebrates of the United States*, 2nd ed. McGraw-Hill, New York.

Bonhote, J. L. 1903. On a collection of birds from the northern islands of the Bahamas group. *Ibis* 8(1):273–315.

Borrero, J. I. 1965. Notas sobre el comportamiento del colibri coli-rojo (*Amazilia tzacatl*) y el mielero (*Coereba flaveola*), en Colombia. *El Hornero* 10:247–250.

———. 1975. Notas sobre el comportamiento reproductivo de *Amazilia tzacatl*. *Ardeola* 21:933–943.

Bowmaker, J. K. 1988. Avian colour vision and the environment. *Acta 19th Cong. Int. Orn., Ottawa* 12184–12194. (UV vision in hummingbirds.)

Brandt, H. 1951. *Arizona and Its Bird Life*. Bird Research Foundation, Cleveland, Ohio.

Brewster, W. 1902. Birds of the Cape region of Lower California. *Bull. Mus. Comp. Zool.* (Harvard University) 41:1–240.

Brown, J. H., and M. A. Bowers, 1985. Community organization in hummingbirds' relationship between morphology and ecology. *Auk* 102:1–69.

Brown, J. H., W. A. Calder, and A. Kodric-Brown. 1978. Correlates and consequences of body size in nectar-feeding birds. *Am. Zool.* 18:687–700.

Brown, J. H., and A. Kodric-Brown. 1979. Convergence, competition and mimicry in a temperate community of hummingbird-pollinated flowers. *Ecology* 60:1022–1035.

Brudenell-Bruce, P.G.C. 1975. *The Birds of New Providence and the Bahama Islands*. Taplinger Publishing, New York.

Brunton, D. F., S. Andrews, and D. G. Paton. 1979. Nesting of the calliope hummingbird in Kananskis Provincial Park, Alberta. *Can. Field-Nat.* 93:449–451.

Butler, A. G. 1932. (Exhibition of a hybrid hummingbird, *Damophila amabilis* × *Amazilia tzacatl*). *Bull. Br. Ornithol. Club* 47:134.

Calder, W. A. 1971. Temperature relationships and mating of the calliope hummingbird. *Condor* 73:314–321.

———. 1973. Microhabitat selection during nesting of hummingbirds in the Rocky Mountains. *Ecology* 54:127–134.

———. 1976. Energy crisis of the hummingbird. *Nat. Hist.* 85(3):24–29.

———. 1993. Rufous hummingbird. In *The Birds of North America*, No. 53 (A. Pool, P. Stettenheim, and F. Gill, eds.). Academy of Natural Sciences, Philadelphia, and American Ornithologists' Union, Washington, D.C.

——— and L. L. Calder. 1992. Broad-tailed hummingbird. In *The Birds of North America*, No. 16 (A. Pool, P. Stettenheim, and F. Gill, eds.).

Academy of Natural Sciences, Philadelphia, and American Ornithologists' Union, Washington, D.C.

———— and ————. 1994. Calliope hummingbird. In *The Birds of North America,* No. 135 (A. Pool, P. Stettenheim, and F. Gill, eds.). Academy of Natural Sciences, Philadelphia, and American Ornithologists' Union, Washington, D.C.

———— and E. G. Jones. 1989. Implications of recapture data for migration of the rufous hummingbird (*Selasphorus rufus*) in the Rocky Mountains. *Auk* 106:488–489.

————, N. M. Waser, S. M. Hiebert, D. W. Inouye, and S. Miller. 1983. Site fidelity, longevity, and population dynamics of broad-tailed humming-birds: a ten-year study. *Oecologia* 56:357–364.

Carpenter, F. L. 1976. Ecology and evolution of an Andean hummingbird (*Oreotrochilus estella*). *Univ. Calif. Pub. Zool.* 106:1–74.

Carriker, M. A., Jr. 1910. An annotated list of the birds of Costa Rica, including Cocos Island. *Ann. Carnegie Mus.* 6:314–915.

Carroll, S. P., and L. Moore. 1993. Hummingbirds take their vitamins. *Anim. Behav.* 46:817–820.

Chapman, F. M. 1902. Notes on birds and mammals observed near Trinidad, Cuba, with remarks on the origin of West Indian bird *life. Am. Mus. Nat. Hist. Bull.* 4:279–330.

Chase, V. C., and P. H. Raven. 1975. Evolutionary and ecological relationships between *Aquilegia formosa* and *A. pubescens* (Ranunculaceae), two perennial plants. *Evolution* 29:474–486.

Clench, M. H., and R. C. Leberman. 1978. Weights of 151 species of Pennsylvania birds analyzed by month, age, and sex. *Bull. Carnegie Mus. Nat. Hist.* (5):1–87.

Clyde, D. P. 1972. Anna's hummingbird in adult male plumage feeds nestling. *Condor* 74:102.

Cogswell, H. L. 1949. Alternate care of two nests in the black-chinned hummingbird. *Condor* 51:176–178.

Cohn, J. M. W. 1968. The convergent flight mecha-nisms of swifts (Apodi) and hummingbirds (Trochili) (Aves). Ph.D. dissertation, University of Michigan, Ann Arbor.

Colwell, R. K. 1985. Stowaways on the Hummingbird Express. *Nat. Hist.* 94(7):56–63.

————. 1995. Effects of nectar consumption by the hummingbird flower mite *Proctolaeleps kirmsi* on

nectar availability in *Hamelia patens. Biotropica* 27:206–217.

Colwell, R. K., B. J. Betts, P. Bunnell, F. L. Carpenter, and P. Feinsinger. 1974. Competition for the nectar of *Centropogon valerii* by the hummingbird *Colibri thalassinus* and the flower-pecker *Diglossa plumbea,* and its evolutionary implications. *Condor* 76:447–452.

Conway, A. E., and S. R. Drennan. 1979. Rufous hummingbirds in eastern North America. *Am. Birds* 33:130–132.

Cook, R. E. 1969. Variation in species density of North American birds. *Syst. Zool.* 18:63–84.

Cottam, C., and P. Knappen. 1939. Food of some uncommon North American birds. *Auk* 49:479–481.

Davis, T. A. W. 1958. The displays and nests of three forest hummingbirds. *Ibis* 100:31–39.

de Schauensee, R. M. 1964. *The Birds of Colombia and Adjacent Areas of South and Central America.* Livingston Publishing, Narberth, Pa.

————. 1967. *Eriocnemis mirabilis,* a new species of hummingbird from Colombia. *Not. Nat.* 402:1–2.

Demaree, S. R. 1970. Nest-building, incubation period and fledging in the black-chinned hummingbird. *Wilson Bull.* 82:225.

Des Granges, J. L. 1979. Organization of a tropical nectar feeding bird guild in a variable environ-ment. *Living Bird* 17:199–236.

Diamond, A. W. 1973. Habitats and feeding stations of St. Lucia forest birds. *Ibis* 115:313–329.

Dickey, D. R., and A. J. van Rossem. 1938. The birds of El Salvador. *Field Mus. Nat. Hist. Zool. Ser.* 23:1–609.

Dorst, R. 1962. Nouvelle recherches biologiques sur les Trochilidés des haut Andes péruviennes (*Oreotrochilus estella*). *Oiseau* 32:95–126.

Dubois, A. D. 1938. Observations at a rufous hummingbird's nest. *Auk* 55:629–641.

Dunning, J. B., Jr. (ed.). 1993. *CRC Handbook of Avian Body Masses.* C.R.C. Press, Baton Rouge, La.

Edgarton, H. E., R. J. Niedrach, and W. van Riper. 1951. Freezing the flight of hummingbirds. *Nat. Geogr.* 100:245–261.

Edwards, E. P. 1973. *A Field Guide to the Birds of Mexico.* E. P. Edwards, Sweet Briar, Va.

Edwards, E. P., and R. E. Tashian. 1959. Avifauna of

the Catemaco Basin of southern Veracruz, Mexico. *Condor* 61:325–337.

Eisenmann, E. 1955. The species of Middle American birds. *Trans. Linnaean Soc. N.Y.* 7:1–127.

Elgar, R. J. 1980. Observations on the display of the blue-tailed emerald. *Avic. Mag.* 86:147–150.

English, T. S. 1928. A diary of the nesting of *Microlyssa exilis*, the crested hummingbird of Montserrat, West Indies. *Ibis* (12th series) 4:13–16.

———. 1934. Notes on a nest of *Microlyssa exilis*. *Ibis* (13th series) 4:838.

Escalante-Pliego, P., and A. T. Peterson. 1992. Geographic variation and species limits in middle American woodnymphs (*Thalurania*). *Wils. Bull.* 104:205–219.

Feinsinger, P. 1976. Organization of a tropical group of nectarivorous birds. *Ecol. Monogr.* 46:257–291.

Feinsinger, P., and R. W. Colwell. 1978. Community organization among neo-tropical nectar-feeding birds. *Am. Zool.* 18:779–795.

Fitzpatrick, A. 1966. My friend rufous. *Fla. Nat.* 39:35–38, 54.

Fitzpatrick, J. W., D. E. Willard, and J. W., Terborgh. 1979. A new species of hummingbird from Peru. *Wilson Bull.* 91:177–186.

Foster, W. L., and J. Tate, Jr. 1966. The activities and coactions of animals at sapsucker trees. *Living Bird* 5:87–114.

Fox, R. P. 1954. Plumages and territorial behavior of the lucifer hummingbird in the Chisos Mountains, Texas. *Auk* 71:465–466.

Fraga, R. M. 1989. Interactions between nectarivorous birds and *Aphelandra sinclariana* in Panama. *J. Trop. Ecol.* 6:19–26.

Friedmann, H., L. Griscom, and R. T. Moore. 1950. Distributional check-list of the birds of Mexico. Part 2. *Pacific Coast Avifauna* (29):1–436.

Gass, C. L. 1979. Territory regulation, tenure and migration in rufous hummingbirds. *Can. J. Zool.* 57:914–923.

Gass, C. L., G. Angehr, and J. Centra. 1976. Regulation of food supply by feeding territoriality in the rufous hummingbird. *Can. J. Zool.* 54:2046–2054.

Gaunt, S. L. L., L. F. Baptista, J. E. Sanchez, and D. Hernandez. 1994. Song learning as evidenced from song sharing in two hummingbird species (*Colibri coruscans* and *C. thalassinus*). *Auk* 111:87–103.

Gill, F. B., A. L. Mark, and R. T. Ray. 1982. Competition between hermit hummingbirds Phaethorninae and insects for nectar in a Costa Rican rain forest. *Ibis* 124:44–49.

Goldsmith, T. H. 1980. Hummingbirds see near ultraviolet light. *Science* 207:786–788.

Grant, K. A., and V. Grant. 1968. *Hummingbirds and Their Flowers*. Columbia University Press, New York.

Grant, V. 1952. Isolation and hybridization between *Aquilegia formosa* and *A. pubescens*. *Aliso* 2:341–360.

Grantsau, R. 1968. Uma nova espécie de *Phaethornis* (Aves, Trochilidae). *Papeis Avulsos de Zoologia* 22 (article 7):57–59.

———. 1969. Uma nova espécie *Threnetes* (Aves, Trochilidae). *Papeis Avulsos de Zoologia* 22 (article 23):246–249.

Graves, G. R. 1980. A new species of metaltail hummingbird from northern Peru. *Wilson Bull.* 92:1–7.

Greenewalt, C. 1960b. The hummingbirds. *Nat. Geogr.* 118:658–679.

———. 1962. Dimensional relationships for flying animals. *Smithsonian Misc. Coll.* 144:1–46.

Grinnell, J., and A. H. Miller. 1944. The distribution of birds of California. *Pacific Coast Avifauna*. (21):1–608.

Gruson, E. S. 1972. *Words for Birds*. Quadrangle Press, New York.

Hartman, F. A. 1954. Cardiac and pectoral muscles of Trochilidae. *Auk* 71:467–469.

Hauser, D. C., and N. Currie, Jr. 1966. Hummingbird survives through December in North Carolina. *Auk* 83:138–139.

Hering, L. 1947. Courtship and mating of broad-tailed hummingbird in Colorado. *Condor* 49:126.

Hinklemann, C. 1988. Comments on some recently described species of hermit hummingbirds. *Bull. Brit. Orn. Club:* 108:159–169.

Hinman, D. A. 1928. Habits of the ruby-throated hummingbird. *Auk* 45:504M–505.

Horvath, O. H. 1964. Seasonal differences in rufous hummingbird nest height and their relation to nest climate. *Ecology* 45:235–241.

Howell, S. N. G. 1993a. Taxonomy and distribution

of the hummingbird genus *Chlorostilbon* in Mexico and northern Central America. *Euphonia* 2:25–27.

———. 1993b. A taxonomic review of the green-fronted hummingbird. *Bull. Brit. Ornith. Club* 113:179–187.

——— and S. Webb. 1989. Notes on the Honduran emerald. *Wils. Bull.* 101:642–643.

Howell, T. R., and W. R. Dawson. 1954. Nest temperatures and attentiveness in the Anna hummingbird. *Condor* 56:93–96.

Hubbard, J. P. 1978. Revised Check-List of the Birds of New Mexico. *New Mexico Ornithological Society Publication* No. 6, Albuquerque.

Hutto, R. L. 1985. Habitat distribution of migrant landbird species in western Mexico. In *Neotropical Ornithology*, pp. 221–239 (P. A. Buckley, M. S. Foster, E. S. Morton, R. S. Ridgley, and F. G. Buckley, eds.). Ornithological Monograph No. 36. American Ornithologists' Union, Washington, D.C.

Ingels, J. 1976. Observations on some humming-birds of Martinique. *Avic. Mag.* 82:98–100.

James, R. L. 1948. Some hummingbird flowers east of the Mississippi. *Castenea* 13:97–109.

Jobling, J. A. 1991. *A Dictionary of Scientific Bird Names*. Oxford University Press, New York.

Johnsgard, P. A. 1994. *Arena Birds: Sexual Selection and Behavior.* Smithsonian Institution Press, Washington, D.C.

Johnson, A. W. 1967. *The Birds of Chile and Adjacent Regions of Argentina, Bolivia, and Peru*, vol. 2. Platt Establecimientos Graficos, Buenos Aires.

Kelly, J. W. 1955. History of the nesting of an Anna hummingbird. *Condor* 57:347–353.

Kobbe, W. H. 1900. The rufous hummingbirds of Cape Disappointment. *Auk* 17:8–15.

Kodric-Brown, A., and J. H. Brown. 1978. Influence of economics, interspecific competition, and sexual dimorphism on territoriality of migrant rufous hummingbirds. *Ecology* 59:285–296.

Kuban, J. F., and R. L. Neill. 1980. Feeding ecology of hummingbirds in the highlands of the Chiso Mountains, Texas. *Condor* 82:180–185.

Lack, D. 1973. The number of species of humming-birds in the West Indies. *Evolution* 27:326–327.

———. 1976. *Island Biology, Illustrated by the Land Birds of Jamaica*. Blackwell, Oxford.

Lamb, C. C. 1925. Observations on the Xantus hummingbird. *Condor* 14:32–40.

Lasiewski, R. C. 1962. Energetics of migrating hummingbirds. *Condor* 64:324.

———. 1964. Body temperature, heart and breathing rate, and evaporative water loss in hummingbirds. *Physiol. Zool.* 37:212–223.

Lasiewski, R. C., W. W. Weathers, and M. H. Bernstein. 1967. Physiological responses of the giant hummingbird, *Patagonia gigas. Comp. Biochem. Physiol.* 23:797–813.

Leberman, R. C. 1972. Identify, sex, and age it. Key to age and sex determination of ruby-throated hummingbirds in autumn. *Inl. Bird-Banding News* 44:197–202.

Leck, C. F. 1973. Dominance relationships in nectar-feeding birds at St. Croix. *Auk* 90:431–432.

Legg, K., and F. A. Pitelka. 1956. Ecological overlap of Anna and Allen hummingbirds nesting at Santa Cruz, California. *Condor* 58:393–405.

Lesson, R. P. 1829. *Histoirie Naturelle des Oiseau-Mouches.* Arthus Bertrand, Paris.

———. 1830–31. *Histoirie Naturelle de Colibris.* Arthus Bertrand, Paris.

———. 1832. *Les Trochilidées on les Colibris et les Oiseaux-Mouches.* Arthus Bertrand, Paris.

Levy, S. H. 1958. A possible United States breeding area for the violet-crowned hummingbird. *Auk* 75:350.

Ligon, J. S. 1961. *New Mexico Birds and Where to Find Them.* University of New Mexico Press, Albuquerque.

Linneaus, C. 1758. *Systema Naturae. Regnum Animale.* (10th ed. tomus I.) L. Salvii, Holminae.

Lodge, G. E. 1896. Notes on some West-Indian hummingbirds. *Ibis* 2(7th ser.):495–519.

Lowery, G. H., Jr. 1974. *Louisiana Birds*, 3rd ed. Louisiana State University Press, Baton Rouge.

Lyerly, S. B., B. F. Riess, and S. Ross. 1950. Color preference in the Mexican violet-eared hummingbird. *Colibri t. thalassinus* (Swainson). *Behaviour* 2:237–248.

Lynch, J. F. 1985. Distribution of overwintering Neotropical migrants in the Yucatan Peninsula. II. Use of native and human-modified habitats. In *Neotropical Ornithology*, pp. 178–195 (P. A. Buckley, M. S. Foster, E. S. Morton, R. S. Ridgley, and F. G. Buckley, eds.).

Ornithological Monograph No. 36. American Ornithologists' Union, Washington, D.C.

Lynch, J., and P. L. Ames. 1970. A new hybrid hummingbird, *Archilochus alexandri* × *Selasphorus sasin. Condor* 72:209–212.

Lyon, D. L. 1973. Territorial and feeding activity of broad-tailed hummingbirds (*Selasphorus platycercus*) in *Iris missouriensis. Condor* 75:346–349.

———. 1976. A montane hummingbird territorial system in Oaxaca, Mexico. *Wilson Bull.* 88:280–299.

Lyon,, D. L., J. Crandall, and M. McKone. 1977. A test of the adaptiveness of interspecific territoriality in the blue-throated hummingbird *Auk* 94:448–454.

MacMillen, R. E., and F. L. Carpenter. 1977. Daily energy costs and body weight in nectarivorous birds. *Comp. Biochem. Physiol. A Comp. Physiol.* 56:439–441.

Marshall, J. T. 1957. Birds of pine-oak woodland in southern Arizona and adjacent Mexico. *Pacific Coast Avifauna* 32:1–125.

Martin, A., and A. Musy. 1959. *La Vie des Colibris.* Editions Delachaux et Niestlé, Neuchâtel, Switzerland.

Mayr, E., and L. L. Short, Jr. 1970. Species taxa of North American birds: a contribution to comparative systematics. *Publ. Nuttall Ornithol. Club* 9:1–127.

Miller, A. H., and R. C. Stebbins. 1964. *The Lives of Desert Animals in Joshua Tree National Monument.* University of California Press, Berkeley.

Miller, J. R. 1978. Notes on birds of San Salvador Island (Watlings), the Bahamas. *Auk* 95:281–288.

Miller, R. S., and R. E. Miller. 1971. Feeding activity and color preference of ruby-throated hummingbirds. *Condor* 73:309–313.

Miller, S. J., and D. W. Inouye. 1983. Role of the wing whistle in the territorial behavior of male broad-tailed hummingbirds. *Anim. Behav.* 31:689–700.

Mirsky, E. N. 1976. Song divergence in hummingbird and junco populations on Guadalupe Island. *Condor* 78:230–235.

Mobbs, A. J. 1971. Stretching attitudes in hummingbirds. *Avic. Mag.* 77:231.

———. 1973. Scratching and preening postures in hummingbirds. *Avic. Mag.* 79:200–204.

———. 1979. Methods used by the Trochilidae when capturing insects. *Avic. Mag.* 85:26–30.

Monroe, B. L., Jr. 1968. A distributional study of the birds of British Honduras. *Am. Ornithol. Union Monogr.* 7.

Montgomerie, R. D. 1979. The energetics of foraging and competition in some Mexican hummingbirds. Ph.D. dissertation, McGill University, Montreal.

———. 1984. Nectar extraction by hummingbirds: Response to different floral characters. *Oecologia* 63:229–236.

Moore, R. T. 1939a. Habits of the white-eared hummingbird in northern Mexico. *Auk* 56:442–445.

———. 1939b. The Arizona broad-billed hummingbird. *Auk* 56:313–319.

———. 1947. Habits of male hummingbirds near their nests. *Wilson Bull.* 59:21–25.

Morony, J. J., Jr., W. J. Bock, and J. Ferrand, Jr. 1975. *Reference List of the Birds of the World.* American Museum of Natural History, New York.

Morrison, P. 1962. Modification of body temperature by activity in Brazilian hummingbirds. *Condor* 64:315–323.

Nelson, R. C. 1970. An additional nesting record for the lucifer hummingbird in the United States. *Southwest. Nat.* 1:513–515.

Newfield, N. L. 1996. Piecing together the hummingbird puzzle. *Living Bird* 15(2):16–21.

Nickell, W. P. 1948. Alternate care of two nests by a ruby-throated hummingbird. *Wilson Bull.* 60:242–243.

Norris, E. A., C. E. Connell, and D. W. Johnston. 1957. Notes on fall plumages, weights and fat condition in the ruby-throated hummingbird. *Wilson Bull.* 69:155–163.

Northrop, J. I. 1891. The birds of Andros Island, Bahamas. *Auk* 8:64–80.

Oberholser, H. C. 1974. *The Bird Life of Texas.* (Ed. E. B. Kincaid.) 2 vols. University of Texas Press, Austin.

Olrog, C. C. 1963. *Lista y Distribución de las Aves Argentinas.* Instituto Miguel Lillo, Tucumán.

———. 1968. *Las Aves Sudamericanas*, vol. 1. Instituto Miguel Lillo, Tucumán.

Oniki, Y. 1970. Nesting behavior of reddish hermits (*Phaethornis ruber*) and occurrence of wasp cells in nests. *Auk* 87:720–728.

Ornelas, J. F. 1987. Rediscovery of the rufous-crested coquette (*Lophornis delattrei brachylopha*) in Guerrero, Mexico. *Wils. Bull.* 99:719–721.

Orr, R. T. 1939. Observations on the nesting of the Allen hummingbird. *Condor* 41:17–24.

Ortiz-Crespo, F. I. 1974. The giant hummingbird Patagonia gigas in Ecuador. *Ibis* 116:347–359.

Owre, O. T. 1976. Bahama woodstar in Florida: first specimen for continental North America. *Auk* 93:837–838.

Paynter, R. J., Jr. 1955. The ornithogeography of the Yucatán Peninsula. *Bull. Peabody Mus.* (9):1–347.

Pearson, O. P. 1954. The daily energy requirements of a wild Anna hummingbird. *Condor* 56:317–322.

———. 1960. Speed of the Allen hummingbird while diving. *Condor* 62:403.

Phillips, A. R. 1964. Notas sistematicas sobre aves Mexicanas. III. *Rev. Soc. Mex. Hist. Nat.* 25:217–242.

———. 1966. Further systematic notes on Mexican birds. *Bull. Brit. Orn. Club* 86:86–94, 103–112, 125–131, 148–159.

———. 1975. The migration of Allen's and other hummingbirds. *Condor* 77:196–205.

Phillips, A. R., J. Marshall, and G. Monson. 1964. *Birds of Arizona*. University of Arizona Press, Tucson.

Primack, R. B., and H. F. Howe. 1975. Interference competition between a hummingbird Amazilia tzacatl and skipper butterflies (Hesperiidae). *Biotropica* 7:55–58.

Pickens, A. L. 1927. Unique method of pollination by the ruby-throat. *Auk* 44:14–17.

———. 1930. Favorite colors of hummingbirds. *Auk* 47:346–352.

———. 1944. Seasonal territory studies of ruby-throats. *Auk* 61:88–92.

Pickens, A. L., and L. P. Garrison. 1931. Two-year record of the ruby-throat's visits to a garden. *Auk* 48:532–537.

Pitelka, F. A. 1942. Territoriality and related problems in North American hummingbirds. *Condor* 44:189–204.

———. 1951a. Breeding seasons of hummingbirds near Santa Barbara, California. *Condor* 53:198–201.

———. 1951b. Ecologic overlap and interspecific strife in breeding populations of Anna and Allen hummingbirds. *Ecology* 32:641–661.

Poley, D. 1976. *Kolibris*. Neue Brehm-Bücherei 484. A. Ziemsen Verlag, Wittenberg Lutherstadt.

Powers, D. R. 1991. Diurnal variation in mass, metabolic rate and respiratory quotient in Anna's and Costa's hummingbirds. *Physiological Zoology* 64:850–870.

Powers, D. R., and K. N. Nagy. 1988. Field metabolic rate and food consumption by free-living Anna's hummingbirds (*Calypte anna*). *Physiological Zoology* 61:500–506.

Pulich, W. M., and W. M. Pulich, Jr. 1963. The nesting of the lucifer hummingbird in the United States. *Auk* 80:370–371.

Pyke, G. H. 1980. The foraging behaviour of Australian honeyeaters: a review and some comparisons with hummingbirds. *Aust. J. Ecol.* 5:343–369.

Rising, J. D. 1965. Notes on behavioral responses of the blue-throated hummingbird. *Condor* 67:352–354.

Robertson, W. B., Jr. 1962. Observations on the birds of St. John, Virgin Islands. *Auk* 79:44–76.

Rowley, J. S. 1962. Nesting of the birds of Morelos, Mexico. *Condor* 64:253–272.

———. 1966. Breeding records of birds of the Sierra Madre del Sur, Oaxaca, Mexico. *Proc. West. Found. Vertebr. Zool.* 1:107–203.

———. 1984. Breeding records of some land birds in Oaxaca, Mexico. *Proc. West. Found. Vert. Zool.* 2:74–224.

Rowley, J. S., and R. Orr. 1964. A new hummingbird from southern Mexico. *Condor* 66:81–83.

Ruschi, O. 1965. [Position assumed by the female during incubation and the warming of the young in Trochilidae.] *Bol. Mus. Biol. Prof. Mello Leitão* (48):1–3.

———. 1967. Some observations on the migration of hummingbirds in Brazil. *Bol. Mus. Biol. Prof. Mello-Leitão* 28:1–5. (In Portuguese.)

———. 1979. *Les Aves do Brasil*. Kosmos, Sao Paolo.

Russell, S. M. 1964. A distributional study of the birds of British Honduras. *Am. Ornithol. Union Monogr.* 1:1–195.

———, J. C. Barlow, and D. W. Lamm. 1979. Status of some birds on Isla San Andres and Isla Providencia, Colombia. *Condor* 81:98–100.

Russell, S. M., and D. W. Lamm. 1978. Notes on the distribution of the birds in Sonora, Mexico. *Wilson Bull.* 90:123–130.

Sargent, R., and M. Sargent. 1996. Spring ruby-throated migration. *Netlines* (Newsletter of Hummer/Bird Study Group, Clay, Ala.) 3(1):5–6.

Schäfer, E. 1952. Sobre la biologie de *Colibri coruscans*. *Bol. Soc. Venez. Cienc. Nat.* 15:153–162.

Schaldach, W. J. 1963. The avifauna of Colima and adjacent Jalisco, Mexico. *Proc. West. Found. Vertebr. Zool.* 1:1–100.

Schmidt-Marloh, D., and K. L. Schuchmann. 1980. Zur Biologie des Blauen Veilchenohr-Kolibris. *Bonn. Zool. Beitr.* 31:61–77.

Schuchmann, F. L. 1979. Okologie und Ethologie des Anna-Kolibris (*Calypte anna*). *Natur Mus.* 109:149–155.

Schuchmann, K. L., D. Schmidt-Marloh, and H. Bell. 1979. Energetische Untersuchengen bei einer tropischen Kolibriart (*Amazilia tzacatl*). *J. Ornithol.* 120:78–85.

Scott, P. E. 1994. Lucifer hummingbird. In *The Birds of North America,* No. 134 (A. Poole and F. Gill, eds.). Academy of Natural Sciences, Philadelphia, and American Ornithologists' Union, Washington, D.C.

Selander, R. K. 1966. Sexual dimorphism and differential niche utilization in birds. *Condor* 68:113–151.

Short, L. L., Jr., and A. R. Phillips. 1966. More hybrid hummingbirds from the United States. *Auk* 83:253–265.

Sibley, C. G., and B. L. Monroe, Jr. 1990. *Distribution and Taxonomy of Birds of the World.* Yale University Press, New Haven, Conn.

Simon, E. 1921. *Histoire Naturelle des Trochilidae (Synopsis et Catalogue).* Mulo, Paris.

Skutch, A. F. 1931. The life history of Reiffer's hummingbird (*Amazilia tzacatl tzacatl*) in Panama and Honduras. *Auk* 48:481–500.

———. 1951. Life history of Longuemare's hermit hummingbird. *Ibis* 93:180–195.

———. 1952. Scarlet passion flower. *Nat. Mag.* 45:523–525, 550.

———. 1961. Life history of the white-crested coquette hummingbird. *Wils. Bull.* 73:5–10.

———. 1964. Life histories of hermit humming-birds. *Auk* 81:5–26.

———. 1967. Life histories of Central American highland birds. *Nuttall Ornithol. Soc. Publ.* 7:1–213.

———. 1972. Studies of tropical American birds. *Nuttall Ornithol. Soc. Publ.* 10:1–228.

———. 1976. *Parent Birds and Their Young.* University of Texas Press, Austin.

———. 1981. New studies of tropical American birds. *Publ. Nuttall Ornithol. Club* 19:37–58.

Slud, P. 1964. The birds of Costa Rica: distribution and ecology. *Am. Mus. Nat. Hist. Bull.* 128:1–430.

Smith, G.T.C. 1949. A high altitude hummingbird on the Volcano Cotapaxi. *Ibis* 111:17–22.

Smith, W. K., S. W. Roberts, and P. C. Miller. 1974. Calculating the nocturnal energy expenditure of a incubating Anna's hummingbird. *Condor* 76:176–183.

Snow, B. K. 1974. Lek behavior and breeding of Guy's hermit hummingbird, *Phaethornis guy. Ibis* 116:278–297.

Snow, B. K., and D. W. Snow. 1972. Feeding niches of hummingbirds in a Trinidad valley. *J. Anim. Ecol.* 41:471–485.

Snow, D. W. 1968. The singing assemblies of little hermits. *Living Bird* 7:47–55.

Snow, D. W., and B. K. Snow. 1980. Relationships between hummingbirds and flowers in the Andes of Colombia. *Bull. Br. Mus. (Nat. Hist.), Zool. Ser.* 38:105–139.

Snyder, D. E. 1966. *The Birds of Guyana.* Peabody Museum, Salem, Mass.

Sprunt, A., Jr. 1954. *Florida Bird Life.* Coward-McCann, New York.

Stewart, R. E. 1975. *Breeding Birds of North Dakota.* Tri-College Center for Environmental Studies, Fargo, N.D.

Stiles, F. G. 1971. On the field identification of California hummingbirds. *Calif. Birds* 2:41–54.

———. 1972a. Age and sex determination in rufous and Allen hummingbirds. *Condor* 74:25–32.

———. 1972b. Food supply and the annual cycle of the Anna hummingbird. *Univ. Calif. Publ. Zool.* 97:1–109.

———. 1975. Ecology, flowering phenology, and hummingbird pollination of some Costa Rica *Heliconia* species. *Ecology* 56:285–301.

———. 1976. Taste preferences, color preferences, and flower choice in hummingbirds. *Condor* 78:10–26.

———. 1978. Ecological and evolutionary implications of bird pollination. *American Zool.* 18:715–727.

———. 1979. Notes on the natural history of *Heliconia* in Costa Rica. *Brenesia* 15(suppl.): 194–210.

———. 1980. The annual cycle in a tropical wet forest community. *Ibis* 122:322–343.

———. 1982. Aggressive and courtship displays of the male Anna's hummingbird. *Condor* 84:208–225.

———. 1985. Seasonal patterns and coevolution in the hummingbird-flower community of a Costa Rica subtropical forest. In *Neotropical Ornithology*, pp. 757–787 (P. A. Buckley, M. S. Foster, E. S. Morton, R. S. Ridgley, and F. G. Buckley, eds.). *Ornithological Monograph No. 36.* American Ornithologists' Union, Washington, D.C.

———. 1995. Behavioral, ecological and morphological correlates of foraging for arthropods by the hummingbirds of a tropical wet forest. *Condor* 97:853–878.

Stiles, F. G., and L. L. Wolf. 1970. Hummingbird territoriality at a tropical flowering tree. *Auk* 87:67–91.

———. 1979. Ecology and evolution of lek mating behavior in the long-tailed hermit hummingbird. *Amer. Ornithol. Union Monogr.* 27:1–78.

Straw, R. M. 1956. Floral isolation in *Penstemon*. *Am. Nat.* 90:47–53.

Sutton, G. M., and T. D. Burleigh. 1940. Birds of Tamazunchale, San Luis Potosi. *Wilson Bull.* 52:221–233.

Sutton, G. M., and O. S. Pettingill. 1942. Birds of the Gomez Farias region, southeastern Tamaulipas. *Auk* 59:1–34.

Tamm, S., D. P. Armstrong, and Z. J. Toose. 1989. Display behavior of the male calliope hummingbird during the breeding season. *Condor* 91:272–279.

Tiebout, H. M., III. 1993. Mechanisms of competition in tropical hummingbirds: metabolic costs for losers and winners. *Ecology* 74:405–418.

Todd, W.E.C. 1942. List of hummingbirds in the collection of the Carnegie Museum. *Ann. Carnegie Mus.* 29:271–370.

Toledo, V. M. 1974. Observations on the relationship between hummingbirds and *Erythrina spp*. *Lloydia.* 37:482–487.

Toops, C. 1992. *Hummingbirds: Jewels in Flight.* Voyaguer Press, Stillwater, Minn.

Trousdale, B. 1954. Copulation of Anna hummingbird. *Condor* 56:110.

True, D. 1993. *Hummingbirds of North America: Attraction, Feeding, and Photographing.* University of New Mexico Press, Albuquerque.

———. 1996. Mighty mite. *Birder's World* 10(2):26–29.

Tyrrell, R. A., and E. Q. Tyrrell. 1984. *Hummingbirds: Their Life and Behavior.* Crown Publishers, New York.

van der Pijl, L., and C. H. Dodson, 1966. *Orchid Flowers: Their Pollination and Evolution.* University of Miami Press, Coral Gables.

van Rossem, A. J. 1945. A distribution survey of the birds of Sonora, Mexico. *Occas. Pap. Mus. of Zool. La. State Univ.* (21):1–379.

Wagner, H. O. 1945. Notes on the life history of the Mexican violet-ear. *Wilson Bull.* 57:165–187.

———. 1946a. Food and feeding habitats of Mexican hummingbirds. *Wilson Bull.* 58:69–93.

———. 1946b. Observaciones sobre la vida de *Calothorax lucifer. An. Inst. Biol. Univ. Nac. Auton. Mex.* 17:289–299.

———. 1948. Die Balz des Kolibris *Selasphorus platycercus. Zool. Jahrb. Abt. Syst. Oekol. Geogr. Tiere* 77:267–278.

———. 1952. Bietrage zur Biologie des Blaukehlkolibris *Lampornis clemenciae* (Lesson). *Veroeff. Mus. Bremen, Reihe A Band* 2:5–44.

———. 1954. Versuch einer Analyse der Kolibribalz. *Z. Tierpsychol.* 11:182–212.

———. 1959. Bietrag zum Verhalten des Weissohrkolibris (*Hylocharis leucotis Vieill.*). *Zoologische Jahrbücher, Abteilung für Systematik Oekologie und Geographie der Tiere* 86:253–302.

Waser, N. M. 1976. Food supply and nest timing of broad-tailed hummingbirds in the Rocky Mountains. *Condor* 78:133–135.

———. 1978. Competition for hummingbird pollination and sequential flowering in two Colorado wildflowers. *Ecology* 59:934–944.

Waser, N. M., and D. W. Inouye. 1977. Implications of recaptures of broad-tailed hummingbirds banded in Colorado. *Auk* 94:393–395.

Wauer, R. H. 1973. *Birds of Big Bend National Park and Vicinity.* University of Texas Press, Austin.

Wells, S., and L. F. Baptista. 1979. Displays and morphology of an Anna × Allen hummingbird hybrid. *Wilson Bull.* 91:524–532.

Wells, S., R. A. Bradley, and L. F. Baptista. 1978. Hybridization in *Calypte* hummingbirds. *Auk* 95:537–549.

Welter, W. A. 1935. Nesting habits of ruby-throated hummingbird. *Auk* 52:88–89.

Welty, C. 1975. *The Life of Birds,* 2nd ed. W. B. Saunders, Philadelphia.

Weske, J. S., and J. W. Terborgh. 1977. *Phaethornis koepkeae,* a new species of hummingbird from Peru. *Condor* 79:143–147.

Wetmore, A. 1946. New birds from Colombia. *Smithson. Misc. Collect.* 106(16):1–14.

———. 1953. Further additions to the birds of Panama and Colombia. *Smithson. Misc. Collect.* 122(8):1–12.

———. 1963. Additions to records of birds known from the Republic of Panama. *Smithson. Misc. Collect.* 145(6):1–11.

———. 1968. The birds of the Republic of Panamá. *Smithson. Misc. Collect.* 150, pt. 2.

Wetmore, A., and W. H. Phelps, Jr. 1956. Further additions to the list of birds of Venezuela. *Proc. Biol. Soc. Wash.* 69:1–10.

Weydemeyer, W. 1927. Notes on the location and construction of the nest of the calliope hummingbird. *Condor* 29:19–24.

———. 1971. Injured calliope hummingbird lifted by another. *Auk* 88:431.

Weymouth, R. D., R. C. Lasiewski, and A. J. Berger. 1964. The tongue apparatus in hummingbirds. *Acta Anat.* 58:252–270.

Whittle, C. L. 1937. A study of hummingbird behavior during a nesting season. *Bird-Banding* 8:170–173.

Wilbur, S. R. 1987. *Birds of Baja California.* University of California Press, Berkeley.

Wiley, R. H. 1971. Song groups in a singing assembly of little hermits. *Condor* 73:28–35.

Williamson, F. 1957. Hybrids of the Anna and Allen hummingbirds. *Condor* 59:118–123.

Winker, K., M. A. Ramos, J. H. Rappole, and D. W. Warner. 1992. A note on *Campylopterus excellens* in southern Veracruz, with a guide to sexing captured individuals. *J. Field Ornith.* 63:339–43.

Witzeman, J. 1979. Plain-capped starthroats in the United States. *Continental Birdlife* 1:1–3.

Wolf, L. 1964. Nesting of the fork-tailed emerald in Oaxaca, Mexico. *Condor* 66:51–55.

Wolf, L., and F. R. Hainsworth. 1971. Environmental influence on regulated body temperatures in torpid hummingbirds. *Comp. Biochem. Physiol.* 41:167–173.

Wolf, L., and F. G. Stiles. 1970. Evolution of pair cooperation in a tropical hummingbird. *Evolution* 24:759–773.

Wolf, L., F. G. Stiles, and F. R. Hainsworth. 1976. Ecological organization of a tropical, highland hummingbird community. *J. Anim. Ecol.* 45:349–379.

Woods, R. S. 1927. The hummingbirds of California. *Auk* 44:297–318.

Wyman, L. E. 1920. Notes on the calliope hummingbird. *Condor* 22:206–207.

Zimmer, J. T. 1950–53. Studies on Peruvian birds, nos. 55-63. *Am. Mus. Novit.* 1449, 1450, 1463, 1475, 1513, 1540, 1595, 1604.

Zimmerman, D. A. 1973. Range expansion of Anna's hummingbird. *Am. Birds* 27:827–835.

Zimmerman, D. A., and S. H. Levy. 1960. Violet-crowned hummingbird nesting in Arizona and New Mexico. *Auk* 77:470.

Zusi, R. L. 1980. On the subfamilies of hummingbirds. Abstract of paper presented at 98th meeting, American Ornithologists' Union, Fort Collins, Colorado, 11–15 August 1980.

Index

Complete indexing is limited to the entries for the vernacular names used in this book; individual descriptive species accounts are shown in italics. Latin names are provided parenthetically for species not thus identified in the text; indexing of genera and species includes text references to such taxa plus the descriptive accounts of each taxon. The appendixes are not indexed.

A

B

C